SOLAR ENERGY

The Awakening Science

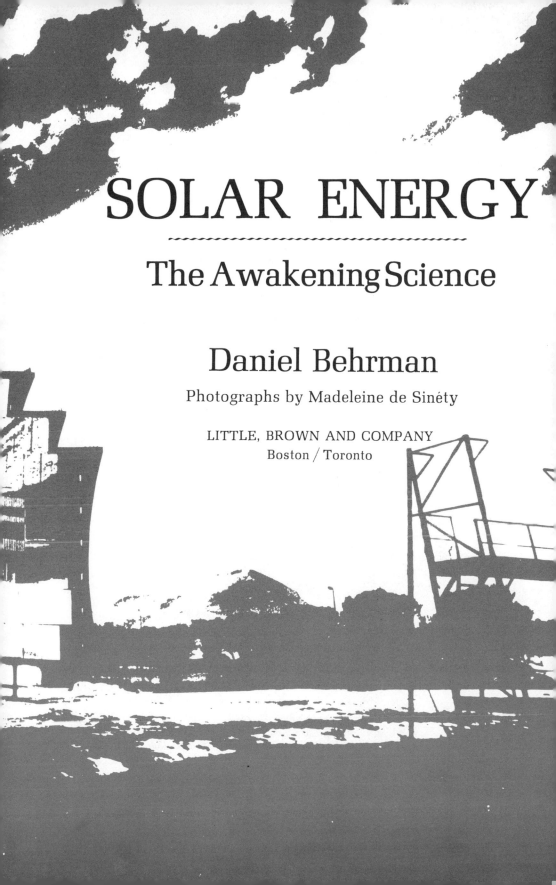

SOLAR ENERGY

The Awakening Science

Daniel Behrman

Photographs by Madeleine de Sinéty

LITTLE, BROWN AND COMPANY
Boston / Toronto

Fourth Printing

T/10/76

For permission to quote previously copyrighted materials,
the author gratefully acknowledges:

Oceanus magazine, Dr. William von Arx, and Dr. William
Heronemus for excerpts from the articles "Energy: Natural
Limits and Abundances" by Dr. William von Arx and "Using
Two Renewables" by Dr. William Heronemus, Summer 1974.
Copyright © 1974 by Woods Hole Oceanographic Institution.

The Bulletin of Atomic Scientists, for excerpts from "Future
Uses of Solar Energy" by Maria Telkes, August 1951. Copy-
right © 1951 by the Educational Foundation for Nuclear
Science.

Allis-Chalmers Corporation, for excerpts from *Power From
the Wind* by Palmer C. Putnam. Copyright © 1948 by S.
Morgan Smith Co.

LIBRARY OF CONGRESS CATALOGING IN PUBLICATION DATA

Behrman, Daniel.
    Solar energy: the awakening science.

    Includes index.
    1. Solar energy. I. Title.
TJ810.B43   333.7   76-19000
ISBN 0-316-08771-8

Designed by Janis Capone

*Published simultaneously in Canada
by Little, Brown & Company (Canada) Limited*

PRINTED IN THE UNITED STATES OF AMERICA

To Jack and Jim

Acknowledgments

The author wishes to express his appreciation to the United Nations Educational, Scientific and Cultural Organization for having enabled him to write this book entirely upon his own responsibility. Thanks are also due to Dr. Harry Lustig who guided him in his research and took a professional look at the manuscript; and to Berol Robinson for his precious advice.

SOLAR ENERGY

The Awakening Science

I.

~~~~~~~~~~~~~~~~~~~~~~~

Solar Reflections

~~~~~~~~~~~~~~~~~~~~~~~

I SIT BY VALAINE POND, fishing for words. It is one of the largest millponds here in eastern Brittany. Shaped roughly like a boomerang, it spreads five acres from the mill dam around the elbow of the boomerang to the upstream end, clogged with tussocks, hemmed on one side by the oak grove where I sit, on the other by a sheep pasture. In winter, when the water table is high, the miller of Valaine gets three-quarters of his power from the pond. He uses a turbine to grind barley into livestock feed. Cannily, he taps the sunshine that drives the hydrological cycle. The sun will evaporate the water that runs over his dam into the Bay of Mont-Saint-Michel twenty miles to the north, lifting it into the atmosphere to condense into rain and drizzle back onto the watershed of Valaine Pond so that his turbine will keep turning. Inefficient though the system may be, the nationalized power company fears it enough to prohibit him from making and selling electricity.

I fish for words, but the three fishermen who were working the pond with their long poles when I arrived have now stopped for lunch. Their poles stand unmanned, planted tall like saplings at the water's edge, ever fishing while the men cook their meal. The wood smoke from their fire is whipped away by the northwest wind. They are downwind from me, so I cannot smell what they are cooking — probably Breton farm sausage to be accompanied by a bottle of local cider.

Around me I see no signs of an energy shortage, an environmental crisis. Across the water, drowsy ewes graze while their lambs caper. Clouds from the English Channel and the Bay of Mont-Saint-Michel drift across the pale blue sky, clean and washed, above the two farmhouses at my end of the pond. The houses are granite and gray, roofed in black slate and incongruous red tile, sheep pastured next to the one, dairy cattle and a work horse in front of the other. It is June, but the air is cool under the oaks and I am grateful to the sun when it can slip between the clouds and warm me through my jeans and wool sweater and heavy jacket and overshoes. God is in His heaven, so is His sun, all is right with His world. When the Other's world goes wrong, perhaps I will be able to return to Valaine Pond. Here I can sense "alternative energy" — the dam, the smoke of the wood fire, the wind, the sun. It is only a matter of semantics. The wind, the sun, the firewood, the water-wheel have been here all along. The alternatives appeared only a few years ago, a matter of decades, hardly half a century since electricity was available for Breton farmhouses, fuel oil for heaters, diesel oil for tractors, gasoline for the fishermen's cars parked near the dam.

Valaine Pond gave me the perspective I had been seeking. It cannot power New York, it could not even spin the dozen-odd electric meters in the village of Valaine, but it serves to recall where we all began not so long ago. Valaine Pond is the appeal of pristine energy, sun and wind, waterpower and workhorses. This is all indigenous energy, as local as a town meeting, as independent of authority as a frontiersman. Small wonder that it keeps haunting a world where it seems to have no business. When heating oil runs short or electricity browns out, the citizens write their congressmen and their newspaper editors about power from the sun. No one wants brimstony coal smoke or a uranium processing plant in the neighborhood. Nobody is against heat or electricity, but breathes there the man willing to see his own backyard stripmined for coal or a breeder reactor on his front lawn? It is like urban freeways: everyone uses cars, no one wants the freeway around his part of town. This is why solar energy, uneconomic and unprofitable as it may have been up to the other day and perhaps today, just will not go away. Of all the remotely possible large-scale replacements for petroleum, it is the only one with grass roots support.

The push is from the bottom, from the concerned citizens who write scientists and engineers and legislators. It is not likely that the chairman of the old Atomic Energy Commission heard from many citizens who wanted a nuclear power plant near their homes. The nuke has a made-in-Hiroshima image. It does not smile like the sun, it does not lend a glow of health; people take their vacations in Acapulco, not in Los Alamos. That is what the sun has going for it, it has a good record. Old Sol beams on the world, he is not Old Overkill.

While my sympathies lie with the concerned citizens, I shall try to remain in a neutral corner. I do not expect to lend my voice to the great energy debate except when my much more authoritative sources wish to have their say. In this debate, the startling disclosure becomes instant history three months later and nothing is more derisory than last year's long-range forecast. It is no place for the book writer who is forced to telegraph his punches a year or two in advance. The forecasts are off because the schedule has gone wrong. In the past, there were no such discontinuities. Old ways did not disappear overnight, they were gradually superseded by new ways that built on them, used them as foundations, before destroying them. No one had to invent the automobile because of a sudden shortfall in the supply of hay. Oil was added to the panoply of technology, it moved in, then it moved the competition out. The displacements were tremendous but not perceptible within the attention span of humanity, that agricultural animal whose biological calendar seldom extends beyond a growing season or a gestation period. It was not an overnight transition, about the only kind that humanity notices.

This time, the change is so far ahead of schedule that it seems unscheduled. Oil has replaced coal; there is no way to move, live, or die except by oil. It might take twenty-five years of lead time to find another way, twenty-five or fifty years, the time that oil needed to drive out coal in this century, that coal needed to displace wood in the last one. We do not seem to have much more than twenty-five or fifty years. The oil is running out fast, the cheap oil has run out forever, we shall never see it again at $3 a barrel. Oil and natural gas were so cheap that cost was no deterrent to their use. Writing in *Science,* Hans Landsberg remarks that a barrel of oil was worth $2.60 in 1948 and $3.39 in 1972. Expressed in constant 1948 dollars, the 1972 price was $1.85. Oil had gone down 30 percent in twenty-five years. No wonder it inhibited the appearance of new sources of energy or of cheaper ways of using the old ones. When the price of oil did go up, nothing remained the same. A Gallup poll taken in mid-1974 found that only 6 percent of the American public regarded energy as the country's biggest problem, compared to nearly 50 percent six months before. It had been replaced by inflation; the name of the game had changed, but the same side was losing.

There was plenty of coal left when oil drove it out, so much that we still have enough for our grandchildren's grandchildren and beyond. It is oil that has let us down. Domestic oil and domesticated oil are running out before the nuclear plants are on line, before sulfur has been cleaned from coal and solar collectors can come from factories. The transition has not been planned the way the others were, at least not by the same people. The future is no longer a well-timed show, rehearsed, marketed, ready to be dropped into our ways of life like color television on home cassettes. Now we are no longer sure. We cannot imagine life without freeways and beltways; blankets the weight of airmail stationery on our beds; kitchens that hum, buzz, and beep like an airliner in landing approach. The economists call it inelasticity of demand. It means we have made no psychological provisions for next year, we have not made our wills, we presume we shall live forever and ever.

Shortly after April Fools' Day in 1974, the gasoline pumps were flowing again all over America. Traffic was backed up for half a mile on Interstate 70 between Stapleton Airport and the Denver business district where we were spurring a rented red Pinto out to Fort Collins, Colorado. It was a remission but we should not fool ourselves, the disease is terminal.

It was an odd moment, that first outbreak of panic after the Mideast war. It will go down in the incidental history of the United States with flagpole-sitting and streaking. We tried unsuccessfully, Madeleine and I, to spare oil by taking a train from Denver to Pittsburgh. There was a shortage of fuel but no lack of empty seats on the DC-10 that we rode from Denver to Chicago. We had two to ourselves, there was a third for Madeleine's overnight bag, another for her camera bag, an acre of void between us and the movie screen.

Yet there had been no room on the Broadway Limited from New York to Chicago, there was no way to get on the Empire Builder from Denver back east. Amtrak was like the French telephone system; nothing worked. I started to call their Denver office at nine in the morning. Either it was busy or it did not answer. Finally, around half-past ten, a gentleman answered. He was kind, but he could not be helpful. He would not have the train diagram until one o'clock, only three hours before departure. His computer was not yet on line. He could not confirm our seats out of Denver, he could not say if we would be able to board the one and only connecting train from Chicago to Pittsburgh which, assuming that we were among the chosen, would dump us at our destination at 1:20 A.M. after a six-hour break in Chicago, time enough to queue up for an Amtrak reservation.

So, once more, we turned in our rented car at another motel near another airport. The renter drove us to the airline gate and, two hours later, we were heading for Chicago on that empty DC-10. There was no way to save any of America's energy without squandering all of my own on futile telephone calls and waiting lines. I was willing to spend a day and a half between Denver and Pittsburgh to get my thoughts together, to slip painlessly through two time zones, to see my country again; instead, I was almost ordered into a wide-bodied Friendship Jet. During the energy panic, it was easier to fly than to walk. Taking a train was like running a maze. Between Denver and Pittsburgh alone, our jumbo jets must have burned fuel amounting to more British Thermal Units than are being produced by all the solar energy collectors on the face of the earth. The *New Yorker* that I bought at Stapleton Airport had an energy cover: twin windmills mounted on the World Trade Center towers in Manhattan. That was in fun; in reality, I had to sail 37,000 feet high over the squared-off farms of Nebraska.

Many in the United States found this period bewildering. In Longmont, Colorado, we spent the weekend in a new house. Ritchie and Alice had been living there for less than a year with their two daughters, one baby son, two quarter horses, one dog, one cat, one ton-and-a-half GMC truck, one light pick-up. Ritchie was a working man and a good one. During the week he followed his trade as a tinsmith, installing heating and air-conditioning systems in homes like his own. On Saturdays he was a farrier, shoeing and trimming the hooves of horses like his own. He had taken the family west from Valley Stream, Long Island, in a camper mounted on the GMC. When they got to Colorado, he went to work. He and his wife and their two daughters lived in the camper, jacked up in a trailer park while he used the truck on his jobs. It is hard to see how Ritchie could have lived the same way or enjoyed the same freedom if he did not have all that power at his beck and call. Countries using less energy per capita may be leading the league in economic growth but they are as far from American egalitarianism as they ever were. If we ever practice conservation as a way of life, we shall have to find a way to distribute it as widely as we have distributed consumption.

"Thank God I wasn't born a hundred years ago," said Alice. "He would have brought us here in one of those covered wagons with me having babies in the back seat." Ritchie's energy was equaled by Alice's beauty. She was of Puerto Rican stock, slim and fair as the upper-class ladies of Spanish America used to be when sunburn was the stigma of the working class. One hundred

years ago she might have ended on a ranch. Now she was in a
ranch house on a new subdivision forty miles north of Denver on
the road to Cheyenne. She, too, had power at her fingertips, the
freezer and the washer, the drier and the range, the garbage disposal unit and, newest wonder of the New World, the trash compactor, built like a bank safe, a lock in front so the children could
not play inside it. The compactor flattened a chestful of garbage
into a shallow slab; it might herald the people compactor we will
need a hundred years from now. The house purred to the rhythm
of its machines; Alice kept busy doing all the bed-making and
cleaning. She had the home of an upper-class lady of South America or West Europe, but she only had herself for help. She told us
that it wasn't right. They had bought all these machines, they
would be paying for them for years, and now they were being told
to save electricity. It was time that the government did something
about it. Ritchie was worried because there would be no more
natural gas for the new homes that he was helping to build —
which meant that fewer new homes would be built. We could be
of no great consolation to Alice. We told her of the house we
rented in Brittany, the wood fireplace that heated our kitchen and
roasted our meats. We spoke of Monsieur Goupil who invited us
for dinner in his home where there was no running water and his
wife saved the dishwater for the pigs. Alice listened in disbelief,
and we found this hard to believe ourselves in Alice's wonderland.

The Mad Hatter's influence was most marked, as one might have
thought, in Washington. Controversy seethed to compensate for
turned-down thermostats. The breeder reactor, nuclear power plant
of the postpetroleum future, took the center of the stage and so did
its detractors. Even the atom's best friends were expressing caution.
Professor David J. Rose of the nuclear engineering department of
the Massachusetts Institute of Technology wrote in *Science*'s
"energy issue" in April 1974: "It is very unsettling that present
reactors contain substantial amounts of plutonium and breeder reactors will contain a huge quantity, close to 10^6 [one million]
curies. The extreme ratio between the resource available and the

allowable body burden emphasizes the necessity of vigilance which must be presumed to exist everywhere, forever."

Ralph Lapp, environmental consultant to the Senate Public Works Committee, was saying in the *Washington Post:*

> The issue centers upon the nagging question about the probability of a major nuclear accident. . . . A modern industrial society demands power and increasingly this power will have to come from central station generating plants.
>
> This means that sites will have to be found for these plants and there will have to be a balancing of risk and reward. Such a balancing act may test the very sinews of our democratic decision-making as the best sites are used up and the remaining sites produce an agony of risk assessment.

There is no denying the argument of the nuclear power advocate when he points to the safety record of the past. When he extrapolates it into the future, he may be on less firm ground. The nuclear power industry is in its infancy. One cannot help but recall where the automobile industry was when Henry Ford brought out the Model T in 1908. If 55,000 Americans had been killed on highways the following year, it is likely that we would have a different transportation system today.

About this time in Washington, Barry Commoner, chairman of the Scientists' Institute for Public Information, was calling the breeder reactor a "disastrous mistake." He charged that the Atomic Energy Commission was covering up a report that solar energy could provide 21 percent of the power that the United States would need by the year 2000 and at a competitive price. The *New York Times* stepped into the argument in a Washington-datelined story that explained:

> On the principle that one cannot beat something with nothing, the antinuclear scientists are cranking up a campaign to tell the public that solar energy is more feasible than is generally understood and will pose much less environmental risk than do reactors.
>
> Disinterested analysts fear that the public will leap to the con-

clusion that sunshine is the solution to all energy problems in much the same way that scientists and engineers led the public 25 years ago to expect early, abundant and cheap nuclear energy — a promise that has continually receded.

The disinterested analysts did not include E. B. White, who took time out from building a wheelbarrow to write his recommendations for 1974 in the *New York Times* a few months earlier. Among them were:

The atom. Nuclear power plants should be phased out, to await the day when someone comes up with a solution to the problem of safety and the problem of disposal.

The sun. Solar heat should be captured and used. A cat dozing in the sunny doorway of a barn knows all about it. Why can't man learn?

Innocently if not disinterestedly, we blundered into this den of bobcats when we arrived in Washington at the start of our sunbeam hunt. I was told that a former classmate of mine was touting a plan to cover the Indian reservations of the Great Southwest with solar cells produced by a new process making them cheaper than throwaway beer cans. I was intrigued; I knew little about solar cells, but I did know that their price had restricted them to generating electricity for use in outer space or on earth only for such tasks as lighting navigation buoys off the Aleutian Islands or charging the batteries of sailboats anchored in lonely coves far from the comforts of the marina. I phoned my former classmate. He did not remember me, but he took me into his confidence.

The oil companies and the Atomic Energy Commission, he said, were hatching a plot to torpedo solar power. How? By pushing solar heating and cooling of individual homes. Now you know, Dan, that no one in his right mind is going to live in a house where he will have to open a window in the morning or flick a switch at noon. The oil companies and the Atomic Energy Commission know this, that's why they're trying to foist these houses on the American public. They know that no one will buy them. Even if they sold 250,000 however, they would cut power consumption by only 1.5 percent.

No, solar cells were the way to go. Weren't they too expensive? Not at all, Dan, not at all. He named a name, that of a researcher who had achieved a breakthrough in producing silicon solar cells. The capital cost was down to 30 cents per watt, only three tenths of a dollar. That was quite a change from their price in outer space, around $100 a watt.

After our conversation, I tried to check his story. I was referred to Dr. Paul Rappaport at the RCA Laboratories in Princeton, New Jersey. Rappaport was responsible for one of the first of these photovoltaic cells that turn sunlight directly into electricity. When I got him on the phone, he said he had no axe to grind because RCA does not make solar cells. I asked him what he thought of the breakthrough that I had been told about in Washington. I named the name that my former classmate had mentioned.

"He's full of it," said Rappaport (that's not *quite* what he said), "and you can quote me."

I was not always so lucky in winging all the rumors flying about in that spring of 1974. Another friend had seen a report that the oil companies were getting a monopoly on solar energy. I could understand this, the American is so used to depending upon his oil company that he regards it as all-powerful. My friend said that he had seen the report in a newspaper but he could not give me a reference. Too bad — perhaps then I might have learned how Exxon would pull clouds over the heads of the nonbelievers, bring rain down on the roofs of the nonsigners, shut off the warming rays from the independent distributors. I do not say that man would not like to get a corner on the sun, I just don't see how it could be done.

Nailing down loose talk was all the harder because I had never seen a science like solar energy. Usually, a body of science has its benchmarks from which one can start prospecting and speculating. Not this one; it was hard to find agreement on anything. The polite language of scientific intercourse did not exist. I heard one eminent figure refer to another as a "snake oil salesman." One source would paint for me a picture of a multibillion-dollar market for sun-heated homes; another quoted Dr. Hoyt Hottel of MIT, a pioneer in solar houses, who is said to have said: "If you wish to lose the least money, get fifty percent of your heat from the sun. If you wish to lose no money, don't get any." (Hottel made his remark during the era of giveaway prices for oil and natural gas; it is understood that his enthusiasm for solar houses has risen with oil prices, but he still believes solar heating is being oversold.)

The trouble with sunshine as a source of power is that it is spread all over, it is hard to concentrate. That is also the trouble with solar energy as a science. There are no centers, no true nuclei. Since it was out of fashion for so long, it is not to be found in fashionable places. Looking for it did not take us to Harvard or Berkeley; instead, we played two- or three-night stands in Newark, Delaware, and Butler, Pennsylvania, and Fort Collins, Colorado, and Odeillo, Pyrénées-Oriéntales, and Aldingen bei Stuttgart. Prior to leaving for Minneapolis, we asked an editor in a Manhattan publishing house if she knew of a good place to stay there. After a moment's reflection . . . "You know, that problem has just never come up."

Solar energy was an intellectual backwater, almost a forgotten subject. Most of its leading practitioners had busied themselves with other things after the solar bubble of the 1950s had burst. I knew something about that early fizzle myself. In 1958 I covered an international symposium on solar energy and I stuck my neck out with these words I wish I could forget: "In lands scattered around the world, solar energy is coming out of research laboratories and into daily life at a price which the average man or his national economy can afford to pay." There were other nuggets, too.

Costs of operating a sun-powered refrigerator have been brought down to the point of competing with classical methods of refrigeration. . . . Only a few years ago, solar energy was the pampered, spoonfed darling of scientists working in laboratories behind closed doors, closed not so much for the sake of secrecy but for protection against the usual form of humor reserved for research off the beaten track. Today, it is a healthy child doing very well in the hard-headed world of industry in the United States. . . . Most important of all, this progress offers new hope to those regions which are the richest in sunshine and, through some quirk of nature, usually the poorest in other power resources: the countries in the arid lands covering one-third of the earth's surface.

Despite such uplifting prose, solar energy could not get airborne. All the predictions were queered by the clearance sale on oil and natural gas, given away at prices never to be seen again. Occasionally, business tactics were borrowed from the world of retail merchandising. In Israel, solar water heaters are being used by 150,000 families. They probably would be used by a good many more if the electric utility had not put a crimp in their growth by offering cheap power at night to heat water, at a price that hardly covered the cost of the fuel the utility was burning.

Solar energy has not benefited from the momentum of money, what might be called the Concorde effect. This supersonic airliner has cost France and the United Kingdom several billion dollars, but development continued long after all hope of recouping the investment from commercial operation had vanished. When that much money says something will occur, it does. It is as if bettors could actually make a horse run faster by putting more money on it. This may not happen on a racetrack but it does in modern economics.

It may have happened, too, with nuclear power which has not been obliged to meet the marketplace test imposed these days upon competing energy sources, solar and others. James T. Ramey, former commissioner of the United States Atomic Energy Commission, has written that the AEC spent $1.5 billion in twenty-five years to develop the light water reactor for civilian power production. Another $2 billion was spent on this type of reactor as a power plant for submarines (the sun never gets spin-off on this scale from military programs). With this kind of money, we are bound to have either an atomic power program or an atomic power scandal.

Solar energy has never received anything comparable. By the start of 1974, the National Science Foundation was allocating about $13 million for solar energy with $50 million expected the following year. This was a big improvement over 1971, when almost nothing was spent, but, even so, it would have taken the slow sun one hundred years at the 1974 rate to get to where the fast breeder reactor had arrived at the beginning of that same year. While barriers to the use of solar energy will not be brought down by massive applications of gigabucks (along the lines of "if we can get to the moon, we can get at the sun"), the minuscule amounts devoted to research and development certainly kept scientists away from it for a number of years. New ones did not enter the field; the old-timers found new fields to enter. Consequently, when attention was refocused on the sun during the first energy worries of the 1970s, activity was not at all proportionate to the publicity. The money is now beginning to trickle but not yet in anything like the amounts needed to make the sun a sure thing.

Solar energy remains a heterogeneous subject, hard to grasp and hard to evaluate. The range of the field is not only geographical; in our own travels, which certainly omitted as much as they covered, we found ourselves talking to welders in a barn, engineers in a factory, physicists in a research laboratory. Occasionally we were wined and dined on expense accounts, but most of the time we chomped Big Macs. Some boosters of solar power spoke to us of a thousand megawatts (a megawatt being a million watts), others were willing to settle for three-quarters of a horsepower to pump water for dying African cattle in the Sahel. We encountered an almost populist belief that good old ingenuity can succeed where the scientists and the bureaucrats only soak up taxes and issue reports. Hippie communes swear by solar energy; so does that establishmentarian pillar, NASA, which has generated more electricity in space with the sun than we have yet to produce on earth. Every science has its fringe; solar energy's is a large one.

It draws scientists with a conscience. We encountered them

widely. Young engineers explained they went into solar energy to be able to do something for society and, at the same time, to avoid a negative tack like antipollution research. They were good to talk to, they were welcome halts in our journey. It's too bad that there are not more of them.

Nearly all the people of all ages involved in solar energy have a quality of the dreamer about them. Professionally, they are realists. They do their homework, their data are sound, they test their assumptions with mathematical models, they do not expect too much from their materials or from the sun itself at their latitudes. Then they emerge from their calculations and their computer runs. They look at the dwindling store of fossil fuels, the mounting level of environmental constraints, and wonder why an infinitely renewable source of clean power should not make at least some contribution.

One of them is Harry Lustig, professor of physics at the City College of the City University of New York. When he was commissioned to write a report for the United Nations Educational, Scientific and Cultural Organization on solar energy, he did not have to look far for inspiration. "As these words are being written, in June 1973, in an air-conditioned room," Lustig stated

> the radio has just announced an 8 per cent reduction in the voltage being supplied at the mains, due to temporary excess demand on the generating capacity of the Eastern Seaboard of the United States. A major oil company has launched an advertising campaign urging people to drive their cars less and more slowly to save gasoline. A few days before, the President of the United States sent another special energy message to the Congress and to the people of the country. The author's desk is piled high with scientific and popular magazines which, over the past few months, have devoted special articles or entire issues to the energy problem.

Lustig reviewed the past, present, and future of energy consumption as so many others were doing at the time. Then he quickly turned to sources of renewable energy, as contrasted to fossil fuels. First, there is sunshine; then geothermal energy "conveyed to the surface of the earth from the interior by the conduction of heat and by convection in hot springs and volcanoes"; finally, the energy of the tides. Lustig concluded:

> The total power influx into the earth's surface environment is almost entirely (99.98 per cent) due to solar radiation; the sun's contribution to the earth's energy income is 5,000 times more than that of the other sources combined. This does not necessarily mean that solar energy is therefore the only or even the major candidate for becoming a source of renewable energy but the latter conclusion does, in fact, appear to be correct.

Nor does this mean that energy from the tides or heat from the center of the earth cannot be used in special cases. In fact, they now generate far more electricity than anyone has ever got from the sun. The tidal plant on the estuary of the Rance River in France has a capacity of 320 megawatts and there are other promising sites, says Lustig, with potential capacities from 2 to 20,000 megawatts. Still, the tides' total potential for power amounts to less than 1 percent of what the world used in 1970. I have seen the figures stated differently: if a dike were built around the entire coastline of the United States, the power generated by the tides could only meet the needs of the city of Boston.

Geothermal energy also looks better locally than globally. In the Lardarello region of Italy, it has been used since 1904 and the plant there has a capacity of 370 megawatts. Two other main areas are the Geysers in northern California and Wairaki in New Zealand. However, the power now stored in the world's major geothermal areas, if converted at 25 percent efficiency, would only amount to what the United States used in 1970. Lustig observes that this is not really renewable energy. If all of it were to be used up in the next fifty years, the contribution to the world's annual power load would be even less than that of the tides. He does allow for the possible success of recent suggestions to tap geothermal heat by drilling deep into the earth's crust instead of sticking to hot springs and volcanic areas where the heat is more or less delivered to the surface.

Now for the sun. We have all seen figures showing just what proportion of the state of Arizona or the Sahara Desert would be needed to provide all the world's power for all time from sunshine. Lustig, in his report, did not quite put it that way. He started with the "solar constant," the flow of solar energy that arrives at the earth outside the atmosphere. Most recent measurements fix it at 1.353 kilowatts on an area of one square meter oriented at a right angle to the incoming sunlight. If all of this incident radiation reached the surface of the earth, one-quarter of the solar constant, or 338 watts per square meter, could be absorbed on an average. The reduction occurs because only half of the earth receives any sunlight at any given time and because, even then, at most places and most times, the light comes in not at a right angle but at a slant. Still, the total for the earth's surface would amount to 172,000,000,000 megawatts. This is more than twenty thousand times the rate at which the world was using energy in 1970 and about one thousand times the power that, according to some estimates, is likely to be needed in the year 2050.

However, not all of the incident radiation reaches the surface of the earth and, needless to say, only a fraction of the energy that does manage to get through can be collected by man and put to work. About 34 percent of the incident solar energy is immediately reflected back to space by the clouds, the atmosphere, and the earth. This leaves us with the often-heard figure of one kilowatt per square meter (one square meter equals 1.2 square yards; a kilowatt is 1,000 watts, the equivalent of ten 100-watt light bulbs, or about one and one-third horsepower), which should be handled with care. It refers to the maximum reaching a spot on the earth when the sun is directly overhead and includes diffuse radiation received from the atmosphere.

Another 19 percent of the incident solar energy is absorbed by the atmosphere, leaving 47 percent to be absorbed by the land and the water on the earth's surface. Of all this absorbed sunlight, about half (or, if you prefer, 22 percent of the total solar constant) fires up the heat engine that runs the hydrological cycle to give us rain, snow, and running rivers. A smaller fraction, less than 5 percent of the solar constant, drives the winds in the atmosphere and the currents in the ocean. They blow and flow because the earth gets more sunshine at the equator than at the poles, thereby creating density differences (expressed in their most familiar form by barometric pressure lines on the meteorologists' charts). An even smaller figure, 0.2 percent, represents what is captured by the chlorophyll of green plants. Through photosynthesis, this quantity becomes "the essential source of energy for the growth of all living matter (as well as, in a minute fraction, the source of fossil fuels)." The remainder of the absorbed sunlight, 20 percent of the solar constant, is reradiated into the atmosphere as infrared waves which we cannot see. These waves get their names because they are longer — the distance from crest to crest is greater — than the wavelength of the color red, the upper bound of the visible band of the spectrum of energy coming from the sun. Still, they are not very long — about .002 to .008 millimeters for infrared radiation from the earth at ordinary temperatures while the wavelength of visible light runs from about .0004 millimeters (violet) to .0007 millimeters (red).

All the components of sunshine can be turned into power. Lustig remarks that the classical approach is to "intercept a tiny fraction of the 47 percent which would otherwise go to heat the land and water surface and the atmosphere and to use it either directly as heat or convert it into mechanical or electrical energy." He gives his own estimate, as good as anyone else's, of how much power is involved here: "With a collection and conversion efficiency of 10 percent, something like 2 percent of the land area of the U.S. would suffice to meet the country's total energy needs into the year 2000."

Next comes the hydrological cycle, particularly the flow of rivers and streams, in scales running from Valaine Pond to the New Aswan Dam. The total potential of waterpower amounts to a full third of the energy the world now consumes. The trouble is that only 8.5 percent is now in use, and the three regions where the rest awaits takers — Africa, South America, and Southeast Asia — are not yet industrialized to the point where they need it. Lustig mentions a variant of hydroelectric power which has recently been mooted. Large bays or inlets could be dammed at their mouths and allowed to evaporate so their surface would drop *below* sea level. Then seawater running over the dam could produce power in a process that we will encounter later on.

Lustig may have overlooked the importance of the wind, certainly an effect of solar heating. "While windmills could be resurrected as a source of small power in certain locations," he said, "wind energy cannot be relied upon on a large or global scale." I think he would find disagreement here in a number of quarters. The wind has a number of things going for it as an energy source. It is cheap, it is widespread, and, principally, it is plausible: we've been using it all along. The Scientists' Institute for Public Information relied heavily upon the wind when they produced their alternative projection for a year 2000 without breeder reactors. They thought that the wind might produce 170 million kilowatts, just about as much as they hoped to get from the sun through photovoltaic cells and steam-driven turbines. One argument against windmills and solar collectors is the aesthetic pollution they would inflict on us if introduced on a large scale. The argument is advanced most often by the wonderful folks who have given us high-tension lines, the Jersey flats, and the Ruhr Valley. I like the wind myself, I think man was making progress when he went from the *Santa Maria* to the *Cutty Sark* which could run 15 knots and develop 3,000 horsepower with no more fuss or stink than a toy sailboat.

The sun drives the winds and, as we have seen, the currents of the oceans. Another source of power lies in what is called the "thermal gradient" of the sea, the difference in temperature in tropical areas between the warm surface and the cold deep water which forms at the poles and creeps along the bottom towards the equator.

Then there is the possibility of using the sun the way we have been using it all along:

Recommendations have been made in several countries for increasing the solar energy harnessed through photosynthesis, by growing suitable trees and plants in special solar plantations and recovering the energy either through burning or the creation of synthetic fuels. . . . It is not yet clear which of the many schemes proposed for the harnessing of solar energy will prove to be the most efficient or how soon solar energy will be exploited on a large scale. But it is clear that among all the renewable sources of energy, the sun is the only serious contender for massive and long-term exploitation and that it is, in fact, the only real alternative or supplement to breeder reactors or fusion for dealing with the energy crisis.

Lustig is among those who believe that government support of some kind (even if orders of magnitude lower than that given other power sources) will be needed for solar energy to become economical.

There is little doubt that further technology developments and mass production will lower the cost of solar energy utilization and even less doubt that there will be a sharp rise in the price of conventional fuels as their supplies run out or their use becomes restricted for political or social reasons.

But it would in this author's opinion be foolhardy to sit back and rely on the latter phenomena, because the crossover point might then be reached at a time and at a price which would have disastrous consequences for the world's standard of living.

It may even be insufficient to limit the positive effort in favor of solar energy to publicly-supported research and development work, no matter how intensively carried out and how generously funded. The reason is that the required major cost reduction may not occur primarily through technological improvements but simply through mass assembly line production. . . .

Since industry is not likely of its own accord to invest in setting up expensive mass production facilities and schedules unless there is an assured market, and since such a market is not likely to materialize until the cost of the solar machinery is brought down, governments may need to intervene to break out of the vicious circle by subsidizing either the production or the consumer acquisition of solar devices.

The economics of solar energy is not just a matter of cents per kilowatt-hour (the amount of energy represented by the consumption of one kilowatt of power during one hour and also the unit used by utilities for billing purposes). On the market, Lustig tells us, this price can run from two-tenths of a cent at a big power plant to $70,000 when the electricity is coming from a watch battery. In other circumstances, it is impossible to set a market price. "There are places where there is at present no fossil fuel or nuclear energy available and where solar energy may be the best or the only hope for relief from suffering. In the drought-stricken zone of West Central Africa where millions of cattle have died and humans are starving to death for lack of water lying underground, fuel and maintenance-free solar pumps might just possibly be more effective and practical than diesel-powered ones, for which the fuel has to be transported over 500 miles of desert and which may become useless at the first breakdown."

Then there is the ultimate argument on the sun's side. Certain forecasters see growth limited not so much by a shortage of energy as by an excess of pollution.

This suggests that, in spite of the fact that the present impetus for solar energy development in the industrialized countries appears to be fed by concern with the exhaustion of the supply of other fuels, a halt to the increase in the consumption of energy will have to occur for other compelling reasons, long before some of these supplies are exhausted. . . .

If that happens, it will not be the renewable aspect of solar energy but its non-polluting nature which will make it an essential source. For it is not only air and water pollution which may call a halt to exponential growth before anything else does, but it may very well be thermal pollution, the disposal of waste heat, which will limit the use of fossil fuels and nuclear energy. The burning of these fuels releases their stored potential energy and converts it into heat. The solar energy which falls on the earth enters into the heat balance whether it is temporarily used by man for work or not. It is the only source of energy which does not contribute thermal pollution.

Such were the dreams of Harry Lustig while he was consultant on solar energy to a United Nations agency. Then he returned to City College where he became dean of the College of Liberal Arts and Sciences as well, a job that brought him up against student energy. His neo-Gothic office was in a state of constant swirl when I saw him in New York but he always had time in between committee meetings and campus crises to talk of the state of solar science and scientists. He is one of several who have become deeply concerned and probably will remain so until the end of their days.

The question now is whether or not solar energy will enter our lives the way it has entered theirs. It is not a simple one to answer. We think about tomorrow but we act today. As long as we are given no new choices, we will keep doing things the same way. We can talk to our heart's content about using the sun, but we will start to use it only if we see it in the window or on the shelf when we go out to buy a heater or an air-conditioner. In these pages, we have tried to concentrate on people with their eyes on today. There are others who speak of the year 2000 or 2025, with solutions to intercept rising curves of energy use and population growth. What is worrying is that others were speaking the same way in 1925 and 1950. We certainly need the visionary scientist who can foresee the day when plants will convert sunshine directly into hydrogen as much as we need the manufacturer trying to sweat a few tenths of a percentage point of efficiency out of a sunshine collector. Yet it is the manufacturer who will come up with a product that can be introduced to start the process of change today.

One forecast can be made safely. We will see the sun warming and perhaps cooling buildings long before it will generate electricity on a large scale. Except for scattered initiatives by individual homeowners, these are likely to be public buildings at first. This is another safe forecast for several reasons. Even before the Mideast war, several such buildings were either being planned or under construction, and the National Science Foundation started to finance experimental heating of schools during the 1973–74 winter. It is much easier for governments at any level than it is for private citizens to pay a higher purchase price for a building in the expectation of lower running costs later on.

It may also be that the rich will follow. In our travels, we came across a manufacturer of solar water heaters and central heating systems in southwestern France. To his astonishment, he had received inquiries from Andorra. Wealthy homeowners living in that tax haven wanted to make certain they would not be at the mercy of a subsequent run on oil. By installing solar central heating, they could keep their tanks full all winter long, buying in the summer and running their solar systems while the less fortunate would be scouting around for oil in cold weather. They also reasoned they could make a profit. By stretching their supply with the sun, they could heat with oil bought at last year's price. In fact, the contents of their tanks would appreciate as the months went by. This is not the way the average man reasons, but this is why the average man is not rich.

Solar energy may appear for the same reason that the automobile did during its early days. It can provide the fortunate individual with a way to liberate himself from the inflexibility of a centralized electric power system, gas network, or oil distribution service. The automobile liberated the rich from public transport until everyone was able to buy a car. One could imagine the rich leading the way with solar homes as they did with Hispano-Suizas and Pierce-Arrows. The sun has a number of things on its side: it is strike-free, it is not unionized. It fits into the behavior of those on the extremes of society, whether the extremely rich or the ecofreaks. Both can serve a social purpose by prying the market open so that the usual incentives will appear and another major change can be made in our way of daily living.

What seems to be established is the growth of the area in which solar energy is applicable. Not so long ago it was limited to Saharan oases, hundreds of miles from the nearest source of fuel. Now the circle is opening. One hears of retired people living in the Rockies and heating their cabins with propane at a cost of over two hundred dollars a month, a whole Social Security check going up the flue. Stunned New Yorkers write to their newspapers about the new bills for running their all-electric homes: three hundred or four hundred dollars a month. These, too, are extremes and they have appeared only recently. They represent the outskirts of the energy market and here the sun has a chance to move in.

There is a surprising community of interests between solar energy and the electric power industry. Nuclear power plants now being installed are to be run at full throttle. There will be no reserve to meet the sudden demands of air-conditioning on a scorching summer day or heating in the winter. Additional plant capacity must be kept in reserve for these peaks, and this is expensive capacity. It is all to the interests of the utility to spread demand in any way that it can. Solar devices are a good way because they always incorporate some sort of storage system. This system can be used when everyone else is drawing on the power company; conversely, it could be charged up with heat or cold during off-peak hours when electricity costs a sliver of the standard rate. Power companies are expressing interest in such systems and it may well be that the *bête noire* of the ecologists, the nuclear reactor, will be instrumental in bringing solar energy within the reach of the homeowner.

Such changes are impossible to grasp, they are beyond our comprehension. I can be an apostle of low-energy living when I am on a bicycle, getting miles from the protein and carbohydrates grown by my neighbors, using firewood to get through the mild winters of the Atlantic coast of western Europe. There I can believe in the power of the sun, a collector on the roof might easily give us the ten or fifteen degrees we wrest from the fireplace. But the process of researching a book like this one discouraged me. We went from motel to hotel, never out of sight of a thermostat. From a bike-riding, wood-heated eccentric, I turned into another particle accelerating through airports, rent-a-car agencies, turnpike tollgates. It was madness to imagine another way to live, I could not move a muscle without jet fuel and high test gurgling into me. Then

I stopped, I ceased to be a privileged particle, I decelerated. I sit in the kitchen in Brittany, the stray cat that has adopted us takes E. B. White's advice as it waits for a saucer of milk in the warmed granite of our doorway. I forget the thermostats and the freeways, there must be other ways.

II.

~~~~~~~~~~~~~~~~~~

The Fiery Furnace

~~~~~~~~~~~~~~~~~~

IN A FOLDER printed in full color and four languages, the town of Font-Romeu in the eastern Pyrenees of southern France tells all that it offers as a vacation resort, summer and winter: fifteen ski lifts, sixteen miles of ski trails, a gambling casino for *après*-skiing, two skating rinks, three heated swimming pools, a riding academy, a forest in the environs, eleven tennis courts . . .

Font-Romeu has sunshine. Twice as much, says the folder where two suns glare out from the logotype on the cover. It lies at the "ideal latitude of 42°30 (the same latitude as ROME and CORSICA)" (and also BOSTON and KALAMAZOO) and at an altitude of 5,400 feet to combine the "blessings of mountain air and Mediterranean weather." In French, Spanish, German and English, we learn that Font-Romeu is the CAPITALE SOLAIRE, CAPITAL SOLAR, SONNEN-HAUPTSTADT, METROPOLIS OF SUNSHINE. It receives three thousand hours of sunshine a year and, to sew up the sales pitch, it is the home of the world's biggest solar furnace, "planted in a site with a maximum of sunshine and a particularly limpid and dry atmosphere." Among the folder's half-dozen illustrations, the furnace gets the same play as the ski runs, the bright night-life lights, and the bikinis around the swimming pools. This bears out our earlier remark: if Font-Romeu had been the home of an oil refinery, no one would have breathed a word about it.

We made our first trip there in mid-March of 1974. We traveled six hundred miles south by car from Paris down the *Autoroute du Sud,* an illustrated lecture on the present-day climate of France and other industrialized countries in these latitudes. There was smog over Paris, rain over Burgundy, stubborn clouds above the Bresse country leading to Lyons; then we crossed a climatic Mason-and-Dixon line at Valence, gateway to the Midi. From here, the road ran down the narrow Rhône Valley, the air whitish-gray with sun and petrochemical haze, the fruit trees still shivering in their winter nudity. We turned east out of the rich valley and over some badlands to Nîmes, old Provence covered with new vineyards to replace the lost wines of North Africa. At Narbonne on the Mediterranean, we veered south again. Now we were on the main route of the great annual summer trek to the Mediterranean beaches of Spain, a solar energy industry if there ever was one. From England over the Channel and through Calais, from Scandinavia and Germany down through Switzerland and the Alps, from Paris and the gray, decaying cities of Belgium, all of northern Europe funnels into this highway. It was lined not by whitened skulls of steers or abandoned Conestoga wagons but by the usual sad milestones of modern travel: service stations and concrete hotels, fruit stands and bars to help the migrant survive a half-hour crawl in the sun through a traffic light at a village crossroads. It was with great relief that we turned east again away from the coast. We were just north of Perpignan and just inside the frontier of Roussillon, the name given to the French side of the land of Catalonia that spreads across from the Spanish border.

The dusty coast changed to a countryside. Cherry orchards took cover behind powerful plane trees that sheltered the road, their trunks permanently bowed by the *tramontane* wind that screams down from the north. In the villages old ladies in black chatted on their doorsteps, old men brought chairs out on the road to sit in the sun, here and there a mule plodded in the shafts of a tank cart.

We followed the Tet River into the town of Prades where Pablo Casals had spent the early years of his exile, out of Spain but still in his Catalonia. Then the road began to climb along the strategic route laid down by Vauban, builder of fortresses to Louis XIV, when he protected the new border with Spain. We rose on a steady grade, never steeper than a team of oxen could handle with a field-piece behind them, but never slacking. At Villefranche-de-Conflent, the first of Vauban's fortified strong points, we were joined by a narrow-gauge railroad that leaped boldly across ravines while the road wound around the contours.

It was getting late on a Sunday evening as we breasted a cease-less stream of skiers coming back from Font-Romeu to Perpignan on their last downhill run of the day, a single file of cars punctu-ated by buses. We passed Mont-Louis, the site of the first solar furnace in the region. The traffic thinned, we were in the forest above Font-Romeu. The road was rough and scarred at the end of the winter, snow banked high on both sides. We cleared a pass and dropped down to the hermitage of Font-Romeu, formerly a monas-tic haven, now secularized with a Hermitage Grill and a Hermitage Nightclub inside its walls. Next came the chalets and the high-rise condominiums of Font-Romeu that had been spewing forth the skiers. Still losing altitude, we eased down to Odeillo, once the main settlement at Font-Romeu but now outnumbered in winter by the skiers and the *après*-skiers. And then, coming out of a hairpin bend in the last flicker of twilight, we saw the solar furnace.

It was a vast silver oyster shell, a parabolic mirror reflecting the mountains upside down at sunset. As we moved, the mountains moved inside the mirror, a psychedelic image for the dead-straight and the cold-sober. I had read up on the statistics. The big oyster shell measured 130 feet high and 175 feet wide, all of silvered glass but not of a piece, no — myriad pieces, 9,500 mirrors each measuring 17.7 by 17.7 inches and bent under the pressure of screws to fit the curve of the parabolic dish. The statistics could not encompass the total effect of that silvered glass standing in the green valley. Why, it took a building nine stories high just to hold the dish upright. The building itself was no great shakes, the usual steel-and-glass exercise in Bureaucrat Baroque, but it looked like someone had gone to work on its nether side with a gigantic ice-cream scoop to make room for the dish.

Nor do the statistics end there. The parabolic mirror covers about 2,400 square yards and focuses the sun down to a spot two feet across. This mirror does not follow the sun across the sky; that would be a trial for such a stretch of vertical real estate. Instead, the big dish stares fixedly at a field of moveable flat mirrors, sixty-three in all, spread out in eight terraced rows on the mountain facing the mirror and the Spanish border just over the next line of peaks to the south. These moveable mirrors add up to 3,100 square yards. They move as the sun moves, catching and reflecting it into the fixed dish, taking the energy falling on the mountainside and concentrating it into 1,000 kilowatts of heat in a space you could easily embrace with your two arms. Not that you would care to: the center of the hot spot goes up to a temperature of 3,800° C., more than half the 5,500° C. on the surface of the sun itself.

The history of the furnace is inextricably linked to the life of Dr. Felix Trombe, director of the solar energy laboratory operated at Odeillo by the French National Center for Scientific Research. I had first met Trombe in 1958 when he was at Mont-Louis, six miles down the mountain. He was working inside a seventeenth-century citadel with a mirror thirty feet in diameter and producing 75 kilowatts of heat, the biggest in the world at the time. He was wearing *espadrilles* and old clothes that hardly went with his status as a Sorbonne professor and a laboratory director in the status-conscious world of French science. When we saw him in Odeillo, he was still in *espadrilles* and his clothes, while not the same, were just as old. He was tall and lean in 1958, agile of mind and body. The intervening years had neither slowed nor thickened him even though he celebrated his sixty-eighth birthday while we were at Odeillo.

Probably the principal reason why the solar furnace is in the Pyrenees is a sentimental one: Trombe himself is from the Pyrenees. His family came from Ganties, a village one hundred and ten miles to the west in a region known as Ariege, not a smiling land of sunshine and vineyards like Catalonia. In his office in the solar laboratory, Trombe let his mind wander back to the mountaineers of Ariege. He told us how a grandfather left for Cuba to seek his fortune in a more likely setting. One of his grandfather's relatives was a bear hunter. In those days, the hunters would kill a mother bear, take a cub and raise it. When the bear was grown, a ring would be placed in its nose and it would set out with its master for a tour of the surrounding towns. The bear would dance and the master would pass the hat. "People gave money," said Trombe; "they were afraid of the bear." He showed us photos of the bear hunters and their dancing wards; he had publications from Pyrenean historical societies with details of local costumes. "I have seen sheep killed by a bear a hundred meters from where we were camped but I have never hunted myself. Over on the border of Andorra, one hundred hunters will go out to kill one bear. It's slaughter. I hate hunters."

He thinks he might have remained on a farm in Ariege like his father if it had not been for a childhood accident. His father raised horses. When Trombe was six, he rode a pony in a race during a Mardi Gras carnival, bearing his father's colors. The pony threw him into the mud. "I can still see my father, beating that pony." After that, the family did not encourage him to stay on the farm. At the same time, Trombe had a cousin who had become a chemist and an example.

The family sent him to Catholic schools, first in the Pyrenees and later in Paris. "I was really impregnated there. Of course, if I had gone to a Protestant school, it would have been just as bad." In Paris, he also came up against classmates who made fun of his southern accent. He may have gained revenge on them later in life by staying in the Pyrenees and spurning the academic honors that Paris could have offered him.

Although he devoted a great deal of time to horses (he still owns and rides them), he did well enough in school to win a scholarship to the Sorbonne where he received his doctorate in chemistry in 1936. In 1932, the French National Center of Scientific Research was created and Trombe had joined its staff the following year. "I've stayed there all my life because I am conservative by nature.

And now I only have two more years to make my colleagues miserable until I retire."

Trombe worked as a young chemist for Georges Urbain who specialized in the so-called rare earths. These are oxides of the rare-earth metals, fifteen in all, that occupy the same row in the periodic table of elements because of their chemical similarities. In the table, they run from lanthanum to lutetium; they are not only relatively rare but they are extremely difficult to separate. Trombe displayed a talent for this work in which he collaborated with his wife. A standard book on the subject, *Chemistry of the Lanthanons* by R. C. Vickery, published in 1953, gives him six references, only two less than Urbain, his patron. The study of the rare earths has often taken on the atmosphere of a race for first place in the discovery of new elements. Credit for isolating one of the rare-earth metals was hotly disputed by workers in Italy and the University of Illinois. Some called it florentium, others illinium, now it is diplomatically known as promethium. The stakes in the race are not merely academic. More and more uses are being found for the rare earths, whether in metallurgy or in the high-temperature applications of jet turbines and rockets.

By 1938, Trombe was working with Marc Foex, his long-time associate who now directs the Laboratory of Ultra-High Temperatures that shares the Odeillo building with solar energy research. The following year war broke out, and Trombe was mobilized into a horse-drawn artillery unit. He saw enough combat duty to be wounded and taken prisoner in 1941. He was invalided out of his *stalag* and back to Paris, his long frame whittled away to not much more than a hundred pounds. There he spent the war years, in science and the Resistance. "Oh, so very little, we just tried to help people, we gave them certificates. I never liked that last-minute Resistance movement."

With the end of the war, new uses appeared for the rare earths. Frédéric Joliot-Curie was interested in dysprosium as a "neutron trap" in nuclear physics research. At that time, the rare-earth and high-temperature laboratory was behind the Sorbonne in Paris. Joliot-Curie moved it to Bellevue-Meudon in the wooded hills west of Paris and asked Trombe to take it over. "He gave us a problem; we had to produce a pure oxide of a rare earth, zirconium oxide. This substance has a very high melting point. In the lab, we used an electric furnace to heat it to 2,600° C. But our oxides were soiled, they were contaminated by the furnace."

Looking for a pure source of heat, Trombe turned to solar energy. He was not the first; Lavoisier had preceded him by nearly two hundred years. Like Lavoisier, Trombe was a chemist. "Physicists made mirrors, but they did not know what to do with them. We did; in fact, as high temperature chemists, we had too much to do. From 1945 to 1949, we worked like slaves at Meudon."

Trombe and his principal collaborators, Foex and Charlotte Henry la Blanchetais, found time to look back into the origins of solar energy as a tool for the chemist. The result was a long contribution by Trombe on solar furnaces to a volume published in Paris on high-temperature chemistry. Even in those days, Trombe was ranging beyond the confines of his discipline. There was a hint of the dreamer in his introduction:

> Solar radiation is at the very origin of most of the sources of energy that man is likely to harness and use on our globe. Coal deposits, the result of the evolution and fossilization of plants, petroleum, hydraulic forces, winds, etc. all depend upon initial activity by the sun. We shall see that the energy radiated over the surface of the Earth by the Sun is infinitely greater than that which one can produce with the help of all the aforementioned means.

Trombe, like many others, traced the first use of solar energy back to 215 B.C. when Archimedes was said to have burned the Roman fleet attacking Syracuse. One version of the story maintains that he used a big hexagonal mirror, another that he lined up soldiers carrying polished shields which they then focused upon the Roman men-of-war. Trombe does not think it could have been done with convex shields; there is no way in which they could have achieved the focal length needed to burn the sails and hulls of the enemy fleet at any distance. He believes that Archimedes could have used brazen mirrors, flat and highly polished just like those of the boudoir. He might have indicated the target to one soldier who then flashed a beam against the side of a ship. Then the others would have zeroed in on the spot with their own mirrors. "Two or three hundred could have done it at a distance of a hundred or a hundred and fifty yards," he said. This is exactly how the flat mirrors on the hill at Odeillo concentrate the sun's image into the big dish which also had its predecessor in antiquity. To concentrate the sun's rays so that they could light their sacred fire, the vestal virgins used a bowl of polished gold.

By the end of the seventeenth century, more profane uses had been found for the sun. In Florence in 1695, an experiment was performed for Cosmo III, the Grand Duke of Tuscany. A diamond, cut and polished, was placed in the focal point of a large burning glass. The sun triumphed. Wrote Trombe: "The diamond, a substance held until then to be immutable, became dull and tarnished, lost its weight, and finally vanished completely."

Concave mirrors were used as well as burning glasses. Trombe describes one of the biggest, the forty-three-inch mirror that Cassini, head of the Paris Observatory, offered his royal master, Louis XIV. The mirror is still in existence and, according to Trombe, it is used to melt a silver coin whenever royalty visits the Paris Observatory (assuming the sun comes along). With this mirror, a contemporary chronicler related, "wrought iron is melted in the time of two seconds and silver becomes so hot that, when dropped into water, it takes on the form of a transparent spider's web." Even bigger mirrors were made by fixing sheets of brass onto a wooden frame. The chronicler continued: "Heat at their focal point is so great that asbestos is turned into yellowish glass and talc into black glass. . . . All minerals and stones undergo vitrification, some quickly, others more slowly."

The fever of experiments in this first solar age did not abate. Emulating Archimedes, Buffon used flat mirrors to burn wood at a distance of two hundred feet. Trombe related another successful attempt to transmit energy in which two huge concave mirrors eighteen feet in diameter were placed face to face. One fired a coal which was then brought up to white heat by a bellows. The other picked up the glow of the coal and reconcentrated it to the point where the experimenters could light tinder or set off a pinch of gunpowder.

The trouble with the concave mirrors of the day, as Trombe explained in his article, was that they had to look up straight at the sun. This was not at all convenient for the scientist trying to see what was going on. By the beginning of the eighteenth century, lenses had come into favor, some as large as three feet in diameter. They enabled the sun's rays to be focused downward, not upward as in the case of the mirrors.

Lavoisier used such a burning glass for the first truly scientific solar furnace. It was a huge lens composed of two hollow halves, fitted together and bound by brass to form an empty sphere four feet in diameter which was then filled with 140 pints of "spirits of wine" (that is, alcohol), cleaner than water and not likely to freeze. Lavoisier tells us that this lens was mounted on "a sort of chariot that turns horizontally around a fixed point in order to follow the sun; a turn of a crank is enough to change its position; another crank, working two long iron screws, raises or lowers the lens as the sun changes height. A single man, without any fatigue, can exert this double movement even when the platform is loaded with eight or ten persons."

In a solar match, Lavoisier was able to test his giant alcohol-filled lens against a rival. Trombe quoted his account fully:

> On the 5th of October, around one o'clock in the afternoon, the sky not being perfectly clear, we placed a 2-*liard* coin on a coal at the focal point of our lens; about half a minute later, it was completely melted and turned to liquid. The same result was obtained with a large copper *sou;* only a little more time was needed to melt it. Never were we able to treat such large quantities with the Academy lens, even in favorable weather and with the aid of a second lens.
>
> The melting of wrought iron requires much more heat than copper. We were not able to produce the required activity with-

out concentrating the rays even further by using a second lens. For this purpose, we used a solid glass lens eight and a half inches in diameter. . . .

We placed shavings of wrought iron in a hollowed coal at the focal point. Almost instantly, they melted into a perfect solution; the iron, rendered molten in this manner, then started to boil and detonated as if it had been nitrate in fusion; a great quantity of sparks shot into the air, producing an effect of fireworks on a small scale. This effect has always occurred every time that we have used the burning glass to melt cast iron, wrought iron or steel.

When we placed platinum in the form of grains at this same focal point, it seemed to cluster together, to shrivel as if about to melt. Shortly afterward, it boiled and smoked; all of the grains came together in a single mass, but without taking on spherical form as all other metals do. After this half-melting, the platinum could no longer be attracted by a magnet, although this had been the case before it was exposed to the sun.

Lavoisier carried out his experiments between 1772 and 1774. He used the sun to obtain temperatures higher than any previously achieved and without contaminating the substances he was studying. At Bellevue-Meudon, Trombe turned to the sun again for the same reasons and installed his first mirror there in 1946. It had come from a German antiaircraft searchlight in a roundabout way. "A merchant in the Paris Flea Market had been using it as a goldfish bowl," Trombe said. "We bought it as junk." It was installed in a shack on wheels that, like Lavoisier's chariot, could be moved by muscle power to follow the sun. The searchlight mirror measured six feet, seven inches in diameter; it could heat rare earths to as high as 3,300° C. and its power was the equivalent of two kilowatts. In this primitive setting, the junk mirror operating inside a shed of corrugated iron, Trombe received many visitors. One was Dr. Chaim Weizmann, the first president of Israel and a noted chemist. Trombe recalls how, in 1949, a procession of black official cars brought Weizmann out to the high-temperature laboratory in Meudon where the two chemists had a chance to talk. This was the start of Trombe's relationship with science in Israel; later, he helped raise money in France for the solar laboratory set up in Israel by Harry Tabor. Tabor and Trombe are charter members of the world's exclusive club of first-rank scientists who have stayed with solar energy even though they have been figuratively burned on several past occasions.

In 1949, Trombe also began his return to the Pyrenees. It was an involved process, the result of still another of his interests. Along with everything else, he had been a confirmed speleologist in his younger days. He only abandoned potholing when he rode a jeep off a mountain road into a ravine, putting a crimp in the steering wheel with his knee. Right after the war, however, he was at the height (perhaps one should say the depth) of his underground career. In 1947, as president of the Speleo Club of Paris, he led a party into a cavern in the Pyrenees known as La Hennemorte (*la femme morte,* the dead woman, in local patois). Enthusiastic help came from General Paul Bergeron, then commanding the Toulouse military region. The army supplied transportation and walkie-talkies to the cave explorers. Trombe and Bergeron remained in touch after the general became chairman of the "Scientific Action Committee" of the French Ministry of National Defense. More than one of Trombe's contemporaries has admired his ability to whet the interest of the military in solar energy and to tap their budgets. To the vast detriment of the science, few others have been able to do so.

Trombe and Bergeron kept up the relationship that had started with the descent at La Hennemorte. The general came to Meudon, he looked at the sky over Paris, and he told Trombe: "You can't stay here, there's no sun. I'll give you a citadel." In 1949, Trombe moved his solar laboratory into the old fortress at Mont-Louis and that was where it remained for nearly twenty years. The setting had to be seen to be believed. Behind the massive stone walls of the citadel, Trombe erected a parabolic mirror thirty feet in diameter on one side of a courtyard. Opposite in a corner stood his "heliostat," a flat mirror forty feet wide that moved like some photophiliac beast to catch the sun and direct it into the paraboloid. The mirror was used to heat zirconium oxide to 3,000° C. and produce zirconia, a refractory material used to line electric furnaces. All this went on inside Vauban's citadel, framed by roofs of round tile, where scientists led the life of a mountain garrison.

Almost from the very start, Mont-Louis was regarded as a temporary expedient and Trombe sought a site for a much bigger plant. The choice narrowed down to two regions, the eastern Pyrenees around Mont-Louis and the southern edge of French Alps. Trombe chose the Pyrenees. "We're further south here, perhaps the sun has more intensity. Perhaps, too, it is because I am from the Pyrenees myself and I wanted to stay close to home."

For seven years, Trombe roamed the region by jeep, looking for a site for his next furnace. Finally, he chose Odeillo. The microclimate there was better than at Mont-Louis and awaiting him was a mountain facing due south where he could install his sixty-three "heliostats" with a minimum of earth-moving. He played his cards close to his vest, but word got out when he made up his mind and the price of land on the site went up a hundredfold. Farmers had been holding onto their pastures in the hope of raising skiers instead of sheep and they needed some persuasion before they would part with it. They were suspicious of the newcomers; they called them "the sorcerors of the sun" and later blamed them for bad weather. Some locals feared the big dish might set fire to forests a mile and a half away. This was not likely because the mirror's focal length is only fifty-nine feet. Nevertheless, the edge of the dish will catch the setting sun between the spring and autumn equinoxes, the 21st of March and the 21st of September. We saw it on a June afternoon around 6 o'clock and we were warned to keep clear of a bright patch near the base of the mirror where the temperature was a good 200° F. Trombe had left standing orders that no cars were to park there in May or June, for the hot spot had already exploded the tires of a car while the dish was being built. Most of the time it only scorched the grass and it was not nearly as troublesome as the mirror inside the old citadel at Mont-Louis. On occasions, its eastern edge would focus the sun down onto a wooden platform, setting the floor on fire and bringing out the staff with buckets and firehoses.

At the time, Trombe had other worries in Mont-Louis. "For mysterious reasons that I knew very well, the military lost interest. The technocrats were against us, the nationalized gas and electric utilities were against us, everyone was against us. It was an un-believable *pastis,* that's how you get buried. If I had not had my reputation as a rare-earth scientist, we never could have built Odeillo. But I protected myself, I stuck to rare earths. And we came here in 1969."

In the United States, too, plans were made to construct a thou-sand-kilowatt solar furnace but they were never carried out. I heard about the American project from Claude Royère, the young chemical engineer on Trombe's staff who is directly concerned with the management of the furnace. Royère, in his mid-thirties, was graduated from the French Higher National School of Chemistry in Bordeaux in 1965 and chose to do research with Trombe instead of taking a better-paying job in industry. Though not a Pyrenean, he liked the idea of living and working in fortified Mont-Louis. After the laboratory moved to Odeillo, he settled there himself, taking a two-hundred-year-old house for his family.

American scientists, Royère told me, decided to operate their thousand-kilowatt solar furnace with a single flat, moveable mirror, a monster that would have measured 160 by 200 feet. A feasibility study was made, the project was deemed too costly and abandoned. "They spent money on a feasibility study, we spent two million dollars on a solar furnace," said Royère with a trace of satisfaction. At Natick, Massachusetts, a furnace about the same size as the French operation at Mont-Louis was built in 1958. But when the time came to go up to a thousand kilowatts, a million watts, a solar megawatt, the French were alone.

Work moved slowly. By 1968, the mountainside at Odeillo had been carved into the eight terraces upon which the sixty-three move-able mirrors would stand opposite the big parabolic dish. In those mirrors there are 11,340 silvered panes and they must be parallel.

Insofar as humanly possible, they act as one, as a single vast sheet when they catch the sun.

We heard how they were lined up from Jean-François Tricaud, a young physicist at Odeillo who is in charge of solar radiation measurements there. Tricaud was almost born into solar energy. Trombe is his mother's brother. His career, however, was his own choice: "Trombe has other nephews, but they're not in solar energy." In 1968, after a year at Niamey in Niger where he worked on an early program to set up solar-driven irrigation pumps in the Sahel, Tricaud came to Odeillo as the moveable mirrors were being installed. It took a year and eight months to do the job with the help of two teams of surveyors who came down from Paris. Each flat mirror consists of 180 individual panes and each pane had to be regulated by the turn of a screw.

Next came the setting of the parabolic dish nine stories high. Here, there are 9,500 panes that had to be bent ever so slightly, again by screws, until they were aimed at the focal point of the furnace. It all had to be done from behind on the terraces of the laboratory building with a light meter at the focal point to act as a target. Once the panes had been bent, the first trials were run. A steel plate three-eighths of an inch thick was placed in the furnace. The heliostats caught the sun and turned it onto the dish. All the energy falling onto the mountainside was sent into the plate, piercing it with a big jagged hole that looked like a shell-burst. The focusing had to be fined down. Gradually, the holes in the steel plates became less jagged, they grew smaller until they were down to thirty centimeters (eleven and three-fourths inches) in diameter. Today, the plates have served their purpose and they are to be found in Trombe's office, mementoes of the early days of the solar furnace when, for the first time since the sun began to shine, its rays were concentrated 16,000 times to the power of a megawatt.

To be sure, this is a thermal megawatt, 1,000 kilowatts of heat not of electrical power. If the mirrors on the mountainside at Odeillo were to be used to generate steam, the end result might be only 100 kilowatts at 10 percent efficiency, say 130 horsepower, hardly what is available under the hood of a compact American car. The lesson here is not the inadequacy of solar energy but the unbelievable flood of power that petroleum represents — and will go on representing until its short life as an earthly energy source is over.

Trombe has mused over the possibilities of Odeillo as a prototype for power generation. "You don't need high-level physics to generate power. We know all we need to know about ways to concentrate solar energy. Odeillo could be repeated. We know it works. With five hundred heliostats, we would have the equivalent of an oil furnace, five hundred mirrors sending five hundred images of the sun onto the same spot. We wouldn't need a parabolic reflector at all; when you want to generate power, you try for six hundred degrees C., not three thousand six hundred degrees.

"The system that steers the mirror is all electronics and that's not expensive. Since the mirrors themselves need not respond quickly, the steering system does not need much energy. We won't use thousand-megawatt plants for solar power. Fifty megawatts will do very well. It's all a matter of mass-producing components. It doesn't require great scientists, just intelligent people. This is not necessarily contradictory.

"You'll see, America will bring out a solar power system for half the price, then they will sell it to the rest of the world. There's no secret, we haven't invented anything. We only introduced the idea of using a number of small heliostats instead of one big one as we had done in Mont-Louis."

Something similar is to be found in one of the oldest suggestions for large-scale power generation. At the World Symposium on Applied Solar Energy held in Phoenix, Arizona, in 1955, the leading solar scientist in the Soviet Union, Dr. Valentin Baum, described his proposal for a solar power plant. Instead of setting the mirrors up on a mountainside, he decided to perch a boiler on a tower 40 meters (131 feet) high. Then he wanted to ring the tower with twenty-three concentric railway lines carrying trains of platforms bearing flat mirrors. Electric motors would keep the trains moving to follow the sun. Baum estimated that with 20,000 square meters (that's almost five acres) of reflectors, he could raise enough steam to generate about 750 kilowatts of electric power and, at the same time, produce nineteen tons of ice an hour in the dry hot climate of Central Asia in the Soviet Union. In winter, the plant could heat a town of nearly twenty thousand.

Twenty years later, the idea has turned up again on a much bigger scale as the "power tower." The version that has been put forth by a group at the University of Houston headed by Dr. Alvin F. Hildebrandt calls for a tower not 40 meters high but 450 meters, 1,500 feet. It would rise from a field of steerable mirrors measuring a square mile in area (in this case, the mirrors would be planted like flowers rather than mounted on railway cars). In effect, they would convert the ground around the tower into a huge lens.

Hildebrandt and his colleagues relied heavily upon cost figures for individual steerable mirrors that had been supplied to them by Trombe. Of particular interest was Trombe's finding that the mechanism guiding the mirrors used only a modicum of the power that the furnace produced. All too often, new energy conversion systems tend to consume more than they generate. Critics of nuclear power enjoy repeating that not until 1971 did the nuclear plants in the United States turn out more electricity than was being consumed by gaseous-diffusion plants used to supply them with their enriched uranium fuel.

In the Hildebrandt project, the mirrors would focus on a boiler on top of the tower standing 145 feet higher than the superslabs of the World Trade Center in Manhattan. This concentrated heat would be turned into electricity either with a steam boiler or a magnetohydrodynamic generator. Like a good many other speculative scientists working in energy, Hildebrandt suggests using this electricity to break water down into oxygen and hydrogen. Then the power can be stored in the form of hydrogen, a handy way to get around the on-again-off-again nature of the sun as a source of electricity. Energy storage is becoming more and more important to the friends of both nuclear and solar power. Sun and wind are too variable, their peaks and valleys must be flattened to fit the curve of consumer power demands. Nuclear power plants, as we have seen, are not variable enough.

Hildebrandt's calculations for a "power tower" economy give an idea of the resources involved in making a major change in the way a modern society gets its energy. He estimates that the power needs of the United States in the year 2000 could be met by mirrors covering between 65,000 and 80,000 square kilometers (25,000 to 30,000 square miles), the equivalent of .86 percent of the land area of the country. As a basis, he assumes a solar energy income of .33 kilowatts per square meter per hour, only half the average sunshine that falls on places like Arizona, southern New Mexico, southern California and west Texas. To cover such an area with mirrors would require 130 million tons of aluminum, well over thirty times America's annual production. To coat them would need four times more glass than the United States produced in 1970; to support them, Hildebrandt calculated, would take only slightly more steel than in all the automobiles on the road in America. Such numbers show, if nothing else, that a shift to solar power certainly would not mean any danger of unemployment.

These are dream figures for a power-tower world. Nothing like them was to be found at Odeillo when Royère took us on a tour of the biggest solar plant that can actually be seen and touched. As often happens in the Pyrenees, it was a rainy day. When we turned up for the tour, Royère was arguing on the telephone with a prospective customer for the furnace. He looked up at us, shielding the transmitter with his hand: "He says he needs forty-eight hours' notice to get his equipment set up. How can I tell what the weather will be like forty-eight hours in advance?"

When Royère completed his call, we left his office and took an elevator down to the ground floor. A huge passage ran from the building proper to the base of the shaft supporting the solar furnace. Most of the passage was filled with experimental gear, crucibles that had been used to manufacture refractory materials. They looked like tall, fat vases. Royère showed us a centrifugal oven that spun at one hundred and twenty revolutions per minute with eighteen hundred pounds of material inside it. Basically, it was a big pot heated from the inside by the sun while water cooled the outside. When it rotated, centrifugal force spread material against the sides of the pot where it cooled and hardened into a glassy crucible. It could not contaminate the material being heated, since it was the same material. Once the contents had been melted, a piston shoved the liquid out of the pot into a water spray that turned it into a powder. One hundred kilograms (two hundred and twenty pounds) an hour could be treated in this manner by the solar furnace which produces very high temperatures in a normal atmosphere and replaces a whole range of electric furnaces. In 1969, a pilot plant was set up at Odeillo to run tests on industrial ceramics and other products used by industry. While Odeillo could turn out one hundred and twenty tons a year of refractory material, its production has been kept down to two and a half tons. "This is a research facility," Royère said; "It's like a racing car. It would be nonsense for us to use it only for manufacturing. No one uses a racing car as a taxi."

Research on thermal shock is an equally valuable use. The solar furnace is a good way to see how the nose cone of a rocket behaves upon reentry because it supplies pure heat while an electric furnace adds the complication of a magnetic field. Then there are atomic explosions. "We can simulate the signal of an explosion at a certain distance. We can open the shutter of the furnace and regulate the exposure time as we go up to 3,000 degrees C. in a few tenths of a second." This work has nonmilitary applications as well. One that Royère mentioned would be the thermal shock effect on the boiler on top of a power tower when a cloud slides across the face of the sun, literally shutting off the heat. The ability of the solar furnace to reproduce such conditions has led to an international research program that is being carried out jointly by Odeillo and the High Temperature Materials Division of the Engineering Experiment Station at Georgia Institute of Technology in Atlanta.

Royère continued his guided tour. The passage from the main building led to the foot of the three-story shaft that supports the furnace proper. We took an elevator to the top, a good 45 feet high. These are tall stories. Each floor of the laboratory building corresponds to a terrace bearing a row of moveable mirrors on the mountainside. Each terrace is separated by 5.25 meters (17.25 feet). This makes for cathedral ceilings in the offices and laboratories of the building where two normal stories could almost be squeezed into one. The building looks cavernous, as if it had been scaled up for a race of giants. On rainy days, it can be gloomy. The gray sky enfolds you through the glass walls as you prowl along high empty corridors. Then the building is like a powerless sunship, adrift in distress, the flat mirrors on the mountainside transmuted from silver to aluminum, a dull sheen on their useless faces staring in every direction. At the end of an experiment, the mirrors are always turned helter-skelter so they will not focus the sun accidentally into the parabolic dish. In this position, they are almost more impressive than when they are on line. To Madeleine, they formed a looking-glass land. They reflected each other, they reflected the big dish and the mountains — the optical permutations were countless. From one angle, they looked like the low houses in the casbah of a North African city; from another, they might have been the glass palaces of Manhattan transplanted on a mountain slope of grass and melting snow. In desperation after a spell of rainy days, Madeleine asked Trombe if he would mind strolling through his house of mirrors under a big black umbrella. Wisely, Trombe refused. It would have been the antisolar picture of the year.

When we left the elevator, Royère took us into a control room where rain drummed on a roof of corrugated steel. It could have been an airport control tower. Big windows of tinted glass faced the moveable mirrors on the hill. On a panel, buttons were arranged in rows corresponding to the terraces on a hill, one set of buttons for each mirror.

"Here is where we tame the mirrors," Royère said. The buttons command hydraulic jacks behind each mirror. They are used in starting the furnace when the mirrors must be turned to catch the sun. Once this is done, an automatic guidance system takes over, locking the mirrors onto the sun. It works with a photoelectric cell divided into quarters. The cell looks at the sun and maneuvers the mirror until each quarter is getting the same amount of light. That is all there is to the system; it is one way in which a solar power plant of the future might track the sun.

Royère apologized for the weather. "We're like farmers, we can only work when the sun is out." In winter, the furnace can be run from nine in the morning until three in the afternoon, in summer from eight to five. It seldom is shut down for maintenance. Of the eleven thousand or so panes that form the moveable mirrors, about one hundred and fifty have to be replaced every year. The air is clean in Odeillo and so are the mirrors, although they had to be hosed down one summer after a red rain. A freak storm had carried dust from the Sahara over the Pyrenees where a cloudburst sent it pouring down.

Since the big parabolic dish is concave, it does not catch the dust. It could be cleaner, though, and that is another problem for the solar power plant of the future. The ideal solution at Odeillo would be a hook-and-ladder truck to hose it down but the local volunteer fire department does not have one. There is one at Toulouse, but the fire department there is understandably reluctant to let it go for a day or two.

Right behind the control room stood the furnace itself behind a pair of big clamshell doors. It was as cold as the rain on the roof. We had to take Royère's word that the doors were the first line of defense against the sun. Once the moveable mirrors are lined up, the parabolic dish concentrates the sun onto the furnace. The mirrors cannot be easily flicked on and off; another way has to be found to stop the solar flux until it is needed. The doors do the job, intercepting the beam from the parabolic reflector before the focal point, the final hot spot. They must still withstand temperatures as high as 600° C. and, from below, one can see the scorch marks on their stainless steel surface.

When it is time to turn the heat on, the doors open. Then the blinding beam enters the furnace, but first it is intercepted by a shutter. Since it had taken that beam no more than a minute to burn a hole in a steel plate half an inch thick, we wondered what exotic alloy had been used in the shutter. None, said Royère; the shutter was nothing but two thin aluminum walls. Water pumped between the walls removed the heat as fast as the sun could deliver it. That was all there was to it. By varying its speed, the shutter could simulate whatever phenomenon had to be studied, whether the split-second exposure of an atomic explosion or the slower heating that would take place in the event of a breakdown in the core of a nuclear reactor.

The solar laboratory has contracts with steelmasters interested in pouring steel into a crucible that need not be heated from the outside. Trombe has spoken of setting up a solar steel industry on the shores of the Chilean desert where there are three hundred sunny days a year. Royère thought that deserts by the sea could also be used to produce hydrogen. Heat water with a solar furnace to 2,000° C. and it breaks down into hydrogen and oxygen in a process known as thermal dissociation. "Any country like Chile or Mauritania could combine sunshine and seawater into the equivalent of an oil field."

The rise in oil price has boosted Royère's morale. "Trombe used to say that solar power was not economic. A solar kilowatt-hour used to cost five or six times more than conventional power. Now it is only two or three times as much. We used to talk about the year 2000 for solar power. We might be getting closer to 1980. I'll be here to see it, I'll be able to work on it. It's not just the price of a barrel of oil, there's the price of pollution, too. People can change; perhaps they will get tired of breathing sulfur."

One way to use an Odeillo-type of installation to generate power would be to triple the focal length of the parabolic mirror to one hundred and seventy-seven feet. At Odeillo, Albert Le Phat Vinh sketched this suggestion for us. Then the mirror would produce the same 1,000 kilowatts but at a lower temperature because the energy would be spread over a much bigger hot spot. Le Phat Vinh envisioned a solar power plant as a series of Odeillos, side by side, rather than as one big complex. "This frightens me a bit. A thousand megawatts would mean one thousand Odeillos . . . why, that's enormous."

Le Phat Vinh is Trombe's oldest associate. Born in Nice of a

Vietnamese father and a Belgian mother, he studied physics in Paris at the Sorbonne. At the outbreak of war in 1939, he was mobilized into the same horse-drawn artillery regiment in which Trombe served. They were taken prisoner and spent a month together before Trombe was freed because of his health. Le Phat Vinh is the quietest and gentlest of men; it was easy to believe him when he remarked: "I had neither the opportunity nor the audacity to escape."

He talked in an office paneled by the photographs that he has taken of his furnace at work. He went back to 1945 when he finally came home to France and saw Trombe again. For the next five years, the two men kept in touch while Le Phat Vinh worked in industry and spent a year in Vietnam. In 1950, Trombe asked him to come to Mont-Louis to help put together the first furnace. "Those were the heroic days of solar energy. Everybody did everything. I and a few others did all the work of setting up and regulating the first parabolic mirror."

His job then was to bounce the sun's energy off the moveable mirror onto the parabolic dish and then into the mouth of the furnace with the least possible loss. He is still doing it. Le Phat Vinh was responsible for the aluminum shutters in front of the furnace. He had to find a way to keep them under control while they opened and closed in three-tenths of a second. To achieve this, he used a linear electric motor; that is, an electric motor with its components "unrolled" flat so that, instead of turning, it moves in a straight line. There is talk of using this technique to run high-speed trains in the future. For the time being, Le Phat Vinh uses it to drive his shutters on a trip sixteen inches long.

"They call me a physicist," Le Phat Vinh said, "but I am really an engineer." He is the most self-effacing of the leading figures at Odeillo. Thin almost to the vanishing point, he wears timeless clothes with no sign of the ski industry's influence. He drives an equally ageless Simca when he comes to work from the wooden chalet where he lives near Mont-Louis. His peers agree unanimously that no one can put the furnace through its paces the way he can.

We had to wait for a chance to see him in action. During our first visit to Odeillo in March, the sun was either absent or hesitant. A cloudy day with intermittent flashes is just as bad as no sun at all. What is needed is a steady flux of energy so that the behavior of materials can be compared against the unyielding standard of the sun. Occasionally, a morning looked promising but one could see the *marin,* the sailor, coming up the valley of the Tet River from the Mediterranean. It was a low cloud that bucked the west wind until it reached Odeillo and sent its gray shoots probing among the moveable mirrors and the interstices in the parabolic dish.

We came back in May for another shot at the furnace. Once more, it was like a vacation on the coast of Brittany or New England. The weather was perfect the week before we arrived and we knew that it would be cold, damp, and drizzling until the day after we left. Reconciled to our luck, we decided to go. It was the Saturday before Whitsunday, the start of a three-day weekend. The solar laboratory was shut tight, the whole staff off on holiday except for a standby crew.

Everyone had left except the sun. On that Saturday morning, we felt it for the first time. The sky almost vibrated as the kilowatts came down through the transparent air. The blue dome overhead was flawless except for a streak of cirrus hugging the peaks to the south. This was why Trombe had put his furnace in Odeillo. He was seeking not only low latitude but high altitude. He had to get rid as much as he could of the atmospheric screen that masks the sun at sea level.

From our hotel in the village of Odeillo, we dropped to the furnace half a mile away. The furnace doors were open for the first time. Perched on top of the elevator shaft, the furnace housing looked a little bit less like a chicken coop displaced by a tornado. Inside the control room, Le Phat Vinh was prodding the mirrors into life. Lights on the control panel went on, one by one, while he activated the mirrors, starting from the bottom terrace. As the mirrors moved, light danced on the hillside facing us through the darkened windows of the control tower. He worked the mirrors a terrace at a time, lining them up like rows of chorus girls. They had not moved for a week, the hydraulic struts that steered them had become sticky.

The trick was to move the mirrors with the buttons until they glinted. That meant that the automatic guidance mechanism could catch the sun and take over. When Le Phat Vinh brought the mirrors under control, they lost the green tones of the late mountain spring and turned silvery white. He hit the buttons on the panel like the flippers on a pinball machine. One rebellious mirror refused to obey; it sent a gleam into the control tower as it swung about, lighting the room as if a lighthouse beam had flashed into it. Mariano Bruffau, one of the crew, rode the elevator down the shaft and then trudged up the hill to adjust its mechanism.

When the mirrors were lined up, the parabolic dish behind us caught the sun. The clamshell doors were shut and so there was no heat at the focal point of the dish — that is, at the mouth of the furnace. Le Phat Vinh said that the sun was putting out 950 watts per square meter, a figure that would remain constant until about two in the afternoon.

Then the day's first experiment began. Slowly, the clamshell doors opened. We wore surgical masks to protect our lungs from the materials that would be charred to dust. Water flowed through the aluminum shutters and dropped to the floor of the furnace housing in a steady cascade. Then the shutters opened. I do not know what the temperature of hell is but we must have been close. For a few seconds the furnace roared like a jet engine. Flames spouted from the sample; its protective coating had been burned away and now it was bearing the full brunt of the sun. Through our welder's glasses, we watched white flame and gray smoke surge out of the furnace. We had been given two sets of glasses to wear. Once we had them on, even the silvery hill covered with moveable mirrors was dark as night, but not the furnace.

The first test ended and the sample was removed at once. It could be handled without gloves. That was the strangest aspect of the experiment, this instant heating and cooling. The power source stayed hot but it was ninety-three million miles away. The experiments continued. From time to time the linear motor gave trouble and the shutter had to be opened by hand. Crewmen in blue overalls, white helmets, and black glasses worked a few feet away from the focal point.

For a change of perspective, I went down to the bottom of the elevator shaft to watch an experiment from the ground. That was almost a disappointment. The clamshell doors opened as if to embrace the sun, a puff of black smoke spurted out, then the doors closed back on the secret recess. It was all over in a few seconds. The hiss of the flaring sample did not even drown out the racket of the one-lung generator that the test crew were using to make sure they had enough current to run their instruments without any peaks and valleys from the Odeillo power plant. The little portable gasoline engine was the noisiest machine in the whole solar plant.

Shortly after three o'clock, the outer mirrors were taken off line. Terrace by terrace, they tilted forward, sending their light patches flashing down the parabolic dish and into the ground. We thought the day's operations were over, but the other mirrors remained in position. It was a test requiring less power. We rode the elevator back up to the furnace and, for the nth time that day, we waited for the sun to make its entrance. It did when the big doors opened: a shower of sparks, a hiss that mounted to a roar, the same acrid odor that had been clutching our throats all day.

Then the other mirrors turned away from the dish. This time, the day was over. The furnace was silent, the little generator had stopped. The setting sun was caught almost in full strength by the edge of the parabolic dish and again, but subdued, by one of the flat mirrors at rest askew on the mountainside. The crew would be back the next day, a Sunday. As long as the sun took no holiday, neither would they.

It was a clean plant. One could well imagine a solar industry of some kind at Odeillo, the workers living in the villas now inhabited only during the ski season and the summer holidays. It could go on as long as the sun lasted; it was not a finite extractive process like stripmining or offshore oil drilling. During the few hours that we had watched the furnace on that Saturday before Whitsunday, it had lived entirely on income. It had not drawn a calorie from the past, it had not borrowed a calorie from the future. The apparatus is vast but only in space, not in time. It sprawls over thousands of square yards instead of drawing on fuel fossilized over millions of years. Solar power is like eating and earning a meal at the same time. One's budget is balanced, there are no debts, no statement from Diners' Club; there is no waste, no garbage. The furnace needed no warming up or cooling off. It stopped when the sun set, it would begin the next morning when the sun rose. The air was just as clean as it had been that morning before the solar megawatt went to work.

III.

~~~~~~~~~~~~~~~~~~~~~~~~~~~~

The Solar Chateau

~~~~~~~~~~~~~~~~~~~~~~~~~~~~

WE SELDOM SAW Trombe by appointment. He would catch us while we were waiting for an elevator (a great deal of time is spent at Odeillo waiting for an elevator) and he would go on talking to us as if our conversation had been uninterrupted since the start of our stay. After he discussed rare earths, horses, potholing, solar power, and Pyrenean folklore, he admitted: "I like to spread myself out." He does, but he manages to do so in depth, quite a neat trick. He stays abreast of rare-earth research through one member of his staff, Germain Malé, who works at Odeillo on ways to prepare the metallic elements of rare earths in their purest forms. This is fine basic science. The intrinsic properties of a material can be learned only if no foreign materials are in the way. Once such purity has been achieved, the chemist can add known quantities of known impurities so that he can achieve new characteristics which he can reproduce at will. These are the lines along which Malé works, somewhat as an extension of the youthful Trombe who had not yet assumed the burdens of running a French national laboratory. Malé has the freedom that the telephone-ridden Trombe gave up long ago. He need not even use solar energy in his work; often he can do it on a rainy day with an electric furnace. When we asked him the applications of these ultrapure rare earths, he was hard put to think of an answer that could be grasped by common earthers. Then his face brightened: "Sometimes, they can be used in lighter flints." We did not waste much of Malé's time, and I am sure that he spent it much more usefully in his frequent consultations with Trombe.

The French Army deserves a great deal of credit for getting Trombe out of his sole role as a chemist. When Bergeron offered him the citadel at Mont-Louis back in 1949, the general tied a string to the gift. He thought that Trombe should not confine himself to solar furnaces but, instead, that he should investigate other uses of the sun, particularly to produce heat and cold. That was the sort of challenge that Trombe enjoyed. "I like to put a group of individuals to work on an idea and watch it blossom. That's the process of discovery. Let everyone participate. Just because I'm the boss, that doesn't mean I'm right. They say in France that youth is skeptical and disgusted. The trouble is that our society today is boring. During the Middle Ages, there was oppression but there was also joy."

From the scientific community that Trombe first set up at the fortress of Mont-Louis and then near the old Font-Romeu hermitage at Odeillo, far from the boulevards and boutiques of Paris, a number of ideas emerged. France is a propitious place to think about solar energy. It was on the short end when the fossil fuels were passed around: no oil and very little coal, not enough to stay with the front-runners when the Industrial Revolution began. It has long lived in an energy crisis, paying prices for electricity and gasoline that would send Americans reaching for their muskets. Nor did France do well with her colonial empire, plentifully endowed with sunshine and little else. Although now independent, the former colonies continue to draw the attention of Frenchmen anxious to be of some help. And then there are the peasant attitudes still ingrained in a country of thrifty farmers who have large capital investments but spend nothing, a contrast to the make-waste way of life. The farmer whose family has worked the same land for centuries is not afraid to acquire capital that can be transmitted to his heirs. Since solar energy is capital-intensive, it may appeal to some atavistic instinct.

At any rate, Trombe's group has been busy. It has studied not only the energy that the earth receives from the sun but also the process in reverse: the radiation at night when the earth loses heat to outer space, thereby balancing its daily heat budget. This is why clear nights are cold: there is less of a water vapor screen to brake the escape of solar heat from the surface of the earth. Trombe tells how he once visited the sultan of Zinder in the old capital of Niger in Africa. The sultan's palace was built around a patio. During the clear nights, the roof radiated heat while cold air funneled into the patio and the rooms under the roof to keep them cool the next day. "There is nothing new about this, Abraham's house was built the same way."

The phenomenon is known as "black body radiation" and it is being applied on the roof of the solar laboratory building. One piece of apparatus up there looks like a square tub, measuring a meter (39 inches) on a side with its walls slightly inclined outward. A sheet of clear plastic film covers the top of the tub; on the bottom is a white aluminum plate. When the temperature at the plate is 32° F. on a clear night, it will drop to −40°F. up at the level of the film as the earth loses its heat to outer space. The laboratory is now trying to learn if this phenomenon could be used to keep food and medicine at 14° F. in a country like Niger where electric power is not always available for refrigeration.

The process developed in France has been used on a bigger scale in Chile to turn brackish water into drinking water. In the northern Chilean desert, water in rivers and springs often contains boron and arsenic. Experimenters in Chile have used "black body radiation" to freeze this water at night in basins about sixty feet long and six feet wide, separating it from the unwanted minerals. They can use it not only for drinking but also to irrigate greenhouse crops, supplying tomatoes, radishes, and cucumbers to copperminers. Under these conditions, natural radiation energy is a good buy because miners pay as much as 230 percent more for fresh vegetables than in the cities of central Chile.

Trombe's system for heating houses is equally simple and contains the same number of working parts: none. For twenty years now he has been working with solar energy at room temperature as well as at 3,000° C. and more. His first patent in the field dates back to 1956 and, since then, solar houses have had their ups and downs in the Pyrenees. They worked but, in the days when heating oil was almost a loss leader, they did not generate much more than an *et alors?* reaction. Trombe had just about given up on solar heating when the energy crisis reached France and the rest of Western Europe in 1973. It was immediately converted into a money crisis, particularly in the pockets of consumers. Interest in Trombe's solar homes rose with the price of fuel oil.

His method has nothing in common with the early solar units developed in the United States after World War II. They used rooftop collectors to heat either air or water which was then distributed in the usual American type of heating system that warms every room in the house. Nearly all these units used water or rocks to store enough heat to get through a series of sunless days. They tended to be complicated and expensive.

Trombe struck out on a different line. He could take advantage of the relatively moderate climate in France where nothing like the cold snaps of a New England or a Midwestern winter ever occurs. Instead of turning on the heat, he acted like a peasant, getting all that he could from the everyday world at the lowest possible cost. That was how the peasants lived during his childhood in the Pyrenees. Tricaud, who comes from the same village, told us how the farmhouses there are built in the shape of an inverted L. Thick walls stand guard against the rain and cold that come on the northwest wind while the yard sheltered by the L is open to winter sunshine on the south.

Living in these high latitudes where the sun never seems to get more than half out of bed in winter, it was inevitable that Trombe should look to the south wall of a house rather than its roof as a suncatcher. Measurements made at Odeillo show that during the five wintry months from September through March, a south wall gets six kilowatt-hours per square meter per day from the sun. In the summer, when heat is not wanted, the south wall obligingly receives only 3.3 kilowatt-hours per day. Sunshine on a wall collector in summer is less of a problem than on a roof collector. An overhanging roof shades the wall in summer, precisely the way Trombe's ancestors solved the problem.

Although the collector is vertical, it works on the same greenhouse principle as the rooftop units. Sunshine strikes the wall of the house, coming in through a pane of glass. The sun brings its energy in over a range of wavelengths running from three-tenths of a micron to three microns (that's not very long: a micron is one-thousandth of a millimeter and a millimeter is but three sixty-fourths of an inch). Right behind the glass pane is the concrete south wall of the house, painted a dark color not necessarily as mournful as black. This surface then reradiates the heat back out to space but in much longer wavelengths, from four to thirty microns, the infrared waves. Glass stops this radiation and traps the heat behind the pane. That is how a greenhouse works; that is how the interior of a parked car heats up on a sunny day if its windows are closed. Even on a cloudy day with the sun obscured, the greenhouse effect applies. Through the glass comes radiation from the entire inverted bowl of the sky. Unlike the burning lens or the parabolic mirror, the flat collector does not need a visible image of the sun, nor must it move to follow the sun through the day or the seasons. On the other hand, it does not concentrate solar energy and, therefore, it produces much less heat.

More heat can be retained by a flat collector if a second layer of glass is placed over the first. Tests have been made with three layers of glass at Odeillo and elsewhere, but there would appear to be no benefit. More energy is prevented from leaving, but too much is lost on the way in.

Trombe uses a concrete wall because this is the building material universally employed for new construction in France, where labor costs are much lower than in the United States and the price of wood is much higher. Concrete may not be beautiful but it suits Trombe's purpose. Used in a wall one foot thick, it will store the sunshine it receives during the day and release it at night. The price of the storage medium is nil since the wall has already paid for itself by holding up the house.

So now we have the start of a solar house, a concrete wall facing south, two panes of glass covering it on a frame. When the sun comes out, the wall will heat up. Next, we have to get the heat into the house. In American solar homes, a pump or a blower is used. It may not take much electricity, but if the power goes out, then the heat goes off. Trombe shares the suspicion of many Frenchmen about modern conveniences and their potential for breaking down. In his system, air behind the glass starts to rise when the wall is heated. Up at the top of the collector, a slot in the wall allows the warm air to enter the room just below the ceiling. It heats the room; then, when it has been cooled, it flows out through a second slot at the bottom of the wall near the floor, to be reheated by the sun. That is all. As long as the sun is out, hot air keeps circulating through the room. After sundown, the concrete wall starts to give off heat accumulated during the day. Experiments show that a wall one foot thick will keep a house warm through the early hours of the morning. Trombe points out that the wall need not be of concrete. The system works with metal walls, if water tanks are placed behind them to serve as heat reservoirs.

What happens when the sun doesn't shine? Trombe decided to solve the problem of long-term heat storage by ignoring it. As soon as the temperature in a Trombe house goes below 68° F., a thermostat turns electric heaters on. He regards electricity as the ideal back-up system for a solar house because its capital cost is so low. The price of hooking the house up to the net is already covered by the normal run of appliances, and the heaters themselves do not run to anything like the cost of the usual central heating system. On his side is Electricité de France, the nationalized electric utility, which is pushing heating systems very hard in "all-electric homes" (do all-electric homes have electric chairs?). These homes must be well insulated, which makes them all the more suitable for solar heating. As an energy conservation measure, standards for all-electric houses have been imposed on every new home built in France. Gone are the days of putting up chilly concrete coops and smothering them in oil when the wind blows in winter.

As a matter of personal taste, Trombe sees no reason why a house should always be kept by a thermostat at 68 or 70° F. "People need temperature contrasts, they shouldn't be too warm at night. If you keep them at the same temperature twenty-four hours a day, you'll turn them into earthworms. They'll become like larvae, they'll lose all their natural defenses."

In a later development, he added ventilation when he worked with Jacques Michel, a French architect who has incorporated the solar wall into houses of his own design. The two men put in a slot in the north wall of the house. In summer, this is opened and cool air is sucked in by the south wall now acting only as a pump and evacuating the air after it has been heated in the room.

None of this would sell very well in Larchmont, but that is not the market Trombe has in mind. He has sought simplicity and low cost. It is cheaper to put a solar collector on a south wall than on a roof because it does not take nearly as much of a beating from wind, rain, hail, and snow. In 1974, Trombe estimated that the cost of his collector would run from $30 to $40 per square meter (about eleven square feet) with a single pane of glass or from $40 to $60 per square meter with double glazing. Stripped to its essentials for the Third World, the system could be cut down to a wooden frame and a sheet of plastic outside a mud-walled house in the cold mountains of the Khyber Pass. No matter what version is used, the cost of the wall collector is the cost of the entire system. The sun pumps the heat and the wall, whether concrete or mud, stores it.

The wall collector has another selling point: it can be used much more easily than rooftop units in a multistory building. It heats the laboratory building itself at Odeillo, although Trombe never mentions this in his contributions to the scientific literature. One can hardly blame him. The architect incorporated the collectors into a standard steel-and-glass curtain wall with the thermal inertia of a pup tent. On a winter morning, scientists come to work in ski parkas at eight o'clock. Since their offices are around 50° F., they keep them on. Then the sun comes out and, by half past nine or so, the temperature is up to 75° F. Many a researcher at Odeillo has regretted that the architect who designed the building has never been invited to spend a year there.

Another heresy was committed when solar walls were installed for symmetry's sake on the east and west sides of the building as well as on the south (the big parabolic dish covers the north wall, remember). They do not get much sun in winter, but they act as greenhouses in summer, making offices on the ends of the building quite unlivable. Trombe is all for putting solar walls on the eastern and western sides of buildings at high altitudes near the equator where heat is needed all year long, but the laboratory at Odeillo could get along without them. Responding as it does to changes in sunshine, the big nine-story building is continually expanding or contracting to the accompaniment of its steelwork cracking and groaning. On a cold wet day with no sunshine, the best place to sit in an office is on top of an electric radiator while you chat with a solar scientist muffled to his ears.

Happily, the big building is not the only live demonstration of solar heating at Odeillo. Trombe and his young associates stayed away from such grand designs in their own work. In 1965, they built their first solar house. It was no bigger than a small apartment, about 860 square feet in area, and it was used only for tests, without anyone ever living in it. Aesthetically, it resembled an overgrown lean-to, its entire south wall blanked out by the solar collector. When we saw it, some ten years later, it had been abandoned but its interior temperature was still around 50° F. despite the cold mountain nights. That is another selling point for the solar wall: while owners are away, it can be used to heat weekend and vacation houses at no extra cost, protecting them from dampness and keeping the plumbing from freezing.

After this successful trial, plans were made for a solar village at Odeillo, thirty houses in all for scientists and technicians. Two were actually built in 1967, more or less along the same design as the first experimental model, but then the big project had to be abandoned for a number of reasons, mostly administrative. Since 1969, these two houses have been inhabited by members of the laboratory staff. One of them is Jean-François Robert, a classmate of Royère's at the Higher National School of Chemistry in Bordeaux. He also started to work on refractory materials and high-temperature solar furnaces. But then he moved into the solar house which had been built about two hundred yards up a hill from the laboratory — "The rent looked good and so did the neighborhood" — and he has never regretted his choice. It changed his career; he now spends his working time on solar housing and it is up to him to reply to queries for help from all over France.

He has learned about solar houses by living through nearly a decade in one with his wife, Laurence, and their three young children. Though his chalet is badly insulated, he can get two-thirds of his heat from the sun. He showed me his figures: if his were an all-electric house, it would need 35,000 kilowatt-hours per year. He buys only 10,500 from the electric utility; the rest come from the sun. Robert, his wife, and their kids were by far the healthiest-looking people around the Odeillo laboratory, but cross-country skiing and mountain-climbing were probably more responsible than solar heating for their physical condition.

The economics look encouraging. Robert said that the solar wall system needs one square meter of collector for 10 cubic meters of space to be heated. Thirty square meters of double-glazed collector could be installed for an extra cost of $1,800 in a new house. At the price of electricity in France, they would save $400 or $500 a year in heating costs, leading to a quick payoff on the original investment.

Such savings mean more in Europe than in the United States. Comparatively speaking, Americans get their energy for nothing. In 1974, when the Environmental Protection Agency first required car manufacturers to state fuel consumption, a sticker on a car we rented explained that it would cost $165 to run a 2,000-pound car ten thousand miles and $500 to cover the same distance with a 5,500-pound car. That was a difference of $335, not even two weeks' pay for the great multitude of car owners. In a European country like France, with gasoline around $1.80 a gallon, the difference would amount to around $1,400, nearly four months' pay for many workers. Economic equality seems to be tied up with access to energy. The Frenchman is much more likely than the American to drive a small car and to look closely at any system that will keep his heating bills from rising at the same rate as oil on the world market.

Robert said that, by the end of 1974, about fifty solar houses were going up in France. One belonged to a young customs officer from a village near Mont-Louis who came to see Robert for advice. He had four children and he was doing everything himself, even digging his own foundations. When we asked why he wanted to install solar heat, he did not talk about ecology or pollution. "It's the price of oil" was the only reason he needed. Nor did he plan to go along with Trombe and architect Michel on a semiautomated solar-electric house. He would put in an oil stove for heat on a cloudy day. He would get his calories more cheaply than from the electric utility and the stove would be no more expensive than an electric heater. He and his wife were willing to go to the trouble of lighting it.

Word-of-mouth advertising of solar homes had reached an Odeillo hotelkeeper, Michel Sageloly. Masons working on new solar houses eat at his place and so do the scientists from time to time. Sageloly got his information from Trombe and now he was doing some serious thinking. His hotel had been built in 1740 as a posting station for stage coaches going over the mountains. The walls were of earth and straw, three feet thick. When the Sageloly family installed central heating forty years ago, they used the cast-iron radiators of the day. Now they muse over the benefits of progress. In 1961, they built another house in the village to provide added rooms for themselves and extra guests. The walls were of hollow concrete block, one foot thick, and the radiators were steel. "The new house has only fifteen rooms," Sageloly said, "but we spend as much on oil there as we do in the old building with twenty-five rooms, two dining-rooms and all the dishwashing for the restaurant." The pendulum has swung back. Sageloly plans to put up another annex, twenty more rooms, but he will build it with the insulation standards of the all-electric house. This should cut his oil bill in half and then he plans to install solar collectors on the south wall.

Ecological purists are tempted to regard the solar-electric house as an unholy alliance between the sun and the devil. If the supplementary heating is electric, they insist that a thermodynamic crime is being committed. One must burn oil at high temperature in a big central plant to raise steam, spin a turbine, drive a dynamo, and put electricity out on the wires and into the home to heat a room to 68° F. In the hearings on solar energy held by the Subcommittee on Energy of the House Committee on Science and Astronautics in June 1973 in Washington, Congressman Paul W. Cronin of Massachusetts remarked: "We have had testimony . . . which indicated that in my area of the country, New England, the normal home oil burner is roughly 65 per cent efficient and electric heat is roughly 25 percent efficient, vis-a-vis total cost to the country in the use of energy. So here we have what was a relatively inefficient way of producing heat being quickly adopted by a large segment of the American public." At those same hearings, the congressman got an answer from Ralph J. Johnson of the National Association of Home Builders Research Foundation: "Of course, they [the electric companies] had the basic incentive, they had a strong demand factor situation, and they didn't have installed load in the winter. Because they had this major problem, they set out on a very aggressive marketing program over a very long period of time, something like fifteen years, and they sold their product; they pounded it home, day after day, bit by bit."

Others seem to be following their lead. The French electric utility started a campaign in Brittany in 1974 to convince farmers to put in electric heating for piglets, vaunting the added pounds to be gained for every added degree in the sty. Posters for farmers went up on village walls; in the glossies, full-page ads told of the need in these parlous times for instantly controllable heat, electric of course. The inconsistency is only apparent. Everywhere, electric companies sell hard when they have to get rid of cheap base load power. Only when it comes to peak time use do they preach conservation.

All this does more good than harm. The power people throw their considerable weight behind the trend towards better-insulated houses and, as we have seen, no one in France is allowed to put up thermal sieves like Sageloly's annex or the Odeillo laboratory building. At the same time, no one can stop the homeowner from insulating his house and then, like the customs man at Mont-Louis, installing another form of heat: oil, coal, even wood. Le Phat Vinh, master of the solar megawatt, uses wood to heat his own house. What is important is that the money and the political clout are coming down on the side of using less energy, not more. Once this happens, the homeowner can decide if he is willing to sacrifice never-lift-a-finger convenience and strike a match from time to time. Not many will make the choice, but it will be there.

Solar heating with the Trombe system in France is not limited to the southernmost latitudes. Michel, the architect, built a solar house in 1972 at Chauvency-le-Chateau in the north, as chill and gray an area as any in France. He could still get half his heat from the sun, although he needed much more wall space for collectors. For my taste, the house had too much of a functional look about it. I do not necessarily want to live in a heating plant.

On the other hand, Michel deserves the highest marks for the new solar condominium that he designed for Odeillo. It consists of three attached houses, three stories high, that stand like a defensive outpost just inside the entrance to the grounds occupied by the solar furnace and its various annexes. From the north the houses do not even look solar — the highest compliment of all. They could be a segment from one of the hill villages that dominate the road up to Font-Romeu through the green valleys of Roussillon. The south side is no longer a glass wall that glares at the world like a drug pusher through anonymous shades. Research by Michel and Trombe has shown that a south wall can provide heat even if it incorporates windows. Michel came up with a collector that works something like a chimney, taking heat from under a window and running it up through flues on both sides. Thus the sun can light as well as heat the solar house.

In passing, one should note that the aesthetic criticisms of the old solar houses at Odeillo are greatly exaggerated. They do not look nearly all that bad, especially after one has been exposed to new housing around big cities in southern France and new resorts on the Mediterranean. Robert's chalet has only a glass door on its southern exposure, but it gives his family plenty of light. As soon as the use of solar energy is mooted, a double standard of aesthetics comes into play. According to the rules of the game, it is permitted to glass in the island of Manhattan and then light and heat it artificially, but you scream landscape-rape if anyone puts a solar collector on his roof to warm a bath. The same holds for money. If solar energy doesn't pay its way from the start, it is unrealistic. Its proponents are well aware of the economic handicaps under which they labor even when the dice are not loaded. If on top of that they can win only when the dice stand on end, they are really not getting a fair chance.

This is why the solar condominium is the boldest experiment of any of Trombe's careers. One of its three houses is his. Not only will he live in it as Robert has been doing in the laboratory's chalet, but he will buy it, too. Dr. Michel Ducarroir and Dr. Benigno Armas, who have the other two houses, are buying theirs. Here we have unsubsidized solar research with the scientist volunteering as his own economic and thermal guinea pig.

Ducarroir is a chemist of the same Bordeaux vintage as Robert and Royère. Originally, Trombe schemed to have him sent to Niger to work on solar energy there during his period of military service. Ducarroir did not pass his physical and Trombe mobilized him instead to conduct high-temperature research in Mont-Louis. Later at Odeillo, Ducarroir worked on a cooling process, using an absorption type of refrigerator with a solar collector to replace the kerosene or gas flame found in the usual household versions. To get the refrigerator up to its working temperature of between 160°

and 190° F., the flat solar collector operating on the greenhouse effect just won't do. A way had to be found to concentrate the sun cheaply, and a parabolic trough was devised that need only be moved once a day to follow the sun. Ducarroir can now make about fifty pounds of ice a day and his goal is to double production so that the solar refrigerator can be used to preserve perishables in tropical countries.

Ducarroir and his family were living in a small rented house in Odeillo when he decided to go into the solar condominium. He will occupy the east end of the building, with Trombe on the west end and, in the middle, Dr. Benigno Armas, a physicist born in the Canary Islands who has worked with Trombe since 1968. His is the smallest of the three row houses and he has high hopes of getting warmth from his neighbors as well as from the sun. He insulated the north side himself, using plastic foam and air space to get the equivalent of four feet of concrete. While the houses form a homogeneous whole, each interior is designed to its owner's taste. Trombe has a number of small rooms as in old French homes while Ducarroir planned his kitchen and living room in a single area much as in countries where household help is not even a memory.

To Trombe, the condominium is the belated triumph of his old plan for a solar village at Odeillo. Bureaucrats stopped that project by decreeing that the National Center for Scientific Research could not build housing for its staff. With all the muleheadedness acquired from his Ariegois ancestors, Trombe kept at it. He saw the mayor of Font-Romeu to try to get help. The mayor needed convincing. Font-Romeu regards itself as the queen city of ski resorts in the Pyrenees and little chalets like Robert's did not fit the image. Trombe agreed that he would do something different, a solar chateau instead of a hamlet of modest huts. The mayor got his municipal council together and they agreed to sell Trombe the land he needed. Half of it was in the hands of a private owner, but he played fair: he said that the mayor's price would be his price. And so the three scientists became the owners of a half-acre of rock. To handle matters, they set up a nonprofit company with Trombe as president, Ducarroir as secretary, and Armas as treasurer. The condominium had just enough members to supply the officers required by law.

"It was a challenge," said Trombe, "we had to accept it." First, the rock had to be flattened to the point where a house could be perched on it. Then the contractor went bankrupt. A second contractor was brought in; he went bankrupt. The whole building industry on the French side of the Pyrenees was in trouble because the Spanish economy had been picking up momentum and luring Spanish workmen back across the border.

The work went on. We could see the progress ourselves. In March of 1974, not much more than the walls were up and we watched Robert as he supervised the installation of a thermocouple to measure how they would transmit heat. He explained that the walls in his small chalet were too thick, almost two feet of concrete. Sunshine received at noon took fifteen hours to get through the wall; it was heating the house after midnight when the family was already in bed. With a one-foot wall, it would take only seven or eight hours for the stored heat, from 20 to 30 percent of the sun's energy received during the day, to be transmitted. If the walls were too thin, they would heat too quickly and simply act as an immense radiator. Robert gave us this lecture from the top of a scaffold where he had scrambled with a mason who was installing the thermocouple. Then he showed us how the collector could be shut off with a crank at night so that cold air would not be drawn into the house. "We could automate this but that would raise our costs. We can't build houses here the way they do in America." While these houses admittedly do not run to current American tastes, Trombe has estimated that they produce solar calories at one-third the cost of other systems.

Two months after we had talked to Robert on top of the scaffold, the small chateau was well along. Trombe's house was the least advanced because, so we had heard, it was he who had changed his plans most often. Without the glass panes of the collectors, the concrete walls looked something like a blockhouse with their floor and ceiling slots for the heating system. The neighbors' houses had made more progress. In the middle, the south side of Armas' house had been painted dark green while Ducarroir's house on the far end was Burgundy red. Sheathed in glass, their houses were a pleasing sight.

Inside Trombe's house, we found Trombe himself prowling along plank walks like a lifelong mountaineer. He showed us his bedroom on the very prow of the castle, almost in the shape of a ship's forecastle. It faced north with a view of the laboratory and the furnace. "When I retire, I shall live here. My collaborators will be afraid, they'll know I'm still here keeping an eye on them."

Like every other veteran of solar energy, Trombe has his moments of caution and even discouragement. Once, he told us: "If in two years from now, thousands of solar houses are not going up, then it will all have been a waste of time." That is not too likely, but one need not share his pessimism. Robert, who will go on working in the future under the eye of his retired director, is less pessimistic. He finds interest in solar energy at the ministerial level in Paris. In a country where industry is nationalized to the point that it is in France, this is tantamount to encouragement from the top strata of the business world. Robert also told us that the publicly owned electric utility is toying with the idea of eliminating meters. The customer would be allowed a given capacity for which he would pay a fixed rate, thereby freeing the power company from reading meters or scaling peaks. If the customer exceeded his capacity by, say, using his electric stove, radiators, color television and washing machine all at once on a cold day, then a circuit-breaker would come into action and black him out, instead of browning out an entire region. Under such a regime, it would be to his interest to invest in solar energy so that he could have extra capacity that could not be shut off.

This is all speculation. Reality is Robert's small home and the solar chateau on the rocky spur just inside the laboratory grounds. The glass collectors look southward at the mountains, they take the sun like the skiers on the slopes. Below the condominium, one can see the low dry stone walls setting off the fields where the Comanges family once grazed their cattle and horses. In the village of Via where the Comanges live, they heat their old house with wood, and a hog-killing on the farm is still a festive occasion. Life moves as disjointedly here as it does everywhere else. Under the jurisdiction of the mayor of the small *commune* of Font-Romeu come the Comanges' on their farm, the ski hotels on the slopes behind the old hermitage, and that near-monastery of the scientists at Odeillo.

The French solar laboratory is hard to understand at first sight. Everyone of any importance there, from Trombe on down, seems harassed by petty problems, and there is obviously no one in Paris trying to make life easier for them so that they can get more work done. But perhaps this is the grain of sand that produces the pearl. With all the incessant telephoning, with the insane building that howls in the wind and protects its inhabitants about as much as a veil, with a budget for operations and equipment so ridiculous that I won't even mention it (it hardly gets into six figures American), these people are producing the only solar megawatt and the only solar house at a truly competitive price in the world.

IV.

~~~~~~~~~~~~~~~

When a Solar House
Is Not a Home

~~~~~~~~~~~~~~~

THE TROMBE COLLECTORS warm the house and nothing else. Water heaters are more in the province of the plumber than of the scientific researcher. In various forms, they have existed for decades, and millions have been produced and sold from Florida to Japan, from southern France to Australia. In his files, Robert had the address of the largest French manufacturer of water heaters, a firm called Héliothermique at Mont-de-Marsan, a largish town in the center of the pine forests that spread over the Landes, the flat "moors" of southwestern France that run from Bordeaux to the Pyrenees. I accepted the address with all the more gratitude because I was curious to see a solar businessman in the flesh.

Our search for solar hot water took us west along the flanks of the Pyrenees. Once we had crossed Puymorens Pass, we saw the last of the ski tows and plunged into a forgotten France of quiet valleys that gradually broadened until we were on the tabletop of the Landes. It is country that looks North American, miles and miles of woods, occasionally a settlement of shanties, nothing in between. When the sun beats down on those pines, as it does a good deal of the time, there is a headiness of resin over the road, the car feels as if you had filled it with bath salts.

The president of Société Générale Héliothermique was Francis Bennavail, a man in his mid-forties who had had the good fortune to make his fortune at an early age, not in oil but in irrigation. He started a firm called Irriland, drilling wells and supplying pumps, pipes, and other hydraulic equipment to farmers in the Landes just when they decided they could make more money growing corn on irrigated soil than pines on dry sand. His business flourished, he expanded it to cover the automatic care and watering of golf courses and racetracks, another growth industry in southwestern France. With seventy-eight on the payroll and $2,600,000 worth of business a year, he decided to try something else. He sold out to concentrate on solar energy. Here, he was running true to form by refusing to run in a rut. As a student from Carcassonne, he had quit engineering school to go to Morocco and Algeria in preindependence days with some friends to seek a change. "We weren't hippies, we were bums. We had no money."

Bennavail's interest in solar energy was purely fortuitous. In the summer of 1971 he was visiting a brother-in-law in a small town near Mont-de-Marsan. On the roof of a nearby house he saw a water tank and a solar collector, although he did not know what they were at the time. He rang the bell and the neighbor, an architect, answered the door and a number of questions. Bennavail learned that the solar water heater had been devised by Auguste Dellac, an engineer at Beziers, the capital of the wine-growing region behind the Mediterranean coast.

Dellac was an inventor of the old school. As I heard the story, his eureka moment came during a rare cold spell in Beziers about twenty years ago. It was no more than 20° F. outside when he left his house to go to work but a strong sun was out. Out of curiosity, he put a recording thermometer in his garden and, to protect it, he housed it in an open shoebox covered with a piece of glass. The greenhouse effect did the rest; the heat that the sun brought into the shoebox accumulated behind the glass. When he came home for lunch, he looked at his recording thermometer. To his astonishment, he saw that the mercury was stuck at the very top of the column at 160° F. on the coldest day of the year.

Dellac designed a solar collector, using the usual glass pane over a black surface but with some added improvements of his own. His collecting surface consisted of two sheets of metal just touching each other. When the sun heated the sheets, water rose up between them by capillary action, eliminating the need for a pump. Since the sheets were flexible, there was no need to worry about a freeze-up on a frosty night. To warm the bathwater, he used a heat exchanger. Water from the solar collectors is piped into a heating element at the bottom of the household tank. There, it "exchanges" heat with the water inside the tank just as if it had been an electric heating element (after two or three sunless days, an electric element does come into play). At night when the solar collector radiates heat to the sky, only two or three gallons in the collector are chilled instead of the whole water tank. Now the reader can understand why Bennavail smiled when a satisfied customer wrote to relate how "solar hot water" had cured his rheumatism.

Dellac started to market water heaters. In his sales literature, he told how they ran on "the new energy — the golden fuel." As the orders came in, he made heaters to fill them. Later, he joined forces with a small aircraft manufacturing concern which went bankrupt (through no fault of the sun). In 1971, when Bennavail found him in Beziers, Dellac was ready to sell his patents and take a less active role in the business. That was how Héliothermique came into being at Mont-de-Marsan.

Bennavail now offers solar water heaters in seven sizes from 300 to 3,000 liters (roughly 70 to 700 gallons). It is instructive to look at his list of satisfied customers. Most of them are in energy-starved corners of what is left of the French empire. By far the biggest is the Kourou Space Center in French Guiana (which I had previously associated only with Devil's Island) where something like 40,000 gallons of hot water *per day* is produced by four hundred and twenty-two solar collectors feeding three hundred and twenty-two apartments and one hundred and forty hotel rooms.

On Réunion Island, another faraway fragment of France, a hospital has a 1,100-gallon plant with a number of smaller collectors working for other institutions and private owners. A similar picture can be seen in New Caledonia, Guadeloupe, Tahiti, and various spots in Africa. In Europe, the Club Méditerranée supplies solar water to three vacation resorts and there are users by the dozens in and around Dellac's hometown of Beziers. They run from the owners of chateaus to the patrons of humble municipal bathhouses. When we saw Bennavail, he was working on plans for a 3,700-gallon plant to serve an apartment house in Nice.

His basic unit is a 70-gallon tank with forty-two square feet of collector. Early in 1974, he was marketing it for $700, twice the price of the equivalent electric water heater (which is included in the solar installation). He calculated that it would save the owner from $100 to $120 a year in electricity bills, so it should be amortized in three years. Since that date the price of electricity has gone up, but so has everything else.

Bennavail uses the same collectors as part of a solar central heating system that he markets almost reluctantly. For a small five-room house, the price of the solar system is $4,500, and that does not include the radiators and the rest of the normal central heating system. This is much too costly; the solar collectors need from eight to ten years to pay themselves off. Bennavail thinks he will be able to cut his prices by 30 to 40 percent. Then, if energy keeps rising, the system could pay its way after only five or six years. Even today he has a few customers for central heating. "If a man has three or four children in his house and he is afraid he might run out of heating oil, he does not worry too much about the cost."

If the collectors are also used to heat a swimming pool, then the payoff is faster. During spring and autumn when the weather is too cold for swimming and too hot to require indoor heat, the collectors are hard at work warming the pool. Roger Bombail, Bennavail's sales director, sees a future here. The more the sea is polluted around the coasts of Europe, the more swimming pools are built in what were once beach resorts. A pool used only two months a year is expensive; solar heating makes the price easier to bear. Solar central heating looks even better if one is up in the bracket of the indoor private pool. Then the lucky houseowner can use the pool to store calories for a cloudy day or, if you prefer, he can swim in his heat reservoir.

About the time that the first big energy scare was tapering off, Bennavail was hesitant. His company was making thirty heaters a month and he was getting thirty to fifty letters a day asking for information. He had two salesmen and a small network of twenty dealers. While he could have easily doubled production, not all requests for information were transformed into firm orders. "I'm afraid, I tend to be pessimistic. If I invest in plant facilities, I'll be in a risky position if the market breaks."

Bennavail's associates think he has nothing to worry about if he can call his shots on solar energy as successfully as he did on irrigation equipment for the corngrowers of the Landes. He is expanding. In his executive Citroen loaded with crates of apples and pears, he drove us to a new industrial park outside Mont-de-Marsan where he was putting up a small factory to replace the old barn where workmen were welding his collectors. It wasn't General Motors but it was a long way from Auguste Dellac's operation in Beziers. There, the solar inventor was all on his own. At one point, so I was told, he had to brush acid on sheet metal so that he could make clean welds. His brushes kept disintegrating until he got the idea of making his own, using hair he pulled from his horse's tail. He had a horse to work his small vineyard where, to quote a local newspaper, he was like all "the happy winegrowers of southern France endowed with the gift of taming and bottling the sun so that it may transmit its warmth and energy to man."

Several weeks later, it was with a trace of wistfulness that we thought back on Bennavail's welders in their barn and the hair of the tail of the horse of Monsieur Dellac. We were standing in the basement of Solar One, another solar house but this time in Newark, Delaware, and the pride of the Institute of Energy Conversion at the University of Delaware. An earnest young man named J. Kevin O'Connor, whose card said he was the institute's program manager, had ridden with us to Solar One in our rented Nova through the drenching rain that always seemed to accompany our first-hand inspection of solar energy projects. He apologized because he could not drive us. He had come to work on a motorcycle.

With O'Connor, we had gone down to inspect the control panel of the house a few minutes after our arrival at the institute. "This is a pneumatic system," he said; "be careful, it'll start with a bang." As we awaited the bang, he went on to explain that it was a "fairly sophisticated" control panel with fourteen different modes of operation. Depending on whether or not the sun was out, whether or not the collectors were above $110°$F., whether or not the house had to be heated or cooled, whether or not batteries had to be charged or tapped, the panel went into action. The cybernetics of the power station had been brought into the home. "This is a laboratory, not a house to live in," said O'Connor, "but we can use it to build simulation models and advise architects."

It was all heady stuff for a pair of travelers suffering from a combined overdose of jet lag and cultural shock. The reader may bristle at this abrupt transition from the *département* of Landes to the state of Delaware, but at least he did not have to suffer the wrench-off of the 747 at Orly, the splashdown into a puddle of heat and smog at Kennedy, the mapless departure by rent-a-car from Manhattan in a lane of trucks oozing slow as slime through the Lincoln Tunnel, the groping along the New Jersey Turnpike through Secaucus, Newark, Elizabeth, New Brunswick, and points south, Interstate 95, a full tank of gas in the car, destination unknown, a hotel-motel somewhere after Wilmington.

We found the motel; it shall remain unnamed but described by Madeleine who, having grown up in a chateau near Tours, had never seen the likes of it before. In her log, she wrote: "The motel was at the intersection of the turnpike and two main highways. Enormous cars were moving all around us. Since the lobby was soundproofed, they seemed to swim as silently as fish but, as soon as one opened a door, the noise took possession of you. The room was immense, two enormous beds, each a double bed, a bathroom with two sinks, a bathtub with two showers, they could spray you with a fine mist, a summer thunderstorm or the stream from a firehose, strong enough to punch a hole in a burning wall. The room was hot, heavy with the odor of the warm bodies of our predecessors. The windows couldn't have been opened for at least six months. Double flowery curtains, color television, eight or ten channels, you could change channels from your bed, infrared lamps in the bathroom to warm your back when you left the shower, a rug as thick as tree moss, a giant closet, about twenty thick white soft towels, cakes of soap everywhere, a hole in the wall from which an endless supply of Kleenex emerged, a screen that opened and closed like a window and, in front of the window, gilded lampposts around a blue swimming pool. The passing cars and trucks left an unbearable stink of gasoline behind and yet a bird was singing in a scraggly bush, the last survivor of the forest that once covered this place and still lined the turnpike and the hill full of the sound of slamming brakes and accelerating motors and the smell of hot tires and gasoline fumes."

Poor Madeleine, her frugal country upbringing in France had not prepared her for the United States wracked by an energy shortage. Her log continues: "Suffocating dry heavy heat everywhere, whether in the motel room or the office of the Institute of Energy Conversion. Outside, we were gasping for breath; in the office we were greeted by a tall slim secretary, an electric radiator next to her legs, its tubes glowing red and purring madly. Dan took his coat off, the girl said, 'Oh' and switched off the electric radiator."

We had made an appointment by telephone from New York with the head of the Institute of Energy Conversion. It was on the strength of that phone call that we had rented the Nova and made our way to Delaware. Later, I was told in Washington that the United States, after having lagged for years in solar energy, had the world's biggest program in 1974 and, within a few years, would probably be running a bigger program than the rest of the world put together. One reason why this will happen is the unbelievable ease of getting through to people who get things done in the United States. The sun may shine with equal intensity along the same latitude in the New and the Old Worlds, but Americans can move with so much more alacrity when the time comes to capture it.

The Institute of Energy Conversion occupied one end of a long two-story building on South Chapel Street in Newark. From the outside, there was little to distinguish it from the Alderman's Court that shared the building. In the lobby, a small exhibit showed various ways to put solar energy to use. The most spectacular was a large plastic model of a World War I German biplane (one of the Red Baron's mounts, no doubt) hanging from the ceiling with a glass bull's-eye on its fuselage. The receptionist explained that if one focused a light beam on the target, then a photovoltaic cell would activate an electric motor, driving a propeller to send the model whirling around on the end of its supporting strut. Too bad, it was out of order. I picked up a folder entitled "The Coming of Age of SOLAR ENERGY." This, too, was low-keyed, much more educational than promotional. In our day and age, it is almost impossible to grasp the insignificant amount of money devoted to developing solar power. Little is left over for advertising and PR, hardly enough for a model airplane and a small folder (nothing like the folder put out by Font-Romeu). In the garage of Solar One, there had been another exhibit, somewhat more ambitious. One panel carried a slogan: "Turn on the Sun — Activate a Real Solar Cell." O'Connor did, he shed some light on the exhibit and a motor on the panel started to run. Then he plucked the motor from the panel to show us there was no connection, no batteries, nothing up his sleeve.

Each cell was but a third of a volt. So much more energy was represented by the five big fuel oil tanks belonging to the Newark Lumber Company just across the street. The Institute of Energy Conversion sits in an industrial area although it is part and parcel of the University of Delaware. But then, there is not much separation between town and gown in Newark. It all looks like part of another age; perhaps the DuPonts reconstructed a small town of the 1940s the way that the Rockefellers restored Williamsburg. The demography is prewar: a town of twenty thousand, a university of twelve thousand students, virtually the same ratio as Ann Arbor when I was an undergraduate at the University of Michigan in 1941. The campus is so small that students can walk or cycle to classes past a distinguished parade of red brick dormitories and halls of learning. Newark (Newark, DELAWARE?) is such an out-of-the-way place that architects seeking a reputation have left it alone. Consequently, there is almost unbroken unity of style from the first building of New Ark College that went up in 1833 to the new dormitories and the student union building of today. There are woods for picnicking and a creek for swimming not much more than a long walk from Main Street and South Chapel Street. One hears a great deal about Boswash and the unbroken megalopolis of the northeastern seaboard. If this is so, then Newark, Delaware, must be an island of countryside entirely surrounded by city. Students adapt to life in the 1940s once they get used to going out at night without worrying if they will get back. After our first night in the neighborhood of Interstate 95, we found new quarters next to the B&O tracks in an old hotel where students came to tank up on beer every Thursday night and the freight trains ran by so close to our windows that our room rocked as if we had been riding in a *wagon-lit*. The lady who owned the hotel also owned a racehorse. She was of great charm, she gave Madeleine a white carnation when we turned up a second time and she offered us breakfast on the house. The world is not bounded by Interstate 95, there are Newarks elsewhere than in New Jersey, there is a New Ark floating somewhere . . .

The one on South Chapel Street in Newark, Delaware, is the achievement of Dr. Karl Böer who qualifies as the sort of maverick personality that one finds in solar science. Böer is a public man, at home in the marketplace and on the board of directors; at the same time, he is capable of private introspection. I had first heard him at the international congress on "The Sun in the Service of Mankind" in Paris in 1973. He appeared to be a young man (he wasn't as young as he appeared; he was born in 1926) extremely convinced of what he was talking about, so much so that he kept talking whether he was understood or not by his entire audience. He has a slight German accent with a heavy American overlayer which can make for unexpected effects when he comes up with a phrase like "I can ballpark it."

From Böer's writings, speeches, and the talks we had with him, it was possible to trace his interest in energy back to June 1945 in his native Berlin. He had lost all his family there, and the city was a shambles. In his home, electric power was available only two hours a day. "I had a small car battery, just six volts, I charged it up when the power was on so that I could have lights at night."

The aftermath of the war shaped his scientific career. While he was studying at Humboldt University in Berlin, nuclear physics was the noblest of the scientific disciplines but it was forbidden to Germans by the Allies. Böer turned to what was to become solid state physics. "We started with a current meter and a few crystals of cadmium sulfide. People didn't think much of our work, they called it sneezing physics. If you sneezed in one corner of the room, you changed the properties of the cadmium sulfide on the other side. You needed the highest purity in your materials and that led to irreproducibility of results. Many serious scientists stayed away."

Böer stuck to it. He got his diploma in physics from Humboldt University in 1949 and published his first paper the following year (for the curious, its title was *Einige zusammenfassende Bermerkungen über Stromgrössenmessungen mittels Elektrönenröhrenvertstärker*). He climbed the rungs of the academic hierarchy in short order. By 1961, he held a professorial chair at Humboldt, a status about on par with God's at a German university; he had founded a new physics department there where he ran a cadmium sulfide program; and he had started a scientific journal, *Physica Status Solidi*. He still serves as one of its editors and he has stuck to his pledge to publish articles within fifty days of accepting them. It was about that time, too, that he came to the United States as a visiting research professor at New York University. He stayed at NYU only a year before coming to Delaware where he was under less pressure in a new country with an unfamiliar language.

This may explain why Böer did not hesitate to go into solar energy, a field with no more prestige than that of solid state physics at the end of the Second World War. "I went through it once, I was a respected physicist in a snug harbor when I left Berlin. After twenty-five years, I still have some confidence left. And solar energy is a challenge. It does not have a perfect reputation. There are low points in every field; perhaps they are more apparent in solar energy."

His present challenge may be a greater one. When Böer came to the United States, he brought with him his prestige in solid state physics and enhanced it. He worked as a consultant to industry, he made his reputation and a good deal more, he became a full professor of physics at Delaware and then, in 1970, the same intellectual restlessness got the better of him. He took a sabbatical and stayed home for six months to think about solar energy, the natural outcome of his work with cadmium sulfide cells converting light into electricity. He was not quite in isolation; his wife, Renate, is an economist and they work together evenings, Saturdays, and Sundays. Suddenly, the titles changed in the flux of papers that he had been emitting since 1950. In 1972, he published on "Future Large Scale Terrestrial Uses of Solar Energy" and he has seldom strayed from the subject since.

He spoke of all this in his office where the temperature was European compared to the incubator in which his secretary dwelt. The decor was practical: greenboards, a model of Solar One, a photo of his wife, two telephones, microphotographs of cadmium sulfide cells on the walls next to a dozen pictures showing the step-by-step construction of Solar One, sparse furnishings to accommodate a small conference like the one with us, a bookshelf where a few titles could be distinguished — *Thermal Physics, Solar and Aeolian Energy,* the *New Cassell's German Dictionary.* In his hand, Böer held what looked like copper foil encased in plastic, about the size of a sheet of typing paper. "That's three watts. There are nine cadmium sulfide cells to a sheet, each cell is one-third of a watt. Seven years from now, they should cost no more than twenty cents a watt." Then he went to his desk and picked up a small plastic box. Madeleine was watching, too, and she noted what she saw. "Böer spoke with a funny accent, round and guttural. Sometimes I understood him more clearly than an American, sometimes it was worse. He got up, he plucked a transparent plastic cube from his desk. Inside was a copper plate. He lifted his arm, he held the cube up to the electric light and now the copper plate spun all by itself. Not bad, this sorcerer."

Böer said: "That cell is six years old and it still runs. I need the presence of hardware. I need to experience it as a system. Then I know how real it is. There are bugs in this solar cell. A few years from now, we'll have the bugs solved."

It would be wrong to regard Böer as an apostle of the cadmium sulfide cell the way others seem ready to stand or fall on the breeder reactor or oil shales. It is a means subordinate to the end that came out of his reflections during sabbatical leave. He got away from solid state physics and, instead, looked at the criteria for solar energy.

"The first is cost, and by that I mean first cost. We can add ten percent to the price of a house. That's all. If it's 30 percent, forget it.

"The second is amortization. Let's not fool ourselves, no one gets five percent mortgages anymore. Interest is over 10 percent but, at the same time, the value of a solar energy system will go up every year. So a mortgage can be an advantage; all the clever people are in debt.

"Third, the system must be absolutely reliable, and fourth, it must be compatible with current equipment. If I have to wait two hours for it to warm up, forget it. You may accept that, I may accept that, but the market will not. Not this market. When the president says to save fuel, people turn their thermostats down to 65 degrees and surround themselves with electric heaters. We need a large market, not just the do-it-yourself pioneers. There are people who build their own television sets but that's not how you sell television.

"Fifth, there is serviceability. Here is the big incentive. A service industry must be created to provide a warranty for the system. Solar energy cannot be self-sufficient. Even if you're rich enough to store heat for five days, you will be frightened, when you watch the meter drop. I want no part of autonomy. We must interact with the current energy market and this is a very powerful industry. If you make enemies there, you will lose the chance to help the energy people enter an otherwise highly fractionated market and to service the product. They need incentives. Let them make a buck."

Böer wants to enlist the sun on the side of the power companies to shave the top off their peak loads. His reasoning is the same that we had heard from Robert in France and from so many others. In any public utility, peak costs are many times higher than normal running costs. That is why long-distance phone calls are so much cheaper at night; that is why railroads and bus companies can only go broke trying to provide commuter service with vehicles that are mobbed two or three hours a day and idle the rest of the time. Building, buying, and owning enough capacity to take care of every possible demand is widely proclaimed to be nonsense . . . except for the consumer. He is inspired to tool up for all contingencies. If he can be convinced to run his own transportation system, then there may be a chance of getting him to buy his own electric utility.

I am not being facetious. I think there is better than an outside chance that Böer may be right. He sees two ways in which the customer can help the power company. First, cadmium sulfide solar cells on the roof can generate electricity and store it in lead-acid batteries. In the event of sudden demand, the solar house-holder would use current from his batteries and, in a real pinch, he might even send electricity back to the power company. Second, he could store heat and particularly cold along with electricity. In summer, he could run his air conditioner at night, lower the temperature of a storage reservoir and then, during the day, coast along without using any power while his neighbors try to cool their homes in the blazing noonday sun.

This is neither idealistic nor revolutionary. Böer makes no such claims. "We must look at how we can make small changes. The *transient* phase is the most difficult of all. That is why it is dangerous to be radical. What is radical is not amusing for long. It's like communal living, it's fun for half a year but when the sex wears out, it's a different story. We must get out of the world of emotion and feeling, we must go to hardware. We must see how we can bring *slight* influence to bear, using available means. We don't try for heaven, we take the wind as it blows. That's what a sailboat does, but it's not rudderless. It goes the way you want it to go."

These were the paths along which Böer's thoughts journeyed during his sabbatical leave. "I'm not more clever than anyone else, just more careful. I'm a conservative man. Solar energy will not come about overnight. It's like a snowball on a mountainside — you don't know what it will be like by the time it gets to the valley. It must meet the demands of different climates, different consumer groups, different income brackets. It represents a wealth of opportunities."

In his blue suit, modish tie and dark-rimmed glasses, Böer is more likely to change the ways of the world than the barefoot protester. He abjures radicalism, yet he is one of the most poetic of the solar prophets on the contemporary scene. Not for him the usual image of how much real estate could supply how much of our energy. Böer has written: "If converted into mechanical energy, sun shining on the United States could lift in one day the entire crust of our nation, 1,000 meters thick, about one meter into the air, and with it all our cities, lakes and forests."

In this article that he wrote for *Chemtech* in July 1973, he let his imagination go:

> Solar energy impinging at high noon on an area of 35 kilometers by 35 kilometers equals the total peak capacity of all existing power plants combined. Even if we continue to increase our demand to the ultimate saturation level estimated at 45 kilowatts per capita (22 times the current level) and if the U.S. population increases to 500 million people, only three-tenths of one percent of the solar energy impinging on the United States would be needed to fill the resulting gigantic demand."

The old problem is how to store all the power. Nature does it.

> Photosynthesis is the classical example: fossil fuels yield the benefit of millions of storage years while about 70 billion tons

of carbon is fixed every year by assimilation in forests. Phytoplankton fix 40 billion tons and other vegetation 10 billion tons per year. These processes use about 10 percent of the carbon dioxide of the atmosphere. . . . This carbon fixation rate corresponds to an average of .5 percent solar conversion efficiency.

Hydroelectric power also exhibits a low solar energy conversion efficiency. A typical hydroelectric plant's water reservoir (Philadelphia Electric Company's 512,000-kilowatt Conowingo Plant) has an area of 13.4 square miles. It produces yearly about 3 billion kilowatt-hours, although the solar radiation on this area is 68 billion kilowatt-hours per year. Furthermore, the water feeding the reservoir is collected from an area estimated to occupy 27,000 square miles. Hence the solar conversion efficiency of such a hydroelectric plant is less than .002 percent.

Since solar energy is already home-delivered, Böer reasons, one might as well use it that way. "It is there in your backyard and in mine," he has said. No big construction job is needed for the plant. "You have a deployment structure on a house: the roof is there. And so you can save some costs, as a matter of fact, considerable costs. You don't have to put the plywood or the shingles on the roof so you can credit this when you have the solar installation there. But more important still, you can do something with what the energy people call waste heat. If you convert solar energy into electricity, using any method, the overall efficiency comes out to be at best 20 percent and, more realistically, somewhat closer to 10 percent. But we can have 50 percent of the energy converted into heat. When this converter sits directly on your house roof, you can process the heat and do something useful with it. In the wintertime, obviously, you can heat your house. In the summertime, hopefully, you can air-condition your house."

The trouble with siting huge solar power plants in the Arizona desert is that they are too far from major population centers, Böer remarks. The trouble with locating them next to these centers is that, in the United States at least, most big cities suffer from cloudy and inclement weather. If the solar plant is installed in individual homes in such cities, it can work handily with existing energy supply systems, supplementing them, taking the peak load off their backs. Again, I am willing to go along with Böer. Once his lab or someone else's comes up with a cheap solar cell and a reliable heat storage medium, then the customer will have a choice. He will be able to run his house with the automation now in the basement of Solar One or, if he is willing to show some of the frontier spirit of his ancestors, he will be able to do it himself like the customs man in Mont-Louis. At least, he will have a choice, which is more than he has now.

Böer's view of energy that emerged from his sabbatical breaks down into a system and its components. It is the system that matters, the components are all replaceable. Even though cadmium sulfide cells are his life's work, it would make no difference to him if another substance should prove more feasible. George Warfield, Böer's executive director, put it nonscientifically: "Once we've got the systems aspect worked out, we won't care what the hell is up on the roof."

At present, the cadmium sulfide cells are basic to the system. They generate electricity through the photovoltaic effect which I did not understand before I became involved with solar energy and which I think I understand now only on the rare sunny days in my mind. It's all well and good to say that cadmium sulfide is the substance that works the light meter in a camera but who really knows (and who ever really cared) how the light meter works in a camera? I sought an explanation because I decided I was tired of being manipulated in my daily life by unseen mysterious forces. I was warned off. Dr. Allen Rothwarf, who was working on the problems of cadmium sulfide cells with Böer, told me gently: "The jargon hides the basic physics which is even more complicated." Up to a point, though, long before the arcana of the inner circles, it is not as bad as all that.

Getting down to basics, we start with the atom: the nucleus in the center, the electrons traveling around it in their circular orbits, the image of the solar system with the sun and its planets. Electrons furthest from the nucleus in their orbits are called valence electrons. They are the least bound to the central nucleus, the easiest to move out of their orbit. This is what happens to them when they are struck by light or, to get into the jargon, when the electron is hit by a photon, one of the little bundles of energy that make up light waves. If the electron in that outermost orbit absorbs enough energy from the photon, it is kicked out into a conduction band. This is what occurs in substances known as photoconductors. The *Encyclopaedia Britannica* puts things so clearly that one need only crib it: "Light striking such crystals as silicon or germanium, in which electrons are usually not free to move from atom to atom within the crystal, provides the energy needed to free some electrons from this bound condition" and an electric current is the result. That is why they are called photoconductors; they conduct electricity when struck by photons. Otherwise, they act as insulators. To convert the photoconductor into a solar cell, a "built-in field" is needed. This can be created by covering the photoconductor with a different material. In cadmium sulfide cells, it is a thin layer of copper sulfide; in silicon cells, it is a layer of silicon with a different impurity. This layer serves as a "hillside" for the electrons knocked loose by light. The electrons slide down the hill to become separated from their place of origin. A current will flow even without an outside battery, as in the light meter. It is the solar cell itself that acts as a battery.

Matters become even more complicated when we get away from silicon, the raw material used in the solar cells that have been sent into space to generate power so many hundreds of thousands of miles away from the nearest wall outlet. The silicon cells have all the virtues save one: they are wildly expensive (in a space program, some hint darkly, this may not be a flaw). Cadmium sulfide, one of several compounds that yield a photoelectric effect, is much cheaper because it can be applied as a thin film on a metal base. It was a starter in the American space program but was soon discarded because it was less efficient in converting sunlight into electricity and it did not stand up. Soon, it was orphaned from NASA-sized budgets and it might have languished completely if it had not been for efforts elsewhere, particularly in France, where money still mattered.

Discussion of the relative merits of photovoltaic conversion by silicon or cadmium sulfide cells lies right in the jungle of darkest controversy that obscures so much of solar energy. Again, I phoned Paul Rappaport at the RCA Laboratories in Princeton for advice and he was quite hopeful about the prospects for cadmium sulfide. Here, one can sense the almost negative influence of money in great chunks. It leads to ways of doing things that work in outer space but not on earth. Yet if one seeks to cut costs and develop a cheaper technique, the money is no longer there. Neither are the laboratories or the scientists or the engineers; they're all in space where the funds are to be found. It is nonsense to say that if we can go to the moon, we should be able to solve Problem X, Y, or Z. It is because we went to the moon that we are having trouble with so many other problems.

Rothwarf, who came to Delaware from a career in academic and industrial research, put it this way: "For space, cost is the last consideration. For terrestrial use, cost is the only consideration." In his bare office, furnished with no more apparatus than a small desk calculator and a large stock of yellow pads, he was trying to get the bugs out of cadmium sulfide cells. Quality variations in the production process must be ironed out; the cells themselves must be rendered less vulnerable to degradation, particularly by heat. Rothwarf was hopeful. "We have a good group here. We are achieving a critical size, we are at a threshold. If it takes too long for his ideas to get a trial, a guy gets bored waiting for feedback. If there are enough people, then they are tried and progress goes up. With two persons instead of one, you don't go twice as fast, you go four times as fast. Soon, we should have thirty or forty people working here on cadmium sulfide. I think that a cell with 7 or 8 percent efficiency and a long lifetime is only one or two years down the line."

Then what will happen? Dr. David B. Miller, a professor of electrical engineering at Purdue University who had taken a one-year leave of absence to work for Böer, was willing to try some forecasts. "If we get an inexpensive photovoltaic cell, then we can look into it for residential use. This is diametrically opposed to the way that electric energy development is now going with plants of one thousand or two thousand megawatts capacity. Since the sun is diffused, let's use it diffused."

Miller was wrestling with the fine points of home power generation at the institute. One is that solar cells deliver not alternating but direct current, good for charging batteries and not too much else. In Böer's first proposal, he suggested running the lights and an electric range on DC with an inverter to provide AC for other appliances. Miller thought the whole house had to be on AC: "Someone will always unscrew a light bulb to plug in an extra radio."

Just as Böer regards solar energy only as a supplemental source of power for the individual home, Miller believes that solar electricity would never be developed by the utilities to the point where it would decrease their base rather than just their peak demand. With fifty thousand homes in the state of Delaware equipped with such rooftop power plants, then the utility would begin to see results. Even without the sun, it would pay to fill batteries with off-peak power at night, then run all day on the charge. There is no need to await new exotic batteries since lead-acid batteries, tried and tested, do the job very well. As long as they are used to store power for a house and not a moving vehicle, their weight is no handicap. Miller did not think there was much likelihood of solar houses feeding power back to a central plant, but the possibility is always there in the case of a sudden peak that might otherwise lead to a brownout. It is doubly wise to keep a reserve on hand because, as anyone who has ever run a car in winter knows, batteries last longer when they are not allowed to discharge completely.

The Delmarva (for Delaware, Maryland, Virginia) Power and Light Company, one of Böer's backers, has calculated the effect of fifty thousand solar homes on its summer load. When you look at Delmarva's graph, you can see that the peak has not only been flattened but also moved into the early hours of night. From eleven in the morning until four in the afternoon, demand for air conditioning is higher but that is precisely when the sun is turning on full peak power.

V.

〜〜〜〜〜〜〜〜〜

The Sun Queen

〜〜〜〜〜〜〜〜〜

GENERATION OF POWER by cadmium sulfide cells fixed to the roof as solar shingles is but one aspect of Böer's system. Like an orchestra conductor, he is trying to get all the components of a modern home, the machines that light, heat, cool, clean, wash, and entertain its dwellers to work in harmony instead of energy-wasting discord. He has calculated that the solar shingles could collect 50 percent of the sunshine falling on the roof — 5 percent as electricity and 45 percent as heat. Some of the heat would have to be vented because cadmium sulfide cells do not resist high temperatures, but the rest could be used.

This is a strong argument on the side of the solar house. As Böer said, it can use waste heat left over from power generation. "Waste heat" is somewhat of a misnomer; it is used to describe the heat left over when big central power plants use coal, oil, or uranium to generate electricity. Since they are too central, too far away from their customers, the heat usually cannot be used in the home. It is dumped into rivers to the howls of environmentalists and fishermen or it goes into the atmosphere. As power plants get bigger and bigger and as they become more and more nuclear, there is less and less likelihood that the customers will want them around in their vicinity. And so the "waste heat" problem is bound to grow. But it is no problem to the man generating his own power. When you drive a car, you don't run an oil burner to keep warm; you use the "waste heat" that the engine has dumped into the radiator. In the solar house, Böer proposed to store the leftover heat from the roof shingles in a basement reservoir. It could either heat the house directly or, if its temperature had dropped too low, it could feed a heat pump to keep the house comfortable.

"Heat pump" is another expression that keeps surfacing in energy matters, raising irritating questions that no one seems to answer. A heat pump is first cousin to a refrigerator and it works on the same principle. In the refrigerator, a gas such as Freon or sulfur dioxide is compressed by an electric motor outside the food compartment until it liquefies. Then it is pumped inside the box where it expands into a gas, picking up heat and carrying it back out to the compressor where the cycle starts again. The mechanical energy of the compressor has been used to cool the inside of the refrigerator or, if you prefer, to warm the kitchen with the heat that has been transferred out of the food compartment. In the case of the heat pump, the primary purpose is not to cool the food but to warm the room.

This can be used to heat houses. A given amount of electricity will produce up to three times as much warmth if it operates a heat pump instead of being run through the coils of a heater. Then why don't we use heat pumps? Why didn't the appliance manufacturers and the power companies get together to bring this boon to the home? Probably for the same reason that Ford and Standard Oil did not team up to put out a Pinto in 1960. Electricity was cheap, it had to be sold.

The story of the heat pump is instructive. In a statement to the House Subcommittee on Energy, Dr. James Comly, manager of thermal research and development at General Electric, summed it up:

> The heat pump may be thought of simply as a reversible air-conditioner. When it is hot outside, it cools the indoors by pump-

ing heat to the hotter outdoor temperature and releasing it to the air just as an air-conditioner does. When it is cold, the cycle is reversed and the heat is removed from the outdoor air and is delivered indoors at higher temperature for heating.

It has the advantage that with one unit of electrical energy it picks up a half to two units of heat outdoors and then delivers the total of 1½ to 3 indoors, thus delivering 1½ to 3 times as much heat to the house as electrical resistance heaters would deliver for the same electrical energy.

Like solar energy for controlled building heat, the heat pump idea has been here for some time. It was first suggested and expanded upon by Lord Kelvin in 1852 and 1853. There were a few ad hoc demonstrations and a number of technical papers in the 1930s and by the late 1940s several hundred special installations were made.

The market really began to open up when GE and others began to sell packaged, mass-produced units through their distribution channels in the early 1950s and the industry market grew rapidly to over 50,000 units per year in the early 1960s.

After 1963, sales leveled off for several years and have never really regained their early momentum. The slackened growth was a direct result of a reliability problem caused by insufficient knowledge of the lifetime stresses faced by heat pump compressor units.

If Dr. Comly had not been talking to a House subcommittee, he probably would have said that the damn things broke down too soon.

"Although the problem was solved by the major manufacturers by 1966, the reputation for poor reliability still lingers." That's true. Every time I came across an energy researcher who touted heat pumps as a way to get more results for the same money, I ran into a critic who told me that they wouldn't last twenty months, they were expensive to buy and, besides, they were no good in small sizes for home use.

Things are now changing. The heat pump is not all that impractical, at least not in the eyes of the American Electric Power Company, Inc., which was running ads in 1974 to get public opinion behind the idea of more coal for power plants. Their line was: "America has more coal than the Middle East has oil. Let's dig it." No one would accuse them of having sold out to the ecology bloc and yet, in their campaign, they came down hard for the forgotten (save to be maligned) heat pump:

> We must heat our homes and offices and there's no more efficient way than with the electric "Heat Pump" system — for it is an energy multiplier that actually delivers up to two times more energy than it takes to power the device. [This sounds like perpetual motion but we are aware of what the copywriter was up against.]
>
> How the remarkable Heat Pump works: All air contains a measure of heat, even in temperatures well below freezing. The heat pump doesn't produce heat, it transfers heat. It absorbs heat from the outside air and discharges it inside the house at the temperature desired. Simple, efficient and a real contribution to energy conservation. That's the heat pump — the energy multiplier.

Böer had incorporated "the remarkable heat pump" into his system without waiting for the electric power industry to revive it. He thought that heat from the solar shingles on a sunny day in winter could heat the house directly or charge up a storage reservoir containing a salt that melts at 120° F. If a day is too hazy or cloudy to permit direct heating, slightly warmed air could charge a "base heat reservoir" up to 55° F., then the heat pump would keep the house as warm as needed. Once the sun comes out again, the brain in the basement could switch the heat pump off.

It would remain off even during the late afternoon when the sun dips and the collectors can supply only the low-temperature reservoir. The house would run on stored heat alone through the early evening. Only after bedtime would the heat pump come on again, charging the main reservoir when few are using electricity and the utility is only too glad to sell it to the solar householder.

Böer says the heat pump will be needed mainly during a spell of nasty weather. Two sunny days allow the collectors to charge up the reservoir enough to get the house through at least one cloudy day. During a really cold cloudy spell, an auxiliary electric heater will be used to work in tandem with the heat pump.

So much for winter. Now we come to summer when demand is for cooling. On a clear summer night, the collectors will radiate heat to outer space just as Trombe's ice-making machine does. Cold air from the collectors will be fed into the base reservoir to bring it down to 70° F. Then the heat pump can take over but, in summer, it will run as a refrigerator, transferring intensified cold to another part of the main reservoir and storing it at around 50° F. This stock of coolness can be used during the day to air-condition the house with no need for the heat pump to operate and add its burden to the power company's load.

It all sounds complex and it is. The control panel in Solar One must operate the system, switching the heat pump on and off, directing heat to the base or the main reservoir, cooling the solar shingles if they get too hot for the comfort of the cadmium sulfide cells. The French criticize this, saying that it turns the house into a factory. It does, but not much more than in any new American home where electronics replaces the labor of the housewife and the building tradesman.

Böer wants it this way so that he can keep solar energy in step with American technology. The simple solution may be right for the home inventor, but it worries Böer. If solar energy sounds too easy, then people will be tempted to build gadgets in their garages and try them on their roofs. If they don't work, then solar energy is likely to get a bad name from the ensuing backlash. "We do not want to create solar junkyards of abandoned collectors and broken-down systems introduced prematurely," Böer said. "We have a gut feeling that we must use sophisticated economic approaches and develop the total system before entering the market."

He did not think that the energy fright of the winter of 1973–74 did any good. "We didn't need an oil crisis to get us into a premature bind. It hurt too many people and we have enough stimulation as it is. We knew before the crisis that we are running out of natural gas. And we know that oil does not have that much of a future. Refineries are not being built today."

To demonstrate his ideas in three dimensions, to get them into the kind of hardware that he likes to see and feel, Böer built Solar One at a cost of about $120,000, financed principally by the University of Delaware and the Delmarva Power and Light Company. The house stands on South Chapel Street, a five-minute walk from the Institute of Energy Conversion. It looks as if it had been sliced neatly along the ridge of its roof with the northern half thrown away. What is left is a roof pitched at forty-five degrees to get the most of the winter sun in the latitude of Newark. Most of it is covered with a skylight to protect the cadmium sulfide cells that gleam expensively behind it. On the ground floor, there are six blind windows that serve as additional solar collectors. Otherwise, Solar One is a frame house built like its neighbors on South Chapel Street. It is smaller than most of them with only 1,400 square feet of living space, enough for two bedrooms, a combined living-dining room, a kitchen, a bath, a garage and a full basement. There is room for two more bedrooms in the attic behind the solar cells but this space is now occupied by laboratory equipment. When we saw the house, the main activity in the attic was aimed at developing a watertight seal so that the precious cadmium sulfide cells would work in an atmosphere free of any moisture that could age them prematurely. "Accelerated life tests" were being conducted on cells to learn if they would last fifteen years without waiting fifteen years to find out. In these tests, the cells were working twenty-four hours a day without a break for darkness. During the first sixteen months of their deployment on the roof, the cells showed no degradation in their performance and, according to Böer, the accelerated tests predicted a possible lifetime for them of more than fifteen years. But only time — real time — can tell if the predictions are right.

The living quarters of the house were fully up to contemporary expectations: a refrigerator-freezer, washer and dryer, an electric range, no penny-pinching on the power. The place had the same stuffy plastic smell of the motel room where we had spent our first night in Newark, but that, too, must have reassured the contemporary visitor. What was surprising to anyone who has lived behind a picture window was the light that got into the living room despite the solar collectors blanking out so much of the south wall.

There have not been too many bugs in Solar One. The institute claims it can get up to 80 percent of the energy it needs from the sun. In July 1973 when it was opened, thirteen hundred people went through it on a 90° day, but the temperature indoors stayed around 78°. Just before it was inaugurated by the governor of Delaware, a slight crisis arose when the mercury switches operating its lights refused to function on the direct current coming from the cadmium sulfide cells. The switches had to be changed in time for the governor to throw the first one.

The house does look better from the inside, although it did not affect the aesthetics of South Chapel Street any more than the three mobile homes in an adjoining lot. Appearances do not worry Böer. "A little controversy isn't a bad thing. Solar One is a silhouette, it was designed by the architect for maximum solar harvesting."

Judged by its value as a symbol, Solar One is successful. It has become known nationwide; its spreading fame has kept pace with the growth of the Institute of Energy Conversion. Böer regards Solar One as the equivalent of the first horseless carriage. He thinks it may be twenty years before he gets solar energy to the mass-market stage of the Model T. Yet he has been moving fast.

In 1971, when he came back from his sabbatical, he worked up a hundred-page proposal and presented it to the University of Delaware. In it, he explained how he planned to stand on two legs: the institute for research; a private company, SES, Inc. (Solar Energy Systems), for the real world. It is SES, where he serves as a consultant, that now seeks to develop cadmium sulfide cells to the point where they can be marketed. Late in 1973, SES was able to announce that the Shell Oil Company had agreed to finance it to the extent of three million dollars.

In February 1972, Böer started with two rooms, a shared secretary, and almost no money. Two months later, he already had 20,000 square feet of office space. Two years later, he was able to announce that nearly a hundred people at the University of Delaware were involved in his Institute of Energy Conversion, forty on a full-time basis and fifty more from sixteen different departments. Eight million dollars had been committed, and his annual budget was running at two and a half million dollars including one and a half million for SES. The money came from government, industry, and private sources, everywhere from Shell and Delmarva to H. Feinberg's, the Wilmington shop that donated the furniture to Solar One. Böer, like many a naturalized American, swears by made-in-U.S.A. free enterprise. He thinks that if his university can help industry and earn royalties, the citizens of the state will be rewarded by reduced taxes and lower tuition fees for their children. He does not believe that any single company should be allowed to carve out a monopoly for itself in solar energy, but it could be given an edge for four or five years if its investments bear fruit.

The university should also get a return on its human investment. "I've been a university teacher for twenty-five years. We have a responsibility to our students, we must give them something relevant. We have been preparing them to be good physicists, but some take jobs as bartenders. The theoretical basic fields of physics and chemistry are on Cloud Nine. People in industry hire only a few students from these fields and even fewer with Ph.D.'s. What is relevant today in physics is energy conversion."

With these attitudes and the money he has to back them up, Böer has been able to put a large group together in a short time. People move in and out; the turnover is high and so is the level of interest. It is all so new, this injection of vitality into solar energy, the sleeping science. Böer has drawn young scientists from near and far. Among those who came the farthest was Dr. Anwar Malik, a Pakistani who had worked at the Brace Research Institute in Quebec for six years until he moved to Newark. Malik had left Pakistan in 1962, then he did his basic studies in heat transfer at Minnesota and Michigan State before he went to Brace with the idea of doing something that might be useful in his own country. Brace is a peculiar institution in an advanced country in that it devotes all of its resources to devising techniques, particularly in wind and solar energy, that are useful in almost every part of the world except its own. Recently, Malik had worked on a solar crop drier with Professor Fred Buelow, chairman of the department of agricultural engineering at the University of Wisconsin. They have shown that air can be heated simply by passing it under a roof of galvanized sheet steel with a piece of plywood nailed over the space between two rafters to form a duct. The added cost over the normal hot tin roof is almost zero. Malik and Buelow worked on this idea to see if it could produce the dozen degrees of heat gain needed to dry a crop.

Malik would like to introduce it into his hometown of Quetta in Pakistan, not to dry crops but to warm people. Almost anything would be better than the way Quetta heats nowadays with coal stoves. "There is a terrible respiratory problem. A yellow sulfurous fog hangs over the city. I'm sure it has one of the highest air-pollution rates in the world; it's worse than Tokyo. And we should be using the coal to fire brick kilns instead of wasting it on heat." Malik goes home every year and works on ways to heat without coal. In his own house, he filled a room with black mattresses and quilts to store heat. "You can use anything, even a waterbed or bricks." He put sliding panels made from bagasse, a by-product of the sugar industry, over his windows to cut outgoing radiation at night. At eight the next morning, his room was still a comfortable 69 or 70 degrees.

In Newark, Malik had left that part of his life behind. He was working on a computer model that could project and predict the performance of a solar-heated home, using any combination of collectors and storage devices at any given latitude. "Until we can do that, we cannot say anything with precision about the economics of solar heating." This work is directly related to his training in heat transfer, a science that has flourished because of its close applicability to the space race and the behavior of nuclear reactors. When we talked to Malik, he had hopes that a solar-heated house might be built in Quetta by the University of Delaware together with the University of Baluchistan. "My interest here is mainly analysis, the physics of the processes of heat transfer, so that we can try to push efficiencies further. I am not interested in putting hardware together. But at home, I want to see solar air heaters developed for existing buildings. You might say that I am working in a number of different streams at the same time."

The streams meet at Newark. Not far from Malik's office, Dr. Tuncer M. Kuzay was putting hardware together and seeking ways to put solar energy to work on existing buildings — "retrofitting" is the word that the trade uses. Böer came from Berlin, Malik from Pakistan, Kuzay used to teach at the Middle East Technical University in Ankara before he left Turkey in 1968 to take his Ph.D. in heat transfer at the University of Minnesota. "The National Science Foundation once told us that we ought to get an American here so that they would know what we were talking about," Kuzay said.

He took us to his laboratory, a standing test bed to see how different materials collected and transferred the heat of the sun. Putting new materials to work on old principles is a modern approach to solar energy. Thin plastic films are an example: used in flat collectors, they are cheaper and lighter than glass; they are easy to frame and mount; but they allow too much heat to radiate back to the sky. Kuzay was trying to build a better heat trap with plastic films. One constantly seeks shortcuts in using the sun, for there is not enough power available to run pumps or other accessories without taking a heavy loss on what can be finally delivered. Perhaps this is premature, nothing more than a foretaste of a remote era when there will be nothing left to burn, nothing surplus to throw away.

Kuzay had more immediate goals. Up on the roof, he was testing a vertical collector to heat either a room or water. In this work, sponsored by SES, he wanted to achieve a collector in kit form that could be installed by drilling two holes in a wall. "It's a relatively new project. We'll keep going until we're satisfied with all aspects: performance, resistance to corrosion, aesthetics, marketability, and cost. If we can meet these criteria, we should have a viable product in a few years."

It was a cool cloudy day on the roof, no sight of the sun, but diffuse radiation was enough to boost the temperature in the collector under test to 100° F., thirty-eight degrees more than the reading on the thermometer, and to heat the small huts built to house the researchers and their instruments. Water from the collector was fed into a radiator running along the base of a window while leftover heat went into a storage reservoir. Thus warmed by the sun, we listened as Kuzay spoke of other projects. Delaware, like other laboratories, was trying to run a refrigerator with the sun. It was the same absorption cycle that was being tried at Odeillo, but there was plenty of scope for research on the best fluid to use as a refrigerant. Kuzay told of efforts to improve glazing, to devise collector surfaces that would absorb more heat and lose less. "When you're working in refrigeration, you cut every corner; you're crawling for every degree you can get."

Kuzay kept an eye out for uses of solar energy outside the home, particularly in agriculture. Crop drying is a promising market because of the high cost of propane, the main fuel now used in the United States. Chicken farmers looking for heat could be helped, and Kuzay saw possibilities in lower Delaware and in Kennett Square, Pennsylvania, where much of the American mushroom crop is raised. Growers have been asking about the chances of using the sun during the one-week period when they must heat their mushrooms to an equatorial temperature. Farmers are good prospects, Kuzay thinks, because they are more interested in performance than looks.

He belongs to the present generation of solar scientists, more interested in physics and the science of materials than in pioneering new principles. Böer has put together a mix of all the approaches to the sun, the new and the classical. One of the brightest stars on his team was Dr. Maria Telkes, who has been twinkling in this galaxy for more time than she cares to remember. Böer considered her the most enthusiastic of his staff, her vitality matching her convictions. She is responsible for a major component of Solar One, the "reservoirs" that store heat and cold collected from the environment or supplied by the electric company during the wee small cheap hours of the morning. It is Telkes's reservoirs along with Böer's cadmium sulfide cells that make Solar One unique among the handful of houses now running on the sun.

Telkes, too, was unique, a milestone in our travels. Our first appointment fell around noon and, upon her invitation, we picnicked in her office on rations that we brought in from a neighboring Burger King. When Telkes heard that Madeleine was born in Touraine, she told us that among her ancestors were French Huguenots who had come to Hungary. Then she recited Racine while the trucks rumbled by outside on South Chapel Street. Madeleine was impressed, as her log bears out: "We visited Maria Telkes, an adorably mad Hungarian, a genius, Professor Tournesol in skirts. [*Translator's note:* Professor Tournesol is a scientist hero of the *Adventures of Tintin,* as every French comic strip reader knows even if you didn't.] With one hand, she started to put on lipstick because I had a camera; with the other, she reached into a steel box and pulled out two, three, four, twenty-four, fifty patents on her own inventions. I liked best the plastic jellyfish that blows up like a balloon and turns seawater into fresh water. It is a friendly jellyfish that rescues shipwrecked sailors or fliers lost on their little rafts. We went out to buy some fishburgers and hamburgers with milk shakes for us and a Seven-Up for Maria. Then we sat down around a table. Dan spilled his sandwich all over his trousers, the table and the rug while Maria discreetly reached into a desk drawer and pulled out a small silver spoon. She ate her sandwich with all the delicacy of a duchess in a tea room."

Madeleine's instinct was right. While Telkes does not like to dwell overlong on the past, she did tell us that she was born in Budapest where she first learned Racine and then got her doctorate in physical chemistry before she came to the United States "some time ago." Elsewhere, a former collaborator remarked: "You can never contradict Telkes. Either she's right or she tells you she's a Hungarian noblewoman."

She is the sun queen, without a doubt. Since 1939 when she worked on the first MIT experiments with solar-heated houses, she has been foremost in the field. She has been an authority on solar stills such as that inflatable jellyfish, she worked on direct generation of power through a "thermoelectric effect," and she went further than any of her contemporaries in endeavoring to store solar energy at various temperatures. What is more, she is keeping this all up. Among her latest patents have been a "solar window" and a "bubble wall," both ways to use the sun without shuttering the windows of a house with solar collectors. "I'm not an anachronism yet," said Telkes, "although I'm not as young as I look." She does look young; she dresses with the care of a prewar European who grew up in an era when the memory of that Austro-Hungarian Empire had not yet faded. Yet the Peruvian sun she wore as a pendant on a dull Delaware day faced the future. Telkes reminds one of the story that John McPhee related in *The Curve of Binding Energy* about the theory developed at Los Alamos that all Hungarians are Martians because of their wanderlust typified by the gypsies, their language akin to no other on earth, and the fact that so many of them are geniuses. It is hard to say whether or not Telkes is a Martian, but she is certainly in a class by herself.

Storage of the sun's heat is the biggest problem she has tried to crack. This is one of our oldest fantasies, it must be ranked with perpetual motion and the transmutation of metals. B. J. Brinkworth, the author of *Solar Energy for Man,* published in the United Kingdom, found a quotation from Swift to describe it: "He has been eight years upon a project for extracting sunbeams out of cucumbers, which were to be put into phials hermetically sealed and let out to warm the air in raw inclement summers."

That states the problem. Since there is too much sun in summer and too little in winter, why not put it aside the way squirrels hoard acorns or farmers make hay? This was tried in the earliest days of MIT's studies of solar heating. Hoyt Hottel, already quoted in his latter-day role as a Cassandra, was chairman of the MIT Committee on Solar Energy Utilization. In a paper he wrote for the Phoenix symposium in 1955, he summed up the results. The first MIT house was built in 1939 as a two-room laboratory intended only to see how well the sun could be collected. Telkes, who had worked with Hottel, called it not much more than "an enlarged version of a solar water heater." It had an insulated storage tank big enough to collect heat in summer for use in winter. The tank was a giant, 17,000 gallons, and it took up the whole basement of the house. It did work. Hottel published figures showing that the collectors could accumulate enough summer heat to bring the water up to 195° F. by Labor Day. At winter's end, the tank was still at 125° F. Hottel was not particularly pleased. In his paper, he stated: "No claim was made that such storage was economically sound, and economic analyses of the time in fact produced a firm conclusion that long-time storage was not practical. The experiment unfortunately served to overemphasize, in the public eye, the importance of heat storage in space heating."

Then a second house was built, first as a laboratory and then remodeled in 1948 to house a student family with one child. It was a modest home with a steeply pitched roof carrying on its south side a collector four hundred square feet in area and tilted at an angle of 57 degrees, the latitude of the house at 42 degrees North plus 15 degrees. Many solar engineers use this rule of thumb for winter heating so they can get the most warmth when the sun is low. Trombe and several others go a step further when they work with a vertical collector that can be installed on a wall.

In this house, the storage tank was cut from 17,000 to 1,200 gallons and moved from the basement into the attic. Only three-quarters of the heat needed in winter came from the sun, but Hottel and his group had already decided that it was cheaper to install an auxiliary heating system than to try for total autonomy.

About this time, Dr. George O. G. Löf, a young engineer and another of Hottel's collaborators, went back to his native Colorado and built a house at Boulder for himself. Hottel described it briefly: "Crushed rock in a storage bin was used to store heat until needed. Like the second MIT solar house, the storage was designed to do little more than carry daytime-collected heat through the night and into the next day." He was equally laconic in his summary of another offshoot from his group.

In 1948 a solar-heated house was built in Dover, Mass., independent of the MIT solar program and engineered, designed and sponsored by the Misses Telkes, Raymond and Peabody respectively. Major engineering objectives were to prove the feasibility of complete solar heating in the Boston area despite the chance of three, four and even five-day periods of no collection, and to demonstrate the merit of using the heat of fusion of Glauber's salt as a means of storing heat.

He did not speak of the results of the work of the three Misses. That was another story that we were to hear from Telkes herself in Newark, after it had been mellowed by time.

She had been an early convert to solar energy and an early proselytizer. In 1951, she was writing for the influential *Bulletin of the Atomic Scientist* on its prospects. The solar houses that MIT had built in the Boston area could collect winter sunshine with an efficiency of 35 to 40 percent. Not bad, thought Telkes, after such a short span of investigation. She wrote:

For thousands of years, the human race shivered in the winter (many still do) using wood-burning fireplaces or stoves with an efficiency of 5 to 10 percent. Considerable research and development were needed to produce the highly efficient furnaces of today. Further research and development work on the solar heating of houses should similarly improve their efficiency and diminish the costs.

She did not see the world of that day as immutable.

There is increasing interest in the sun, as an untapped and potentially unlimited natural resource, for heat and for power. Conservative engineers treat this subject with near derision, pointing to the fuel reserves in oil, natural gas and coal. On the basis of use at the present rate and of use at a progressively greater rate, the estimated fuel reserves of the United States should last for 100 to 500 years.

A goal definitely more remote than a life-span is of secondary interest to most people, in particular to many engineers who prefer to engage in the pursuit of tangible results, achievable within a short time. However, technological changes that may occur during a century may be vividly imagined by comparing the technology of 1851 with that of today. The past fifty years alone saw the automobile and the airplane emerging from the "impossible" category. The horse and buggy appeared as the best solution of the problem of transportation and agriculture fifty years ago. The rapid development of the automobile did more than relegate the horse to the racetrack: it created entirely new problems and techniques, such as the assembly line, highway construction and gasoline stations, Suburbia, Motels and drive-in movies.

The direct utilization of solar energy at the present time may be compared to the automobile in its infancy. Like those who preferred the horse, there are those who insist that with domestic fuel reserves assured for another 100, possibly 500 years, there is no need to indulge in a search for substitutes. . . .

With the exception of the polar regions, solar energy is available everywhere; it does not need transportation facilities and cannot be cut off by sudden emergencies created by man. Solar energy cannot produce soot, fire hazards and mining accidents. It is the cleanest and healthiest fuel.

Telkes's presentation may not sound overly sophisticated but anybody who could have foreseen in 1951 the environmental and energy concerns of today deserves high marks. She concluded her article with words that she would still use:

The total research and development expenditures made thus far in solar energy utilization are infinitesimal when compared with the expenditures made in the development of other natural resources. Sunlight will be used as a source of energy sooner or later anyway. Why wait?

Telkes didn't. At that time, she had been looking for better ways to store energy. Water and rocks, she thought, are inadequate. They take up too much space. An exhibit in the lobby of the Institute of Energy Conversion, right under the wings of the Red Baron's biplane, showed what she meant. In the latitude of Boston during the winter months, it takes 300,000 British Thermal Units (one BTU is the heat needed to raise the temperature of one pound of water one degree Fahrenheit and its metric equivalent is 251.9 calories) to keep a house comfortable on an average winter day. Roughly a million BTUs are needed to heat the house for three days, the amount that a solar house would have to store to survive three days in a row without sun. The exhibit displays the sums that Telkes has worked out: to do it with stones, you will need 2,100 cubic feet weighing one hundred and twenty-five tons; with water, you're somewhat better off — 1,000 cubic feet and twenty-five tons. Long ago, Telkes considered this was too much space for the homeowner accustomed to a furnace room that takes up no more than four percent of the volume of the house that he is paying for. That could hold enough rocks to store heat for one and a third days, she estimated, and only for two days if water were used.

She decided to use heat of fusion instead. The physics of that is easy, she explained as we munched our Whoppers and our Whalers. Any substance acts in a straightforward way when one begins to heat it, its temperature rising in direct proportion to the amount of BTUs applied. Then its behavior changes when it reaches its melting or fusion point. No matter how much heat is added, it remains at the same temperature, the melting point, until it has changed its state completely from a solid to a liquid. "As a physical chemist, this is one of the first experiments that anyone performs," she said. "You melt things, you melt ice." She showed us the ice turning to water in the plastic cup holding her Seven-Up.

This requires a lot of heat. Back in 1949 ("I must have been ten years old at the time"), Telkes remarked that it takes 144 BTUs to turn a pound of ice into a pound of ice-cold water. If one then applies this same amount of heat to the ice water, its temperature will shoot up from 32° to 176° F. There had to be some way to use this effect in a solar house, and Telkes did not have to look very far. In Europe towards the end of the nineteenth century, inventors were seeking something better than hot bricks to warm beds or keep passengers from freezing in railway trains. They tried salt hydrates, inexpensive chemicals consisting of a salt and water in a crystalline combination. At a given temperature, the crystals would melt, absorbing heat. Packed in rubber bags, such a substance could be dumped into a kettle of water and heated until its melting point. Then it could be used to warm beds or feet, giving off heat until it froze back into a solid. Nothing more than simple chemical cookbookery was needed to get these freezing and melting points within a useful range. However, the salt hydrates did not turn out to be practical. After only a few cycles of liquefying and solidifying, they would "supercool"; that is, they would stay liquid below their freezing point. This happens with water, too, when it loses the nuclei around which crystals must form to start the freezing process.

Telkes thought something could be done about this. It was just a matter of adding nuclei to the salt hydrates — "nucleating" is the proper word — so that they would keep solidifying. The expected returns made the effort worthwhile. Instead of one hundred and twenty-five tons of rocks or twenty-five tons of water, Telkes calculated that she could store a million BTUs, three days of heat, with only five tons of salts taking up 135 cubic feet, not more than a large closet.

She had an early opportunity to try her idea in 1948 in the house at Dover that Hottel mentioned. Eleanor Raymond, a Boston architect, had designed the place for Amelia Peabody, a sculptor with the means to sponsor the experiment on her own with no help from MIT. To the credit of Misses Raymond and Peabody, it must be said that their solar house was a beauty. Above the ground floor, a great wall collector stared south into the sun with 720 square feet of glass. Telkes sought enough storage to get through five to seven cloudy days, a big change from the second MIT house where storage, according to one report, had been "variable, including none." Sun-heated air from the collector was blown by a fan into heat storage bins compact enough to fit into interior walls. In those heroic days, Telkes had to take what she could get in the way of materials. She needed twenty-one tons of chemicals and they were packed into seven hundred five-gallon drums.

The building of the house drew the curious. Telkes remembered how one wealthy lady came to her with the idea of storing heat in a greenhouse. Telkes gave her the name of a chemical and then, a few weeks later, met her again. "She was cool to me. She said she had sent her chauffeur to the druggist to buy a few canisters of the chemical. Then she put them into her greenhouse and nothing happened. She didn't need a few canisters, she needed a few tons."

An economy-minded New Englander filled a five-gallon drum with Telkes's salt hydrates and put it in front of a small dormer window in his attic. "He expected that it would heat the whole house. It was like using a cigarette lighter to run an oil burner."

The experiment at the Peabody house itself was inconclusive. It ran on the solar collector for five years, and then the owner converted it to ordinary heating. Telkes kept working on her salt hydrates. She was using a very ordinary substance, Glauber's salt, found in dyeing, glovemaking, and the manufacture of detergents (it is also a laxative). It melts at 89° F., a good temperature for house-heating purposes. She added a dash of borax to stop it from supercooling and a thickening agent to prevent it from settling into layers.

Since no one wanted the stuff for heating houses, Telkes found a military application. For four years during the early 1960s, she was director of research and development for a firm called Cryo-Therm. When Polaris, Minuteman, and Apollo guidance systems were shipped, they had to be kept at the same temperature for a hundred hours, even in subfreezing weather. It was Telkes who devised their "temperature-controlled shipping and storage containers." This was a way for her to stay in an offshoot of solar energy after she left MIT. A previous post as research director for Curtiss-Wright's solar energy laboratory in Princeton had ended in 1961 when the company's building was bought and razed by the same Shell Oil Company that is now serving as a Maecenas to Böer in Newark. Telkes had been working at the University of Pennsylvania when she encountered Böer in 1971 and moved to Newark.

Böer was not interested in long-term energy storage, in putting sunbeams away in pickle jars. He needed Telkes's salts to get over winter and summer humps in power demand so that he could make

Solar One look attractive to the electric power industry. She is using sodium sulfate decahydrate to fill two of the heat bins in the house. It's cheap; she can buy sodium sulfate for thirty dollars a ton and mix it with water to form Glauber's salt with a final cost of less than twenty dollars per ton or one cent per pound. Instead of using steel drums as in the old Dover house of 1948, she seals it in plastic tubes or pans to eliminate any danger of corrosion. By altering its composition, she can vary the melting point of her material. While Glauber's salt itself goes from solid to liquid at 89° F., the two heat reservoirs in Solar One contain salts that melt at 120° and 50° F. respectively. Telkes has stated that the material melting at 120° is sodium thiosulfate pentahydrate, $Na_2S_2O_3 5H_2O$, which can store nearly 10,000 BTUs per cubic foot.

To operate Solar One in the summer, Telkes cooked up a "coolness storage material." This is Glauber's salt again with what she guardedly describes as "suitable added salts, including sodium chloride and ammonium chloride." The melting point of this concoction lies down at 50° F. "In some respects, coolness storage materials act like ice, which absorbs heat at the rate of 7,900 BTU's per cubic foot of ice as it changes to water at 32° F. Coolness storage material melting at 50° F. can store nearly the same amount of cooling as ice does [7,200 BTUs per cubic foot], and in addition its volume does not expand during freezing, a great advantage as compared to ice which involves a volume change of 9 percent and the danger of breaking pipes and heat exchangers."

Just as in the heating system, these salts are sealed in plastic tubes and kept in a storage bin where air is blown over them and into the house. During the summer of 1973, she reported, they could maintain a temperature of 70° F. while outside temperature was ranging from 60° to 87°. The test was made with a standard air-conditioner switched on after 10 P.M. to freeze the salts which then melted the following day.

It would appear that such a system could be of great value, yet every other solar house I know uses water or rocks to store heat and cold. Some people say that the salts are corrosive, others say that they supercool. Telkes says: "I've written many articles about all this, but few people read them." As far as I can gather, very little money has been spent either on testing or improving heat storage materials, certainly not enough to conclude who is right in another fuzzy debate in this unfortunate science.

Telkes goes on inventing, her mind as fertile as it was the day she went sailing with a friend and saw a big jellyfish in the water. That gave her the idea for the solar still she devised during World War II at MIT. It could be carried in a package the size of a paper cup, then blown up to distill a quart a day of drinking water from seawater let in to soak a black absorbing pad. Later she worked on solar distillation at NYU and she wrote the main paper on the subject for the 1955 symposium in Phoenix. She traced its progress, starting with the great solar still built at Las Salinas, Chile, in 1872. It covered 51,000 square feet, yielded up to six thousand gallons per day and it was still working in a picture taken in 1908 and published with Telkes's article. She herself put together a 200-square-foot still that ran in Cohasset, Massachusetts, not nearly as sunny a climate as Chile's, in 1951. In producing fresh water as in everything else, solar energy lost out to then-cheap fossil fuels, except in places like a life raft or a lonely island where demand is not too high. An example of the latter is the solar still on the Greek island of Nisyros which supplies twenty-five hundred gallons a day for the population of seven hundred. The place is inaccessible enough. When Harry Lustig visited it in 1971 on his solar-energy survey for UNESCO, he reported that he took "a small commercial airliner to the neighboring, larger island of Kos, where an ancient taxi awaited to take us over backcountry trails to a small fishing port. There, we boarded a specially chartered old ferry boat which, after an hour and a half on the Aegean, landed us at Nisyros. From there, it was but a short ride on the back of a truck to the site of the installation."

Telkes did not limit herself to distilling water. She was among those who worked on a solar cooker for India, a Ford Foundation project. She spent some time studying the possibilities of solar thermoelectric generators. These are based on the principle of the thermocouple. If two wires of different metals are joined in a circuit and one is heated while the other stays cold, electricity will start to flow. The principle is widely used in thermometers and it looked attractive in the early 1950s as a way to get power from the sun without bothering with a steam engine. Telkes learned that it did not work that well. The more one heated the hot wire, the harder it was to keep the cold wire from warming up. She could not get more than 5 or 6 percent efficiency from the device and dropped it.

She kept innovating elsewhere, always leading, never following. "If you do something, everyone tries to imitate you. Then some smart aleck gets a government contract to do what you did on a shoestring. What hurts is not that he's using your idea but that he's doing it wrong. These days, everyone is making breakthroughs in solar energy, they reinvent everything. They never bother to look at what has been done. They should. If you didn't read Racine, how could you ever know he existed?"

With fifty patents to her name, she still keeps busy in her own laboratory at the institute. It is filled with old wooden workbenches; it must look much as laboratories looked during her days as a gymnasium student in Budapest. The contents are not period. She showed us drawers and drawers filled with heat storage materials in lightweight plastic tubes. Cold-storage tubes were all melted; the temperature in the laboratory was well over 50° F. The high-temperature storage material was solid and hard, since it was a long way from 120° F.

"Rocks and water aren't so inexpensive," she said; "stones take up space. You can't just fill a cellar with stones, you need air between them. Even if water costs nothing, you need an insulated tank and an expensive foundation. You may save money on a secondhand tank, but when it starts to leak you are in trouble. Many people are naïve, they do not count occupied space. Even in a basement that's expensive, and many modern houses have no basements."

Telkes has never stopped seeking new ways to collect the sun. One that she displayed resembles a section of Venetian blind in aluminum: eighty slats about half an inch apart and set to reflect the high summer sun and keep a room cool. In winter, they could act as a solar air heater. "Aluminum is a perfect selective surface when multiple reflections are used. Each reflecting step absorbs a little energy; if the number of reflections is large enough, practically all impinging energy is absorbed. Then, like a greenhouse, it absorbs energy, but it does not emit long-wave radiation." She could foresee a solar window heater with two positions for the aluminum slats, one for summer and the other for winter.

Like Kuzay, she used the institute's crowded roof to test devices that could be adapted to existing houses, obviously the biggest market. One is her "bubble wall" based on an invention she patented in 1969 to cut heat loss in glass-walled buildings. No one was really worried by the problem during the innocent 1960s, but Telkes had observed that conventional approaches hardly worked at all. Double-glazing was expensive and not as good a defense against cold as a foot-thick brick wall or the average frame house with no insulation at all. She set out to find a way to let light through a wall without allowing heat to escape. What she did was to improve on plastic foam using a transparent plastic. By blowing bigger bubbles in the foam, she could get more transparency without losing more heat. The best diameter for the bubbles was about half an inch. If they were too large, then heat would be lost through convection just as in the double panes of a greenhouse when they are set too far apart. Unlike the greenhouse panes, the bubble wall contains no parallel surfaces. This optical trick enables it to transmit 70 percent of the sun's light while allowing only 4 percent of the heat to escape. Glass lets twice as much heat out.

On the roof, Telkes had another wonder from her bag of tricks, a solar window. It was made of a new plastic, two panes of very thin coated Plexiglas, that bore up better under the weather than the old plastics and did its work three times as well as double-glazing. "This is the key to solar heating," said Telkes with her eternal enthusiasm. "It is so much cheaper than two panes of glass." It may also solve an embarrassment that has long beset builders of solar houses. During the 1930s, early experimenters found that a solar house with big windows facing south actually cost more to heat than an ordinary house during a winter in the northern United States, but that was before we all started to live in glass houses.

Telkes had something else in mind when she created the solar window and the bubble wall. Unlike so much of solar hardware, they are not hard to look at. In her patent application for the bubble wall, she remarked that colored plastics could be used as well. If so desired, one could have a stained glass window effect along with the solar heat.

Madeleine tried to take a picture of Telkes and her bubble wall but, as usual, the weather was against us. She could store heat but she had no way to put sunlight aside for a dark day. We had to wait until we returned from our further travels along Interstate 95 to Washington. Then we had a bright day and a solar panel of bubble wall on hand but Telkes was working at home. An executive assistant, Carol Wooley, volunteered to take the panel up on the roof for a photo. The result showed that solar can be beautiful.

VI.

~~~~~~~~~~~~~~~~~~~~~~

The Political Sun

~~~~~~~~~~~~~~~~~~~~~~

IT IS IN WASHINGTON that several of the streams in solar science meet in turbulent confluence. To the orthodox in the laboratories of universities and industry, Washington is a fount of funds, if not wisdom. Their science is like a premature child thrust into the world before it has grown to viability. It must be protected if it is to survive; only public money can provide the incubator. It is in Washington that the paying public finds its outlet, bringing pressure to bear upon its elected representatives so that they may speed the course of scientific inquiry and, hopefully, hasten the advent of its benefits. Then there are the heterodox, those who claim that the time for solar energy will not come five years from today nor in the year 2020 but now. They are heard in the halls of power, all the more clearly because the most prominent of their number, Dr. Harry Thomason, lives in a Washington suburb, occupying a solar-heated house for all to see, note, admire, or criticize. He has come under heavy attack from certain members of the scientific community but he needs no sympathy; he has repaid them many times over. In his public statements as well as in his way of life, Thomason is the leading folk figure in solar energy, a pole of attraction for all those who want to believe that ingenuity in a home work-shop can achieve results faster than Ph.D.'s on government grants.

When we phoned Thomason at his home in District Heights, Maryland, we were disappointed. He answered in a soft southern accent and politely told us that he would be unable to see us. Ten people were in his house, they were learning about solar energy, and he had neither the time nor the personal energy to devote to an interview. He recommended that we buy the plans for his solar house, then available for ten dollars from the Edmund Scientific Company in Barrington, New Jersey. We said with some exaggeration that we had come from Paris, France, to talk to him. "I've been getting phone calls from Little Rock, Arkansas," he replied, with no exaggeration.

There was no dearth of material about Thomason and his house. His renown had spread to Europe with the help of an Associated Press story that ran in the *International Herald Tribune* when the price of heating oil started to gain altitude early in 1974. The AP writer described how the Thomason family had paid only $15 in heating bills for the winter through February 1 while at the same time enjoying a heated indoor swimming pool and room temperatures well above the thermostat settings recommended by the government. But they were not squandering fuel oil; their heat was coming from the sun.

The newspaper story described how he did it. So did the special energy issue of *Science* that appeared about this time, as reputable a source as anyone would care to quote:

Harry Thomason's house near Washington, D.C. is a good example of what can be done. . . . A key to the system's success is its use of large volumes of low velocity low-temperature air (temperatures as low as 75° to 80° F) compared to conventional heating systems (140° to 160° F). Low-temperature operation means fewer heat losses, more efficient operation of the solar collectors, and more hours of use. Backup heat is provided by an oil furnace which, according to Thomason, is needed only a few hours per week.

The account of Thomason's home that appeared in *Mechanix Illustrated* under the finger-licking title "Can You Heat a House for $12 a Month?" related how he made his discovery:

> Thomason's breakthrough came during a hot, sunny afternoon at his wife's parents' home in North Carolina, about 15 years ago, when a sudden thunderstorm blew up. As he dashed into a barn for shelter, drops of warm water dribbling off the sheet-metal roof pelted his bare head. Like lightning from a cloud came the concept: a metal roof is a heat collector. Just run water down its sun-baked surface into a container and you've captured solar energy. When the sun reappeared, Thomason held a thermometer under the hot metal roof. Ping! It broke as the mercury passed 140°.
>
> Thomason, then 35, was supporting his wife and five children with a full-time job in the U.S. Patent Office, going to law school five nights a week and in the summer building houses to rent. Patent Office employees aren't permitted to submit a patent and he wanted one. So he quit his job, became a patent adviser at the Pentagon and spent evenings and weekends building a house to be heated by solar energy. Everyone in the family, even the 6-year-old twins, pitched in.

By far the best source for Thomason is Thomason. While we did not manage to buy his book and his plans, we found that he had given an extensive statement to the House Subcommittee on Energy in 1973. From page 274 to page 291 in the subcommittee report, it ran on in dense body type relieved only by headlines and italic lines such as "LEARN FROM PAST FAILURES" or "3. To Go Ahead Or To Continue To Stall For 5 Years, That Is The Question."

There was also a *Correction:*

> In the Congressional Record of December 11, 1971, Vol. 117, No. 194, pages S. 21460-2, an article appears about the Thomason Solar Houses. The following error appears: "There are about *two dozen* houses in the United States which use sun-

power for a *major* part of their heat requirements." (Italics added) It should have read, "There are about *four*. . . . three belonging to the Thomasons near Washington, D.C. and the fourth being in the severe cold climate of Coos Bay, Oregon, that one also using the Thomason 'Solaris System'." Although (in 1973) there are several houses under construction that should obtain more than 50% of their heat from the sun (most of them using the 'Solaris System'), only one or two others have met this difficult task and have remained in operation. The Thomasons teach that the house heating system must obtain the major part of its heat from the sun to be worthy of the title of a *solar-heated* house. One that obtains 75% of its heat from gas, and only 25% of its heat from the sun, for example (The Colorado House) is a *gas* heated house with some assistance from the sun. The house built in 1959 by the Massachusetts Institute of Technology obtained 45% one winter and 55% one winter. However, it failed completely within three years. Errors such as that in the Congressional Record have appeared in print many times.

This is good polemics but it is not quite fair to MIT where, as we noted, experimenters decided early in the game that total solar heating with long-term storage was possible but not as economical as a system using an oil furnace as an immediate alternative.

For the *Record,* Thomason struck out again at his critics:

Even today, some express disbelief and skepticism about the "Solaris System." Comments have been made such as: As far as we are concerned that house just doesn't exist, we refuse to recognize it! Is Thomason the first man in history to defeat the Second Law of Thermodynamics? . . . But the facts are: The Thomason houses are in existence, the Thomasons are not defeating the Second Law of Thermodynamics."

Then he went on to explain how his system operates:

A large but low-cost, open-flow solar heat collector on the roof faces slightly *west of south*. A heat storage bin and domestic water heater occupy about ⅓ of the space in the basement. A

simple 1,600 gallon water tank in the bin has three truckloads
of stones poured in around it. On a winter day an automatic
control starts a small, low-power pump ($\frac{1}{6}$ HP to $\frac{1}{2}$ HP) to
draw cool water from the tank and send it to the top of the collector. From there it flows down the many valleys of a simple
blackened sheet of low-cost corrugated aluminum under a glass
cover and with low-cost insulation underneath. The sun warms
the aluminum which, in turn, warms the water.

At the bottom of the collector a simple gutter collects the
warm streams from the valleys and returns the water to the tank,
passing through a simple low-cost heat exchanger to warm the
domestic water for baths, dishwashing, etc. The warm tank of
water warms the stones. When the thermostat in the living room
calls for heat it automatically turns on a very low-power blower
($\frac{1}{6}$ HP) to withdraw the chilly air from the living quarters, filter
it and send it through the warm stones, and around the warm
tank of water, where the air is warmed and returned to the living
quarters. The stored solar heat has kept the home warm in the
cold, half-cloudy climate of Washington, D.C. as long as 4
cloudy days and nights, without sunshine and without auxiliary
heat, in December. Even then all of the stored solar heat had
not been used. There was enough left in the warm water to fill a
1,600 gallon outdoor pool at a temperature of 78°. So, the
United Nations filmed scenes of children playing in the warm
pool on December 13, 1961 while the surrounding air was a
freezing cold 32°, for their official movie, "Power On The Doorstep."

In summer, Thomason uses a simple air-conditioner to chill the
water tank and his fifty tons of rocks instead of heating them,
thereby storing coolness and dryness. Like Böer and others, he
runs his plant at night on off-peak rates, thereby cutting his own
bill and not using the electricity that his neighbors need by day.
"Members of the President's Council on Environmental Quality
who visited the Thomason home and invited the Thomasons to
brief others from various other Agencies in June, 1973, seemed
impressed when the system was explained to them, with its unique
capability for keeping the home both cool and dry on hot, muggy

days without running the compressor. . . . A member of the President's C.E.Q. commented, 'The power companies should love you.' "

Thomason reminded the House subcommittee of the family's other achievements.

Thomason children, Teresa and Mary Ellen, became the youngest inventors when, prior to the age of *10* they invented the "WIGWARM," a solar heated tepee. Their invention has taught thousands of youngsters the principles of simple solar heating. . . .
The "Sunny South Model" (patents pending) looks promising for southern states. . . . This model is intended for low-cost housing. It includes a flat-roof design as illustrated at page 153 of *Popular Mechanics* magazine, June 1973. The system also includes "Pancake" heat (cold) storage under the floor. Water is pumped up to the roof in the morning where the sun warms it, and is allowed to drain back at night to warm the floor and home. Full heating and air-conditioning are believed to be obtainable in many states by other features not illustrated in the Popular Mechanics article. (Unfortunately the article was not submitted to the Thomasons for correction of errors in the various systems after the article and sketches were completely reprepared by *Popular Mechanics*. The Thomason systems are much more simple than illustrated and described by *Popular Mechanics*).

Thomason remarked that Harold Hay has used a water-on-the-roof system to heat a building in Phoenix, Arizona. Full details of this system are to be found in the same report of the House Subcommittee on Energy whose chairman, Mike McCormack, had asked Hay as well for his views on solar energy. Hay is a scientist-inventor who has successfully tapped the outgoing radiation from the earth to the stars at night, as Trombe has been doing in the Pyrenees. He wrote McCormack:

It is not recognized that about as much electricity and fuel can be saved by using the lack of solar energy at night as by daytime use of solar radiation. The Sky Therm system is a successful and economic heating and cooling method based upon selecting a comfortable interior climate from the diurnal energy flux; solar heat is used in winter and cooling to the night sky provides summer comfort. . . . Only Sky Therm uses the solar energy collector as an equally efficient heat dissipator for cooling. Only Sky Therm has the undisputed claim which appeared at the lead of a January 1970 cover-story in the official journal of the American Society of Mechanical Engineers which read in full:

"An air-conditioning system using solar radiation as a heat source, and the atmosphere as a heat sink, has been tested throughout an entire year in Phoenix. A movable insulated roof permits water ponds, in thermal contact with a metallic ceiling, to absorb and retain heat in winter and to dissipate summer heat gain to the night sky. With ambient temperatures ranging from subfreezing to 115°F, the system maintained room temperature between 68°F and 82°F throughout a normal year of Phoenix weather without supplementary heating or cooling."

Or, in nonengineering words, a roof panel slides away by day in winter to let water on the roof take up the sun's heat, then slides back at night to keep the calories from escaping. In summer, it goes into reverse: the panel stops the sun from warming the water during the day, then opens at night so that heat built up inside the house can be released. It is a beautifully uncomplicated system because the collector and the storage medium are one and the same, but it requires clear skies.

Hay writes widely and persuasively on energy and conservation. A member of the board of directors of the International Solar Energy Society, he does not have many kind words for Johnny-come-latelies who are now leaping on the bandwagon. In his reply to Congressman McCormack, he said:

Solar energy (direct use) and solar power (indirect use by conversion to electricity) are not synonymous. Solar energy can be economically used now; solar power will be uneconomic for many years — too late to meet the onrush of present and near-future needs. By confusing solar energy and solar power, the false belief is propagated that solar energy use is for the future. Statements terming solar energy use as "capital intensive, requiring large land areas, and needing large research and appropriations to develop an economic system" are applicable to solar power but not to direct use of solar energy! Such statements, by seekers of large research budgets, appear to reflect wishes for long-term sinecures and they create a credibility gap for science since most often there is premeditated avoidance of any mention of systems already developed to the stage of commercial availability.

Thomason took up the same cudgel in his statement to Congress:

[T]he Thomasons asked how so many respected scientists and engineers could be so far wrong after so many years of building and testing. They have spent millions of dollars from Government contracts, and money from institutions, private sources and corporate firms. They have devoted hundreds of thousands of man-hours to these various projects. The answer was, they pretty much have to promise more than they can deliver. Otherwise someone else will promise more and they will get the contract. Often it seems the Government money goes to those who promise the most. . . .

When the Wright Brothers invented the successful airplane the world used it instead of waiting 60 years for the Boeing "747." The world did not wait around for years to see if some of the lumbering giants would fly — many didn't, they never even got off the ground. . . . Why then is our Government waiting to try to develop the "Boeing 747," complete with solar air conditioning, which may never fly, where there is already a system in operation, that has been in operation 14 years, investigated 10 years ago and found to be worthy by the Federal Housing Administration? The Thomasons can find no rational answer to this question.

The Thomasons have not been feeding at the public trough.

For the record, the Thomason family has never received one dollar as a grant from the Government, or anyone else, for their Solar Energy work. However, they have voluntarily given the Government royalty-free licenses to use some of the $25 \pm$ inventions they have patented, or have patents pending on. They have also donated thousands of dollars worth of Solar Energy apparatus to public schools to help students learn, and set up a Solar Energy Scholarship to help a needy student in college Solar Energy studies.

When Dr. Thomason was offered a position as Consultant to NSF at full pay he counter-offered to serve to help his Government as best he can at no charge, or as a "dollar-a-year" man. The Thomasons have made no charge to Universities, Colleges and Schools that have brought large classes from near and far to see their solar houses. At the request of the National Academy of Sciences and State Department they hosted the Chief Solar Scientist of the Soviet Union, Dr. Valentin Baum, for three days and three nights in their home, at their own expense. They have acted as Consultants, at no charge, to a former Vice President of the United States, and a number of branches of the Government. They have made no charge to the United Nations, British Broadcasting Company, Yorkshire TV of London, American Broadcasting Company, Mutual Broadcasting Company, and others, for days of filming and recording in and around the Thomason solar homes. Mrs. Hattie D. Thomason has prepared dozens of fine meals for visitors from foreign countries and out-of-state visitors, at no charge, and has invited many to stay overnight.

Subsequently, Thomason served as instructor for a course in "Solar House Heating and Cooling" offered in Washington by George Washington University in its Continuing Engineering Education Program. A prospectus from the university said that the course was designed for "company executives, builders, developers, investors, construction foremen, heating/air conditioning engineers, plumbers, electricians and others." It lasted five days, from March 4 to 8, 1974, starting with the United Nations movie, *Power on the Doorstep,* and ending with a study of the latest versions of the Thomason's Solaris System. The university announced: "The fee for the course is $425. This includes a copy of *Solar House Plans,* and its latest supplement, the booklet *Solar Houses* and *Solar House Model,* lecture notes, supplies, parking, and tour of three solar homes. . . . Housing and meals are not provided. However, there is a wide variety of hotels, motels and restaurants nearby."

Thomason attracts students when he talks in terms of the present. In 1973, he told the House subcommittee:

> There are conflicting views among Solar Energy experts as to solar heating and cooling of buildings as the nation heads into a crash program to utilize Solar Energy to lower costs of living, to abate the Pollution Crisis and to meet the Energy Crisis. . . . On the one hand, *idealists* say we should attempt to develop solar house *heating* systems within five years. Those who want five more years for solar heating experimentation say we possibly could move on to *solar* air conditioning for houses within 6 to 10 years. Specifically, the *idealists* maintain: "*If* solar development programs are successful, building *heating could reach* public use within five years. . . ." (Quoted from page 5, "Solar Energy As A National Energy Resource," NSF/NASA Solar Energy Panel, dated December 1972, released February 1973, underlining added). On the other hand, the Thomasons who are *realists* first and idealists second, believe it is not necessary to wait five years. For the past 14 years they have had low-cost solar heating systems in continuous use that have supplied most of the heat for the homes from sunshine.

Hay was not as generous as Thomason to the NSF/NASA panel in the remarks he made to the House subcommittee. Two of his criticisms were:

> The panel had only 20 per cent of its members who were active in the International Solar Energy Society and none who

had produced a commercial result for terrestrial use of solar energy — in fact, it would be worth determining what percentage of the members ever had an original, commercially accepted solution in any field. . . .

The panel had too few impartial and authoritative members; it was predominantly composed of those persons interested in receiving funding for their own projects which they hoped to receive by not criticizing any other project. The attitude suggests one of "pork-barreling."

As oil prospects darkened, the future of solar energy brightened and recriminatory tones started to disappear from intramural discussions. At the Round-Table on Solar Energy held in January 1974 by the New York Scientists' Committee for Public Information in conjunction with the Energy Institute in New York, Thomason could not be blamed for playing the prophet who was no longer crying out in the wilderness:

> You might guess I'm pleased to see the presidential energy advisor, William Simon, order home and office thermostats lowered 6 to 10 degrees. It is true that he cannot order the thermostats turned down in solar heated homes.
>
> But I cannot find happiness in knowing that my neighbors and their children are shivering in their non-solar heated homes. You can be sure that Noah was not happy to see his neighbors drown just because he and his family were safe inside the ark which he had the foresightedness to build. My family and I are sorry that our neighbors have their difficulties in getting oil, keeping warm, and might lose their jobs due to the energy shortage.

Then Thomason raised a point we had heard Trombe and Robert make in Odeillo when they looked to new standards of insulation to make it easier to introduce solar heating. Thomason put it in his own way:

> The energy crisis has forced the electric power companies to admit these truths: there is just not enough electrical production capacity to switch to electric heat. In the first place, the electric

heat is terribly expensive; in the second place, extremely heavy insulation is necessary to conserve every precious BTU of electrically produced heat.

I do want to acknowledge that the electrical power companies have performed a very valuable public service. The price for heat produced by electricity is very high so in order to keep that high cost from becoming astronomical, they had to sell the public on the idea of extremely heavy insulation for electrically heated homes. That is good, and solar house builders do not have to do the selling job.

Thomason agreed with Trombe about the virtues of lower temperatures in the home.

When we had to heat our homes as warm as toast to satisfy the Americans' wasteful desires, we did it primarily with solar heat, but we used a little oil to help out. For those warm homes we could obtain only 65 to 95% of our heat from the sun. By lowering the thermostat . . . , we can obtain substantially 100% of our heat from the sun free . . . and the system can be smaller and less expensive. First, the solar heating system actually obtains considerably more free heat from the sun when operating at the lower temperatures and, second, less heat is needed when the home is kept cooler. We just completely shut our furnace off for the entire first week and for the entire last week of December. We obtained 100% of our heat from the sun, as you will see in our new book that is due off the press a month from now.

Thomason's works get around. By as early as 1973, he was able to report that he had sold 6,600 of his books on solar houses and 934 sets of plans. His methods have spread as far as the hills of Wales where a Thomason type of roof was installed as a solar collector to provide hot water in summer to the farm home of a commune known as Biotechnic Research and Development. A member of the commune, Philip Brachi, described the installation in an article he wrote for *New Scientist,* published in London:

The basic corrugated aluminum roof, anodised dark grey, went up and on in a single memorable morning; nailed direct to the rafters. . . . The corrugations run down the roof's slope, each valley being fed a trickle of water from a perforated pipe running the length of the roof's ridge. . . . The water, warmed in its passage down the aluminium, collects in a plastic gutter and gravity-feeds indoors. Flowing through a copper spiral within the hot-water cylinder of the domestic plumbing system, it yields up its heat, and is then recycled by a small pump back up to the ridge.

Brachi told how the commune, using its own labor, was able to build the solar collector for little more than the cost of ordinary roofing in Wales. Then came the test:

A midsummer day's bottle of Guiness we crack, as the first water flows up the ridge pole . . . and returns too hot to touch. An awesome heat from so soft and gentle a sun. The first solar shower. . . . Even grey-long days were worth a couple of showers. With its present small storage (soon to be increased eight-fold) the system must work when we adjust our day: afternoon bathing as well as by evening. One's life is gently changed, it seems, by a more honest link with the font of our Earth's well-being than that purveyed by the energy moguls.

With all this, one is tempted to ask: what are we waiting for? Why not listen to the Thomasons and the Hays and convert the temperate zone to solar heating, using their techniques? An answer was given at that same round table in New York by Dr. Alfred J. Eggers, assistant director for research applications at the National Science Foundation in Washington. He said:

I have been to Dr. Thomason's home and I was there with Congressman McCormack and his committee. There is no doubt that it is an operating system. It is very suitable for the needs of Dr. Thomason and his family. I think there are very fundamental questions, though, in terms of the widespread applications of the technology. Such widespread application will be necessary if the costs are to be brought down to a level to make it economically competitive with other modes for providing heating and cooling capability.

The more we read about Thomason, the more our curiosity grew. As he had explained, his valuable time was all taken up but we reasoned that we could look at the house, if nothing more. An opportunity soon arose when we were returning from an appointment on the edge of Washington. District Heights was only a few exits east and south on U.S. 495, the beltway that circumnavigates Washington so that the sites of history are never more than a few exits removed from the seats of government. If the Exxon map of Washington that we brought back to Brittany as a memory-refresher is correct, we left the beltway at the Highway 214 junction, then snaked along Walker Mill Road running into District Heights.

One could not miss the Thomason house. We liked it at once; in that stretch of suburban Maryland, it was the only home for miles around with its lawn inhabited by ducks, geese, and guinea hens. There was no one home, the doors were locked. Next to the garage on the north side of the house stood two cars and a cabin cruiser roosting on a trailer. Over the south roof stretched the solar collector, a broad sheet of water moving slowly down a slope of darkened aluminum faced with glass. The house was big, low, and rambling; except for the collector, not too different from its neighbors on Walker Mill Road.

We waited for a member of the family. We were not alone as we waited. A lady from a farm near Annapolis had driven up. We talked next to the garage; she was an oysterman's wife and she ran their eight acres while her husband was out working in Chesapeake Bay. Following directions, she had come here to buy goose eggs. While she had never heard of the solar house, she was impressed by the way its owner had raised guinea hens. Madeleine and I wandered about, taking pictures until we encountered another photographer. He was Bruce Young, a thirteen-year-old pupil at the Sheridan School in Washington, and his mother had driven him to District Heights to carry out a project on the solar house. "The teacher makes us write three million reports and then she approves one," said Bruce. But he liked what he saw at the Thomasons'. "The sun is the ultimate public utility," Bruce continued, sounding something like a speaker at a round table; "I'd like to build a solar roof of my own and live under it."

We did manage to get under Thomason's solar roof by an un- witting subterfuge. While we were taking notes and pictures, two friends of the family drove up. The ladies took pity on us, took us to their homes, plied us with food and drink and then, once they knew that the TV news program that absorbed Thomason's atten- tion every evening was over, they took us back to the solar house. Almost ashamedly, we entered inside this Trojan horse. We found Thomason in the flesh much as we had found him on the telephone, a hurried, harried man. He was slender and almost hollow-eyed, somewhat the prisoner of his success. Yet his manner was courtly and we had time to pay a short visit in his living room while he awaited yet another phone call due at eight o'clock from a seeker of solar knowledge. It was a lived-in room, not a laboratory; it was filled with the objects that a family acquires and allows to accumu- late. Thomason talked pleasantly about this and that until I asked him how he had first got the idea of installing solar heat in his home. I had read the story but I wanted to hear it in his own words.

His tone changed. He repeated what he had said on the tele- phone: he had been talking ten hours a day for two days running, he was expecting another phone call. He was losing his time and strength replying to hundreds of letters the world over. To raise his morale, I told him how we had seen young Bruce Young taking pictures of the house as a school project. "He's just one of thou- sands," said Thomason.

The telephone rang once, he started, the telephone stopped. "They must be trying to reach me." The big grandfather clock in the living room started to chime, its gilded hands read eight o'clock. Before the bells died away, they were echoed by the telephone. Thomason rose and went into an adjoining room. As he disappeared, he turned and thanked us for the visit. "Sorry I couldn't spend more time with you," he said.

We went out, closing the door behind us and then, as we reached the car parked on Walker Mill Road, Madeleine realized she had left a bagful of lenses in the house, Thomason's house that stayed locked and shuttered all day long, its telephone ringing unanswered except by appointment. What new stratagem would we need to get back into the house and recover the lenses? We tried the screen door, it was unlocked. We tried the front door. Unlocked as well. We entered the living room, picked up the lenses and left. Thomason, talking in a low voice in the next room, did not hear us as we came and went. After the sun goes down, his front door is unlocked.

He has been living in the house on Walker Mill Road now for some fifteen years, most of them in peace under the steady cascade of water that comes sheeting down his roof over the black collectors. I have never seen an evaluation of the house to show how much capital must be invested and amortized to produce its solar heat. Nor do I know the costs of maintaining the systems. Two of the collector panes were out at the end of a long winter and they must have represented a loss of heat, though not enough to prevent the others from keeping the house comfortable on a clear cold April day after a frosty night. Not only does Thomason run seminars, he teaches by example. Schoolchildren look at his house and write projects about it. One of his neighbors had been thinking of trying a little solar energy experiment of his own, perhaps improving on the Thomason system with a roof that could turn to face the sun. That is precisely the sort of complication that Thomason avoids but, at least, a process of inquiry has started. Thomason may not have published data in scientific journals, but he has been obliged to let his house stand up to the scrutiny of his neighbors who go through the same winters that he goes through and who know how much heating oil is being delivered on Walker Mill Road. Throughout the world, the man who builds his own solar heater often builds it the Thomason way which lends itself more perhaps to individual initiative than to mass marketing.

Before the oil scare of 1974, Thomason's phone seldom summoned him to supply his solar answering service. Now it seldom stops, it has invaded his professional and private lives. In the winter of 1973–74, his neighbors in District Heights cut their oil consumption by a good 20 percent, so we were told by a local dealer. They lowered their thermostats, they shut off rooms while the dealer's supplies fluctuated in price according to mysterious influences much higher than District Heights. The price remained well within the range of the neighborhood; the oil dealer's average customer was spending four hundred dollars a year for heat. Some of the unaverage customers could spend much more, as much as twenty-one hundred dollars a year in the case of a lady we met. When I asked her how a solar heating plant would look to her if it would save a thousand dollars a year, she answered: "What's a thousand dollars?" She was straight out of the era of wonderful nonsense that was ending; the era when the ladies of suburban Maryland flew along the beltway in their Thunderbirds to borrow a cup of sugar or exchange some conversation, strolling down the beltway in their V-8 slippers, ten miles to the gallon.

I must not knock the beltway too zealously; it enabled us to compress the solar systems in and around Washington into a few afternoons, to run the spectrum from Thomason in District Heights to the authors of that report on solar energy produced by NSF and NASA. We were obliged to use public transportation only once and then we longed for the beltway as we jolted down Connecticut Avenue with a cargo of senior citizens riding on reduced fares. Every aspect of that bus, from the schedule to the vandalproof plastic seats, had been intended to drive passengers to drive themselves.

Our search for the source of the report by the NSF/NASA panel led to the University of Maryland at College Park where the panel's "technical coordinator," Dr. Frederick H. Morse, serves on the faculty of the mechanical engineering department. We spent only an afternoon at College Park, hardly time to acquire many more impressions than the sound of "Maryland, My Maryland" chimed when the campus carillon struck the hours, and the sight of students surfing about on skateboards once they had parked their cars. I remember, too, a conversation with a young man who was working on a scheme to generate power with windmills contrived from helicopter blades declared surplus by the U.S. Navy (apparently he's wrong; they won't work vertically).

Morse had been no less imaginative in his activities both as an engineering professor with the university and as a "program manager" on loan to the National Science Foundation. In a piece that he wrote for the *Chronicle,* a quarterly published by the university's graduate school, he remarked that the Potomac Electric Power Company which serves 425,000 customers over an area of 643 square miles could generate all the power it needs from the sunshine falling on only 27 square miles, 4 percent of the area it covers. In 1972 his department received an NSF grant to organize the solar energy panel and Morse's career took a new turn.

Previously he had been teaching mechanical engineering at Maryland with heat transfer and thermodynamics as his main interests. Then a friend, John Fairbanks, working at the NASA/Goddard Space Flight Center just north of Washington over the Maryland line, told him of a problem that had been contracted out to the Lockheed Missiles and Space Center. Arrays of solar cells mounted on spacecraft would turn out three times as much electricity if a way could be found to keep them facing the sun as the spacecraft moved across the cosmos. This could be done by electric motors and mechanical linkages, but they represented something else to go wrong ten thousand miles from the nearest screwdriver. Lockheed and NASA were working on the idea of a "thermal heliotrope," a device that would always turn to the sun like the flower whose name it bore.

Morse went to work on it as well; his students produced analyses of the system, and support came from NASA. The mechanical flower is based on a helix, a spiral composed of two different metals. One end of the spiral is fixed to a spacecraft or, on earth, a roof and the other might carry an array of solar cells or perhaps a small parabolic collector. When the sun heats the spiral, the two metals expand at different rates, twisting the spiral. It keeps on turning until it falls into the shadow of a small shutter mounted on its free end. Through this feedback, it controls itself.

It looks like one way to keep a solar collector facing the sun without either an expensive tracking system or a cheap laborer to move it every so often. This makes it attractive in solar cooling where, as one is told time and again, much higher temperatures are needed than in a heating system. When we saw Morse, he was trying to spray some of the bugs out of the heliotrope. A way had to be found to protect it against high winds, something that NASA did not have to worry about in outer space. Morse thought that it might be incorporated into a new form of collector, a tube or half a cylinder, that would be less likely to blow away.

The heliotrope could help provide heat for a solar air-conditioning system that Dr. Redfield Allen, another Maryland engineer, was studying. He had been examining all the variants on the absorption refrigerator that have been used. It is hard to buy appliances on the market and adapt them to the sun because they are built to run on the much higher temperature supplied by a gas or a kerosene flame. It's not that the problem is impossible or even difficult, it is just that there has never been any great incentive for anyone to solve it. One cannot pour government money in at the top and wait for results to come out at the bottom. The air-conditioning industry is not interested in this sort of aid, Morse observed. "It's a competitive industry, they don't want government-funded research which they cannot patent. We have started to work closely with them so that what we do will be meaningful." Allen was not ready to predict what system would finally be adopted. He was studying not only absorption coolers using various fluids but also devices running on a solar-driven heat engine. Morse was optimistic: "I believe that, five years from now, we will see experimental homes cooled by solar energy using any one of these approaches."

In the meanwhile, they helped spread the word. Allen edited the proceedings of a "Solar Heating and Cooling for Buildings Workshop" that was held in Washington in March 1973 while Morse served with the National Science Foundation during the 1972–73 academic year. There he could witness the takeoff of solar energy in government circles. When he arrived at NSF, there was only one worker with Dr. Lloyd Herwig, then director of Advanced Solar Energy Research and Technology in NSF's program of "Research Applied to National Needs." When he left a year later, there were twelve, not a Pentagon-sized operation but at least a start toward putting some reality behind the conclusions of the NSF/NASA Solar Energy Panel on which Morse had worked. The panel, scourged by Thomason for namby-pamby caution, made some forecasts that must have sounded like wishful thinking:

A substantial development program can achieve the necessary technical and economic objectives by the year 2020. Then solar energy could economically provide up to 35 percent of the total building and cooling load, 30 percent of the nation's gaseous fuel, 10 percent of the liquid fuel, and 20 percent of the electric energy requirements.

If solar development programs are successful, building heating could reach public use within 5 years, building cooling in 6 to 10 years, synthetic fuels from organic materials in 5 to 8 years, and electricity production in 10 to 15 years.

Conservative though this may appear to some, it was a bold statement by a bureaucratic body at the time. For twenty years, up until 1971, the United States government had been spending no more than $100,000 a year to support research on solar energy. "Congress just wasn't there with the support," Herwig recalled in his office at the National Science Foundation in downtown Washington. In the 1971 fiscal year, NSF took a million dollars out of the hide of other programs and put the money into solar energy. They doubled the figure the next year, then it went up to $13 million for the 1974 fiscal year. Herwig accounted for the scanty budget with all the care of a housewife counting her pennies: in 1974, the sun also got $900,000 from NASA, $600,000 from the Atomic Energy Commission, $200,000 from the Defense Department, and $200,000 from the Postal Service. That was the last complete fiscal year before the birth of the Energy Research and Development Administration along with the Nuclear Regulatory Commission in January 1975 as the old AEC faded away.

NSF does not spend its share of the sunshine fund to put up a front. It hides its headquarters in an office building; the only sign in sight from the street says that this is the home of the Madison National Bank. There are no security guards, there is no parking lot. NSF operates in the front lines, right behind the trenches of the subway that Washington has been digging for itself, the rapid transit system of the year 2000 that is to be born only seventy years after the IRT in New York, a century after the Underground in London.

Herwig's office was in the Advanced Technology Section of NSF. He excused himself because he was late. His telephone had been ringing when we turned up and it could not be turned off. He hardly had time to lead us through the legislative thicket of solar energy in Washington when the phone rang again. We tried not to listen, we concentrated on the decor of his office: a potted plant growing like a beanstalk, a painting of a solar farm, two models of reflectors, a blackboard. Herwig's voice rose slightly: "If it's fifteen to twenty-five million dollars, NSF doesn't have that kind of money." Of all the objects in his office, the one that awed us the most was a calendar showing the three forthcoming months at one glance.

The phone call ended; it was obviously from another department wondering how to get money from NSF. Before he had begun to oil the wheels of government, Herwig had been a physicist. He had worked on nuclear reactors for Westinghouse and laser technology for United Aircraft before he joined NSF in 1964. In 1970 he was drawn into solar energy. "It was a little visionary then." He was a down-to-earth man; it was here, in his office, that the high hopes and the bright prospects had to be worked into budgets and schedules. He listed the six fields where NSF was putting its money down: solar energy for buildings, solar thermal conversion (such as the power tower), photovoltaic conversion (the silicon or the cadmium sulfide solar cells, for example), bioconversion for fuels (you grow it and you burn it), wind conversion, and ocean thermal difference conversion (the former is self-explanatory, the latter means trying to generate power with the difference in temperature between the top and the bottom of the sea in the tropics). Along all these lines, NSF hopes to achieve "proof-of-concept" experiments which could lead to reliable systems that will work in the outside world. NSF does not conduct this research itself but gives money to other agencies or to industrial and university laboratories to do it. The sums are not great but they are big enough to steer work along these lines because there is very little money around from other sources.

It was not lack of money that worried NSF. "Congress is now pushing us hard," Herwig said. "They want to see hundreds of thousands of homes using solar energy by next year. But the technology is primitive. We need another two years before we can come to any conclusions." Talking about the more distant future, Herwig was equally circumspect. He hoped to see the government's solar research budget rise above $50 million a year, as it did with no great trouble in fiscal 1976, and then to several hundred million dollars. "There are so many variables. We do not know how long gas and oil supplies will stretch, nor whether nuclear systems will be shown safe and acceptable. If they are thrown out the window for any reason, then solar energy will develop a hell of a lot faster."

While most of NSF's programs looked five years ahead for results, some had already taken shape. Solar heating systems had been funded for four schools at a cost of $1.3 million and they went into operation in the late winter of 1973–74. "Schools are excellent as a first application of solar energy," Herwig said. "Some have been shutting down for lack of fuel. They are public institutions and they need large installations. We can use schools in different parts of the country as test beds."

Then NSF has helped finance a "mobile research laboratory" to be housed aboard two big trailers. The laboratory had been built by Honeywell out in Minneapolis and it was to move all around the United States to judge the performance of solar equipment in various climates and, at the same time, to let what Herwig listed as "designers, architects, building contractors, zoning and building code officials, mortgage lenders, and others" see solar energy in action for themselves.

The laboratory was making its first stop in the Washington area at the National Bureau of Standards in Gaithersburg, Maryland. On Herwig's advice, we decided to inspect it for our first look at NSF in action. A phone call to the Bureau of Standards confirmed that the mobile laboratory was out in the parking lot. The next morning, we pointed the Nova north to Gaithersburg after making a panic stop on Connecticut Avenue for gasoline at the one station that was open. We drove miles until we got to the Bureau of Standards, then we drove miles through its parking lot in search of the mobile laboratory. The girl at the reception desk said the laboratory was out there somewhere. We wandered over acres, a police cruiser turned up when we overlooked a stop sign and came to our rescue. The laboratory had left two days before, said the cruiser crew. They didn't know where it had gone to; neither did the receptionist in the lobby of the Bureau of Standards. One of NSF's solar energy projects was missing.

We found it, with the help of the public relations officer at the Bureau of Standards. He had our answer in half an hour: the mobile laboratory had moved down to Fauquier High School at Warrenton, Virginia. This was also the site of one of the four school-heating experiments that NSF was undertaking.

VII.

~~~~~~~~~~~~~~~~~~~

Carry Us Back to the
Old Energy Plantation

~~~~~~~~~~~~~~~~~~~

WE FLED WASHINGTON. Like a stone whirling at the end of a sling, we spun off the beltway, out of the dormitories and the preserves of bureaucracy, south and west towards the Blue Ridge Mountains. Warrenton lies forty-five miles from Washington, close enough to reach it, far enough to be left in peace. A turn off U.S. 29 took us straight into town, bypassing the bypass with its new motor city. We had no idea where Fauquier High School might lie, but we wanted to sneak a look at Warrenton before we hunted it. A hill led up towards the white spire of the Fauquier County courthouse and a free municipal parking lot. We got out, walked around and looked around. To our regret, we had only a few moments for a glimpse of the old hotel, once a tavern where General McClellan took leave of his officers during the Civil War, and the statue of Chief Justice John Marshall, a native son of Fauquier County. We had not been long in Warrenton before we were told, and told again, that the government had made a survey a few years back to pick the most pleasant places to live in the United States. Of the seven that were found, Warrenton was the only one east of the Mississippi River.

Gaithersburg and the parking lot of the National Bureau of Standards, big enough to lose a whole solar energy program, were far behind. So was breakfast; it was time for lunch. As usual, the food started Madeleine's literary juices flowing and she put everything down: "Warrenton was a tiny village, all pink, blue and white with shop signs bursting out against the hard blue sky. We ate in a real drugstore: on one side, there was a pharmacy, a sweet shop, a *parfumerie;* on the other, a plastic counter. Under each plate, a small paper tablecloth with a drawing of two clasped hands next to a prayer of grace for Jews, Catholics and Protestants. A little elderly lady, all pink and smiles, her hair white and fluffy, her eyes a transparent blue, was the cook. She went to work before our eyes, before the mirrors that reflected her face back at us, that reflected the glass cases full of cakes, the colored plastic signs each displaying the name of a different flavor of the ice cream that she served us, enormous creamy spongy, sprinkled with walnuts, almonds, peanuts, striated like marble with pink and green veins, brimming on all sides over the sweet brown cone that became its handle. Then we walked through the village, the white quarter and the black quarter where a big beautiful girl asked me if I wanted to take her picture. She laughed, I laughed, we winked at each other. I would like to live here the way I live in Brittany, I know I could. We walked a little further, then the village fire alarm began to shriek. We ran up to the yellow and white firehouse in time to see the firemen racing out in their beautiful red gleaming chrome trucks, the last man sitting on the rear of the hook-and-ladder, holding his steering wheel, his yellow hat jammed over his ears."

The firemen roared down Main Street but it must have been a false alarm because they soon came back, meek as lambs. They backed their trucks into the firehouse and the bystanders returned to their banana splits at Rhodes' Drugstore. And we could get back to our solar collectors. That day, with the Honeywell trailer and the school heating plant all within the village limits, Warrenton must have been gathering more solar BTUs than any other community in America. All we had to do was to take a right turn at the courthouse, go out to the bypass and keep going a few hundred yards beyond it to reach the Fauquier High School grounds. Oblivious to the scientific history that was being made around them, kids were shooting field goals on an outdoor basketball court where the lost solar trailer was parked. Behind it on an embankment, the collector of the school's heating plant was taking the afternoon sun.

We were given a tour of the plant by Douglas Eaton who worked for the InterTechnology Corporation in Warrenton, the builder of the "world's largest single planar solar collector." Eaton, who described himself as a "staff engineer and half-baked designer," had been given the job of getting the collector and the heating system up and into action within eight weeks. He did, with the help of a willing crew of building tradespersons (they included a girl laborer hired in the name of equal opportunity employment). The collector faced south and it was tilted at a 53-degree angle to get the most that could be gotten from the sun in January. In a single span of glass panels, it spread over 2,400 square feet, held up by the girders of an Eiffel Tower-like framework. The collector heated ten thousand gallons of hot water stored in an underground tank, a concrete transformer vault that happened to be around. From the tank, water was being piped to five mobile classrooms that had been parked behind the main building of the school. They had been there on a temporary basis since a junior high school had burned down, but they seemed to have settled in for good. They were only mobile homes, "protected" from the weather by aluminum panels. One hundred and fifty children occupying 4,000 square feet of classroom space were being heated there by the sun.

Herwig was right, schools are good candidates for solar heat. Storage is less of a bother because the heating system can be shut down at night. Eaton said the classrooms were kept at 60° F. at night. Then a pump went to work at 7:30 A.M. to bring them up to 70° when school opened. The ten-thousand-gallon tank stored enough hot water to last through ten overcast days. "We're helped by the sensible heat that we get from the children. I suppose that's the only time that high school kids are sensible. Even when he sits still, each kid is giving us three hundred and fifty BTUs and he never sits still."

The heating system went into operation at the end of March so as to catch the tail end of the 1973–74 heating season for data collection purposes. It worked too well. The collector kept the water so hot that it went on circulating even after the pump was shut off. The temperature in the classrooms went up to 95° and one pupil prayed: "May the Lord save us from solar energy." Eaton consoled him — "Don't worry, son, you're sweating for science" — but he installed valves just the same to shut the heat off.

Eaton said the principal of the school had been cooperative from the start. "We told him that we could heat his gym and that would be a good thing for his football players. He's gung ho on football." He also must have been running a happy school. One of the questions that is often raised about solar heating is the vulnerability of glass collectors in what is euphemistically called the urban environment. At Warrenton, no one had to ask such questions. Even the classrooms in the flimsy mobile homes were well ordered, without a sign of the blackboard jungle. Only one kid had given way to the temptation to shy a stone at those 2,400 square feet of glass shedding their warmth over the temporary buildings. No other form of heat had been used since the start of the test, Eaton said. When the school's main heating system broke down during the 1974–75 winter, only the solar classrooms remained warm.

His employer, Dr. George C. Szego, the president of the Inter-Technology Corporation, was much more affirmative: "Our school heating system is roughly 50 percent better than Brand X. And it costs 50 percent less than Brand X." We met Szego in the offices that InterTechnology occupied out on the bypass next to the Warrenton Bank and across the road from Howard Johnson's. The tone of the place was set by a row of yellow hard hats hanging on hooks inside the door. Szego is a chemical engineer with a curriculum vitae that reads like three lifetimes. He was born in Hungary and came to the United States for a coast-to-coast education, starting with a bachelor's degree at City College of New York and working west through the University of Denver to a doctorate in chemical engineering at the University of Washington. He admitted speaking German, English, Hungarian, Romansch, Russian, and Spanish with a smattering of Turkish, Chinese, Arabic, the Scandinavian tongues, "and a little French." All this had been placed at the disposal of various employers like the Institute for Defense Analyses where he worked on the Manned Orbiting Laboratory, the Space Technology Laboratories for "systems engineering, space programs and weapon systems of a wide variety," the General Electric Company as manager of space power and space propulsion operations, and at Seattle University as head of the department of chemical engineering. Then Szego made a successful reentry from outer space. In 1968, he founded InterTechnology. "We decided to work on the energy and power problems of this world, rather than the other world. You might say our theme has been: The Resource Crisis, or Is There Intelligent Life on Earth? Or . . . On Account of Circumstances under Our Control, Will Tomorrow Be Canceled?"

InterTechnology did a three-volume, 1,500-page study for the National Science Foundation in 1971 on energy in the United States. It had made Szego a formidable interlocutor in any discussion of the problem. He was worried by it before such concerns came up. While his ideas do not ring as new as they did in 1972, they are worth hearing. What bothers Szego is the rate at which energy use has been going up, not only in the United States but in the rest of the world. He notes that the rate is growing three times faster in the world as a whole than in the United States, 4.5 percent versus 1.5 percent. "It's not just American wastefulness. Per capita consumption may be less in the rest of the world, but there are so many more people involved . . . three and a half billion of them." Szego does not argue the justice of fair shares for all; he is more interested in what there is to share. He talks in Q's, one Q being 10^{18} (that is, one followed by eighteen zeroes) British Thermal Units, the equivalent of thirty-eight billion tons of bituminous coal, if that is any help. It becomes easier to grasp when he explains that from the dawn of history until today, we have used fourteen Q's worth of energy — and we will be using another fourteen Q's in the next twenty-five years.

How much is there left to use? Szego thinks something like two hundred and thirty Q's. Thirty are represented by oil and gas reserves, including shale, tar sands, and new sources such as the North Slope of Alaska. Energy is being used up at a rate between one-third and one-half a Q every year, which means that we do not have much left to use up.

Then there are 200 Q's available in the form of coal. "Oh boy, that's three hundred years of coal . . . but we can't burn it. Its sulfur content averages 3 to 3.5 percent. Over two hundred years, we would be releasing twenty-five times as much sulfur as we do now. The world can't handle sulfur dioxide; it's a toxic substance. It adds eighty-five dollars a kilowatt to capital costs to treat sulfur in an electric power plant. That's a lot of capital; don't forget, capital is getting as scarce as oil.

"But the real stumbling block, the brick wall, is carbon dioxide. The present content in the atmosphere is three hundred parts per million. In seventy-five years, it will go to six hundred parts per million. As we all know by now, it creates a greenhouse effect by trapping infrared rays within the earth's atmosphere. If the average temperature of the planet were to go up by one degree centigrade, we would begin to get into climatic changes beyond our present experience. We might see sustained droughts. A rise of three degrees centigrade will lead to the irreversible melting of the polar icecaps. So we don't have two hundred years in which to burn coal.

"What about nuclear energy? There are only two long-term hopes, the breeder reactor and fusion. I do not think that the deployment of breeder reactors will ever be permitted on a large scale. The public is now too cognizant of the dangers of a plutonium economy. The value of plutonium is to heroin as heroin is to gold. Plutonium is by far the most potent hazard introduced by man. The half-life of plutonium-239 is twenty-four thousand, nine hundred years. If you spill it, it will decay so slowly as to be forever.

"As for fusion, no one has yet demonstrated the technical soundness of a fusion reactor. It is a long way off, one hundred years or never."

That leaves the sun as a source of power. Szego has done some thinking about that. He sees an immediate use for low-temperature collectors such as the one heating the Fauquier County High

School. He was much less optimistic about photovoltaic conversion, electricity from solar cells. "Photovoltaics are nowhere. The capital costs for a power plant are three hundred dollars a kilowatt for fossil fuels, six hundred dollars a kilowatt for nuclear, and three hundred thousand dollars a kilowatt for photovoltaics. I agree that the costs will come down, but not enough in the next fifteen years to make any difference." While not on the same plane, costs also bar the way to "high-temperature photothermal use," engineer's shorthand for big collectors that concentrate the sun. "The price is now twelve thousand dollars a kilowatt without storage — and the sun does not shine all the time."

Instead, Szego proposes we use the sun the way nature does: to grow plants which can then be burned to produce power. Wood fueled the United States until long after the Civil War and, even today, Szego claims that it generates four times as much power as all nuclear sources put together. Pulp mills burn their wood waste to make their own steam and electricity. At current prices, Szego calculates that it costs them $1.50 per million British Thermal Units, as compared to $3.50 if they burned oil. And this is being achieved with crops and trees used "fortuitously" for fuel. He and Dr. Clinton C. Kemp, another chemical engineer who is a vice-president of InterTechnology, have calculated that corn grown in the Midwest for silage can convert almost .7 percent of the sun's energy falling on a field every year to fuel, but they do not propose to burn corn. About the same values can be achieved with sycamore, and they are convinced that the performance will go up to 1 percent for a crop of trees raised solely for fuel.

This is the "energy plantation." It has a number of advantages. Perhaps the biggest is its capacity to store solar energy. The sun falling on the field or the forest is converted into plant matter which can be burned on demand, like coal or oil. But, unlike the fossil fuels, the energy plantation's crop will be almost sulfur-free because plant matter usually contains less than .1 percent of sulfur. Ash from the furnaces would be used as fertilizer, an added benefit rather than a waste disposal problem.

What the energy plantation does is to extend the daily use of solar income to a seasonal or an annual use. In a way, this is the best of all possible worlds. The summer's sunshine can be stored for winter use but neither the heat nor the carbon dioxide balance of the earth will be tilted as they are now. Energy released when wood is burned is energy that has been supplied by the sun only recently; it is not the unlocked accumulation of hundreds of millions of years. Kemp and Szego remark that the forest consumes as much carbon dioxide as it will ultimately release when it is burned.

Then there is the use of land. Writing in *Chemtech,* the two engineers describe it in almost idyllic terms:

> Consider a tract 15 to 30 miles on an edge. In its center is a power plant. One area of the tract is being harvested. In another area crops harvested earlier are curing in the sun. In still another area cured product is being taken to the power plant to be burned . . . because this is an energy plantation. It is grown for the main purpose of fixing solar energy. Of course during its maturation period it has all the other uses to which people put green areas.

To get an idea of the amount of land involved, Szego and Kemp studied the paper industry in the south of the United States. By the end of 1971, there were twenty-six pulp mills there producing a thousand tons a day or more of pulp. To support each mill, about three hundred and fifty square miles of pulpwood forest were required. They think that a forest this size is enough to feed a 400-megawatt electric generating plant even if only .4 percent of the sunshine falling on it is converted to fuel. That is enough electricity to meet the needs of a city of two hundred thousand at current consumption rates.

When it was first mooted, the energy plantation drew scorn and snickers. There were those who said that it would take more energy to haul and harvest the crop than it would ever produce. But now others have joined InterTechnology in suggesting the use of photosynthesis as the easiest and quickest way to get power out of the sun.

During Kemp's stay in Brazil from 1958 to 1962 with a paper

company, he came across the Paulista Railroad that had been built in 1903 to serve São Paulo and its hinterland. The railroad ran for two hundred miles and six plantations were set up to supply wood along the way to fire its steam locomotives. Kemp thinks that this was the first time a crop was actually planted for energy. In 1900, the railroad sought a suitable tree for its plantations: it had to be straight and without branches so that it could be chopped into five-foot sticks that a fireman could handle. Eucalyptus was the species chosen. Five trees can be sprouted from the same stump, the wood is so dense that it floats with only the top of a log breaking the surface, and it grows quickly in the Brazilian climate. The line operated on its self-renewing fuel supply until 1958 when the antiquated steam engines were replaced by diesels. Kemp learned of it when he worked on plans to use the redundant "energy plantations" as a source of pulpwood for a paper mill.

This is technological history perhaps about to repeat itself. While Kemp has found no other case of crops actually planted for fuel, it was not at all uncommon in the preoil era to see steam tractors devouring corncobs in their fireboxes. As late as the Second World War, farmers in Brittany were threshing wheat with machinery driven by steam engines running on homegrown wood, branches lopped off trees growing in the hedgerows. Kemp said that Inter-Technology is involved in three experiments in Missouri, Georgia, and eastern North Carolina to select crops best suited to local soils and climates. Attempts have also been made to grow hybrid poplars on unused farmland in central Pennsylvania. Kemp has reported that one hybrid can produce a million BTUs at a cost between $1.25 and $1.45, compared to about $2 for oil and $1.31 for coal. He proposes to grow these "BTU Bushes" on marginal land unfit for food production and he maintains that he could generate all the power that the United States uses with 160 million acres, assuming that the trees turn only .4 percent of the sunshine into energy. "If we could get a 1 percent efficiency, we could be exporting the stuff," he has said.

Szego himself lives on what with a bit of hyperbole might be called an energy plantation. Outside Warrenton, he raises thoroughbreds in the sweet-smelling foothills of the Blue Ridge Mountains, converting six hundred acres of green grass into forty-eight race horses about which he can wax as enthusiastic as he does about solar collectors. His prize stallion, he told us, once beat a Kentucky Derby winner when both horses were in their racing prime. This is more than a part-time occupation for the Szegos. During the foaling season, his wife stays up all night monitoring mares on a closed-circuit television set (the only television program that she seems to watch). The family's home, Oakwood Farm, is a showplace for miles around, but they live in a small wing where Mrs. Szego cooks for her husband and their four adopted daughters, two from Vietnam and two Algonquin Indians, to say nothing of Gog, a heavy-pawed Great Dane pup when we met him but Gog only knows what size by now. His predecessor could put his paws on Szego's shoulders.

As if this was not enough to absorb the president of Inter-Technology, he takes pictures with his three Nikons and makes his own transparencies to show his projects' progress. He has any number going at a given time. He and his staff had no sooner got the kinks out of the collector at the Fauquier County High School when he started work on a new corporate headquarters in the center of Warrenton, right across Main Street from a defunct filling station and the First Presbyterian Church, very much alive. He had taken over a marginal supermarket and he was converting it into a solar-heated home for InterTechnology with offices, display space, a cafeteria, and a solarium beneath 3,500 square feet of collectors arranged in rows rather than in the single sheet out at the school. It looked promising, the air is clear under the skies over Virginia and there is nothing higher than the spires of the church and the courthouse to put his sun roof in the shade.

Szego's building should bring solar energy home to Warrenton. To demonstrate it to the rest of the country, InterTechnology has come up with an "Eccovan" at the request of one of its clients. Kemp has described it: "It is a modern American house with all of the modern American features in it, like a 21-inch color television set, self-cleaning oven and a few other fancy things, with the requirement that it only has two utility connections, one of which was to be the telephone. We chose to allow the other to be water. After we had designed this thing, we were then asked to package it so it could be moved around the country, so that you could walk in it and bask in its comfort no matter what the weather was outside. It is built in a trailer, eight feet by thirty feet. The unit has a solar collector and a storage system built into the frame for convenience so that we could move it around. We are going to move it around the countryside in a climate like Washington, D.C. Most winters the solar collector will be adequate for the heating. About one winter in three or four, we will have several days when the solar collector will not be adequate. In the summer, we will be able to maintain 75° indoors under the worst conditions, this is, temperatures about 95° outdoors. If you are in the same place that we have our little van wandering around, you will be able to walk around a solar-cooled and solar-heated house on wheels."

The Eccovan is not nearly as large as the mobile solar research laboratory that had led us to Warrenton but it does not have the combined resources of the National Science Foundation and Honeywell, Inc., behind it. In October 1973, Honeywell started to put the laboratory together in Minneapolis, using its own money and a $225,000 grant from NSF. It was intended to be something like a traveling solar yardstick against which the performance of various collectors can be compared in sites throughout the United States. When we caught up with it in Warrenton, it was on its first job, checking to see how much Szego's collectors were getting from the available sunshine. Roger Carlson, a young Honeywell engineer, had flown in from Minneapolis the day before to oversee the operation. He and Harold Meagher, who had driven the rig east from Minnesota, were housed in a motel on the bypass. While there was plenty of room in the laboratory, it had not been designed for living.

It consisted of two units: a forty-five-foot semitrailer that carried the solar collector and all the plumbing, refrigeration, and storage tanks that came with it; and a fifty-foot trailer which went along ostensibly as a display room but, from the engineering standpoint, acted mainly as something to be heated or cooled. The collector was deployed on the roof of the first trailer, all six hundred and thirty square feet of it hard at work. The panels were heating a mixture of water and ethylene glycol antifreeze, 1,050 gallons in all divided between two storage tanks. Carlson explained that the water was flowing through the collectors where, that day, it was being heated to 155° F. Then it ran down into a heat exchanger coil where some of its warmth was removed by a stream of air and distributed throughout the trailer. Any heat not needed during the day was stored in the tanks for use at night.

The collectors were covered with a "selective surface" developed at Honeywell so that they could retain more heat than ordinary black-painted metal. Carlson said they had already brought the temperature to 204° F. in the storage tanks, as close as anyone cared to come to boiling point. Three days of storage had been incorporated in the system with a propane-fired boiler as an auxiliary. Although the trailer had reached the Washington area on the first of March, no propane had been purchased. The heat was stifling in the "furnace room" where the two insulated storage tanks were housed along with a small 80-gallon tank to supply hot water for visitors who wanted to wash their hands.

A good deal of space in the trailer was devoted to data recording equipment and controls that could command twenty different modes for heating and thirteen more for cooling. A data logger was picking up measurements from a hundred different points and putting them on tape for transmission later that day by telephone to a computer back at headquarters in Minneapolis. Once this shakedown cruise was over, Honeywell planned to send the laboratory to places like Denver or Key West to test solar air-conditioning. The trailer had two systems on board, one running on the absorption principle, the other a vapor turbine operating a compressor and much more closely related to the usual household air-conditioner.

When we visited the laboratory in Warrenton, it was functioning well, measuring wind speed and solar radiation so that background data would be on hand when the performance of the collectors at Fauquier County High School was evaluated. Not many hitches had occurred, except the day when water came down from the collector so hot that it blew off a connecting hose. The biggest problem of all was moving the trailers east from Minneapolis. Meagher, the driver, had taken six days to make the trip. "Those loads are overweight and overheight. They're fourteen feet three inches high and the legal limit is thirteen feet six. I had to bypass four underpasses in Wisconsin. The shortest route through Ohio is two hundred and sixty-four miles but we did four hundred and fourteen." Meagher said the temperature in the storage tanks was 137° when he left Minneapolis. Seven days later, it still stood at 107°. While the trailers are intended to test rather than demonstrate solar equipment, they had already sold Meagher. "I've got a sixteen by thirty-two foot swimming pool at home. I'd like to install some solar collectors to heat it."

FELIX TROMBE

CLAUDE ROYÈRE

The sixty-three steerable flat mirrors that focus onto the big parabolic dish at Odeillo take their rest, turned off from the sun, reflecting only their optical guidance mechanisms and the backdrop of the Pyrenees.

Under a cloudy sky at sunrise, the Odeillo laboratory stands as a nine-story support for the great parabolic mirror with the tower housing the solar furnace in front of it.

The sun painted this picture on the parabolic mirror while the solar furnace was hard at work.

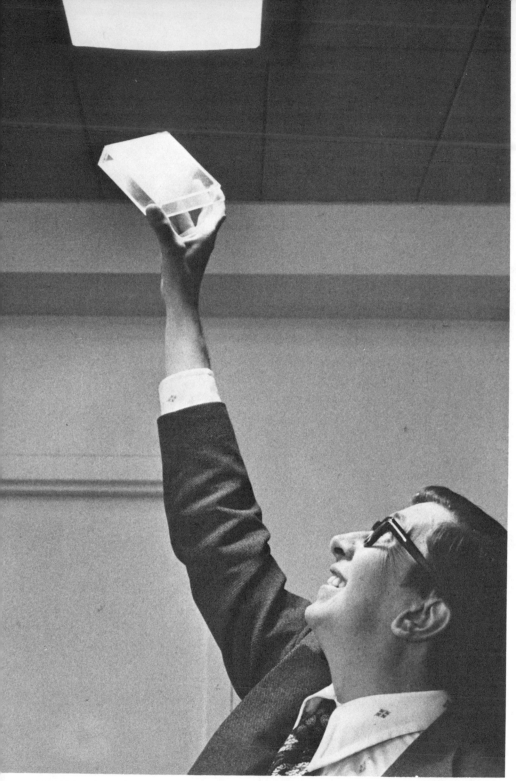
Karl Boer demonstrates a cadmium sulfide cell under the light in his office.

Solar One: The solar shingles containing the cadmium sulfide cells are on the steep roof. Six vertical collectors face out from the ground floor.

CdS/Cu$_x$S Solar Cell
Cross Section

Outer glazing

Gas

Polymer
Epoxy
Grid
Cu$_x$S

CdS

Contact

Base Metal

Cutaway view of a cadmium sulfide solar cell showing the protective outer cover and running down through the layers of copper sulfide and cadmium sulfide where the photovoltaic effect takes place.

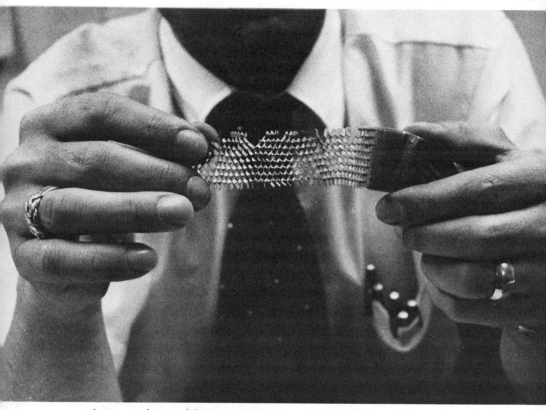

An experimental honeycomb material is under test at the University of Delaware as a sunshine trap.

MARIA TELKES

A solar room tests collectors on the roof of the Institute of Energy Conversion.

Solar collectors face up to the weather on the roof in Newark.

Harry Thomason's house in District Heights, Maryland. Water sheeting down the roof picks up the sun's heat and transfers it into a basement storage tank.

CLINTON KEMP

GEORGE SZEGO

VIII.

~~~~~~~~~~~~~~~~~~

Mr. Sunshine

~~~~~~~~~~~~~~~~~~

OVERSIZED AND OVERWEIGHT though they may have been, the two big trailers were but the easternmost tip of what was being done for the sun in Minneapolis by Honeywell and the University of Minnesota, working together. For several years Honeywell had made the biggest commitment to solar energy of any major American company while the university was contributing the talents of its heat transfer group. While the arguments raged over the pros and cons of solar concentrators as a power source, Honeywell and the university were testing solar absorbing and reflecting surfaces in the desert of Arizona, the damp of Florida, and the extremes of Minnesota. Costs per square foot of flat collectors were another salient in the war of words that was being waged over heating and cooling systems in the halls of Congress and on editorial pages; in Minnesota, Honeywell was building collectors to see if it could make a living making them. In the persons of Roger Schmidt at Honeywell and Richard Jordan at the university, Minneapolis had two well-known solar apostles. Another school was being heated there with the blessings and the budget of the National Science Foundation. It was probably the best place to see how solar energy shaped up in the eyes of what is usually called Middle America.

Although I had never been there before, I felt as if I had never left it. The University of Minnesota is the mirror image of its rival, my own University of Michigan where thirty years before I had spent my student years. I knew I was home when I heard the accents of the people and the music of their speech, flat undisguised Midwestern. I had not heard it for years; it is seldom exported. Once, I picked it out of the BBC on a taped program relating, in the tones of an announcer from Duluth or Kalamazoo, the life of Elvis Presley. I heard it again when I picked up the phone at the Minneapolis–Saint Paul International Airport to rent a car. The man at the local rental agency spoke the same way; not only that but he drove out and picked us up himself, then put us into a magnificent Malibu without a hitch, even though I have never bothered to acquire a credit card. I listened to the man in the car rental agency, he sounded like the people I worked for on my first newspaper job with a county weekly in Ann Arbor. I was the staff reporter, bill collector, and advertising salesman, although I never sold enough space to cover the price of the hamburgers the editor and publisher bought me when we drove to the printers to put the paper out and bring it back in our panel truck, the three of us crowded into the front seat and five thousand copies baled and riding behind us. We worked all Thursday night, and I slept all day Friday. On Saturdays, when I had a class, I threw pants and shirt over my pajamas and grabbed some toast and coffee at the Colonnade before I faced it. The Colonnade was a parody of the Greco-Roman administration building on State Street. The University of Minnesota was built in the same way — none of the ivied Georgian of the East Coast; instead Banker's Classic, top-heavy pediments above useless pillars slapped against a no-nonsense wall of brick. It had come from the same mold and minds that had created the Michigan campus in Ann Arbor.

As for the Colonnade, it had grown over the years almost malignantly. In Minneapolis, it was King's, Food Host USA, an island in a parking lot, a big glassed-in lighthouse on University Avenue

with the grain silos over by the Mississippi almost blocking the view
to the east. King's, Food Host USA was ballroom-sized, but at
seven in the morning there was only one girl serving, a student like
nearly all the labor around campus, and not more than three or
four tables were occupied. We all could have fitted into the Colonnade in Ann Arbor. There was no jukebox in King's, music came
out of a speaker. I could not identify it. If there had been a jukebox,
I would have played *American Patrol*. Every time I hear it on the
BBC, it reminds me of breakfast at the Colonnade, my unwashed
body inside my pajamas and still warm from my bed in Mrs. James's
rooming house on North University as it was warm that morning
from my bed in Room 241 at the Imperial 400 Motel on University
Avenue. Campuses change but little; I think they can grow by
organic accretion, they need not destroy themselves to expand as
American cities must do. That must be why the University of
Minnesota looked so familiar to me. This moment, too, in the mid-
Seventies was somewhat like the early Forties. The age-spread of
the students appeared almost the same; the ex-GIs were gone, so
were the Yippies and the hippies; fraternity row and campus
carnivals were back. If I return to the University of Minnesota in
another decade, perhaps I will find King's, Food Host USA shrunk
to the size of the Colonnade, its parking lot a lawn, a terrace, a
playground. Perhaps I could return by train on the line that seems
to run right alongside the campus on buckled zigzagging rails, relics
of the day when one did not need to move by Malibu.

Madeleine traveled back with me to the life of a midwestern
campus. She guided me with the road map in her lap as we ap-
proached the campus. Her journal has it all: "We had to keep
turning right until we reached Erie Street but it was the wrong way.
We turned around, went back to Huron Street and now we drove
through a charming quarter, small wooden houses on shaded
streets. Nearly all of them had signs: 'Rooms' or 'Room for Rent.'
These were rooming houses for students. What if we tried to stay in
one? I had to twist Dan's arm, it was impossible, impractical, etc.,
I wanted so badly to live here as if I had been living here a long
time. We knocked on a door after climbing the four wormeaten

steps of a wooden porch, painted in a pale green that was flaking away. No answer to our knock except the slow ticking of a clock back where the kitchen must have been. The door was open, a shiny wooden staircase with a worn carpet led upstairs. The glass doors of the rooms on the ground floor were all curtained, there was a somewhat musty odor in the house. Behind the front door stood a brass umbrella stand where a few canes had been planted. It was all old, worn and peaceful, the engravings on the wooden walls were askew. I wanted to stay there so badly, I imagined a nice old American lady, dressed in pink and baby blue under white hair, standing quietly at the bottom of the stairs with love in her eyes as she watched a huge American student running downstairs, glasses on his face, legs so much longer than we ever see in France, a few books under his arm. It was he who shook the quiet house that was too small and frail for him, who knocked all the engravings askew on the walls, who wound up the clock in the kitchen as a favor to his old landlady who treated him as if he was her grandson.

"Another try, another wormeaten porch. We rang the bell, we opened the screen door that protected the front door. Yes, it was a lady who answered, but she was young. She was very sorry but she only took in boys, she couldn't put up a couple.

"And the third time, the same wooden house. An old gentleman, bent in two, white hair and slippers, answered the door. He was sorry, too, he had no rooms, but since we had come from France, we had to go back into his kitchen. He liked France, he even spoke a little French because he had rented rooms to the parents of a five-year-old French girl who had been treated here in Minneapolis in the beginning of the Fifties. She was a blue baby, a team of American surgeons had operated on her heart. The operation was a success, the child was saved. He showed us clippings from French and American papers of the day. He had kept them all in a thick album with a brown plastic cover. Up on the refrigerator, a bird sang in a cage. The kitchen was tiny, cluttered with objects, souvenirs from the man's life. It was a kitchen-office, a telephone sat on the table next to stacks of papers and old newspapers. The flowers on the linoleum were fading under his footsteps that set the wood floor creaking and singing. The walls and windows had been painted and repainted, all the corners and angles were rounded off, plugged with thick navy-blue lacquer, shining like the deck of a ship constantly being painted, one coat laid on top of another. As he told us the story of Jacqueline and her blue baby, he leafed through the pages of the telephone book. He called three or four friends who ran rooming houses, but the answer was always no. We had some coffee, we told him how we had come to Minneapolis. We decided we would sleep in the motel we had seen on University Avenue. We said goodbye, we started the car, then the little old gentleman rushed out of his house to give us directions once more. We lost sight of him as he stood on his porch, waving goodbye.

"The motel was a fine place except for that same strange smell of a room where the windows had not been opened that I found in every new motel room when I came in for the first time. Later on, I guess I got used to it or else perhaps my own smell replaced my predecessors'.

"We left the room to walk around the campus. It was the end of a magnificent day. The sun outdid itself, it showered down on the stone fraternity houses silhouetted against the blue sky. A terrible din was coming out of the open windows of the fraternity houses. Students were straddling the windowsills, hammering nails, drinking beer, calling out from lawn to lawn. Others were carrying immense frames of wood and canvas covered with strange paintings. Tomorrow was the campus carnival, each fraternity had to put on a show. A big black student wearing a red vest and a gray top hat waved his cigar at us from the door of his fraternity house where he took his ease, his eyes laughing. Round and red, the sun dropped behind the horizon of the red sky as we reached the metal bridge that leaped the Mississippi to connect the old university campus with the new one. It was a double-decked bridge; cars on the lower level; up on top, pedestrians, bicycles and a covered passageway for rainy days and the winter cold, glassed-in and pasted up with posters and notices."

We walked by the fraternity houses and the Field House, all the appurtenances of the mind-broadening process at Minnesota or Michigan. Then we went into the Union, not the Michigan Union, the Coffman Memorial Union. A medical student asked Madeleine if she was taking pictures for the *Daily*, not the *Michigan Daily*, the *Minnesota Daily*. We sat down to dinner with him and his girl friend. He was young-old, he would not tell us his age, he had started to study medicine in Zurich in 1961. He was old enough to take Madeleine for a student, not old enough to take me for a contemporary. He was an unusual young-old man, the son of a carpenter's apprentice who had come from Sweden. He had the stiff bearing of the Swede and the kindness of the Minnesotan. He sneaked us into a closed-off dining room where we all ate our fried shrimp and fried chicken in lonely splendor. I found my way through Coffman Memorial Union though I had never set foot in it before. The cafeteria was in the basement just as it was in the Michigan Union, nothing had been moved since 1943 when I left it to go to war. I came back, all of us came back except one night editor of the *Daily*, not the *Minnesota Daily*, the *Michigan Daily*, who died on a bomber. Before coming to Minneapolis, I saw the yearbook of the Class of 1943 at a friend's home in Washington. The night editor's photo was there, sharp-faced, old-young, never to decay like the rest of us.

Every mirror image is necessarily deformed. The Ann Arbor of the early Forties remained aloof from industry, it was a refuge from Detroit. The University of Minnesota was no such thing; there seemed to be more factories on its edges than around Honeywell's Systems and Research Center on Ridgeway Road in a spread-out residential sector of Minneapolis, far from the bustle of the campus. It was a three-mile trip that took us in steps across the map, each turn a right angle. Minneapolis was laid out like the rest of the Midwest, not like Washington's concentric circles that L'Enfant brought over from Paris. But the movement was easy, like the flow between Honeywell and the university's engineering department. It is hard to tell where it begins; there are three main nuclei. There is Schmidt, whose business card says that he is technical manager of solar energy for Honeywell although everyone refers to him, in and out of his presence, as Sunny Schmidt or Mr. Sunshine. There is Jordan, head of the department of mechanical engineering at the university, who has watched solar energy come the full circle from the Fifties through its subsequent eclipse to the present. There is a center of excellence, the heat transfer group in Jordan's department under Dr. E. R. G. (for Ernst Rudolf Georg) Eckert who was among the Germans spirited out of Europe at the end of the war for the brain race into space. As a result, Minnesota was not as isolated as so many other solar energy centers. Roger Schmidt learned his profession from Eckert and worked on ways to heat and cool space vehicles for Honeywell, bringing his influence to bear on the course of in-house research. All the elements were present for a critical mass to come together when the space program started to taper off and, at the same time, solar energy showed signs of stirring. Honeywell and the university worked on a way to generate power with high-temperature collectors planted on a solar farm; then Honeywell struck out on its own in low-temperature applications to heat and cool buildings, starting with the trailer we had seen in Warrenton and continuing with a school on the outskirts of Minneapolis.

Circumstances more or less dictated our itinerary through the nearest thing to an industrial-academic solar complex that the United States had to offer. Schmidt was away from the plant, shuttling in and out of his home between speaking engagements. He could find a moment to talk only later at home. At Honeywell's research center, we fell into the clutches of a public relations man. We did not protest, for without him we never could have gotten beyond the security guards at the entrance to the plant (which, it goes without saying, must research other subjects besides solar energy to pay its way). Once we were past the gate and Madeleine's cameras had been checked, recorded, and registered, we found Honeywell a tight ship but relaxed. People talked easily and even the public relations man, a refugee from aerospace, felt better when we ran into a project to run something like a gas turbine with a solar collector. He called it a "sun-powered jet engine" and that made him a friend of ours for life. He was not quite as happy later on when we visited a solar collector under test in what looked to us like a shed but in what he insisted on calling an "assembly facility." His services had been supplied by a large PR facility, one of the largest, and he gave Honeywell its money's worth.

At the research center, John Kopecky supplied details of the solar collector that Honeywell had installed for the National Science Foundation under a $358,000 contract at the North View Junior High School in Brooklyn Park, a suburb lying northwest of Minneapolis. The site was chosen because the school is close to Honeywell and there was plenty of room to put up a collector between the school and its football field. The Honeywell collector measures 5,000 square feet but, unlike the InterTechnology giant at Warrenton, it had been mounted in three separate rows and not as a single sweep of glass so that maintenance would be easier. Kopecky said the collector was intended to heat only 3 percent of the big school's 165,000 square feet of floor space. Most of the solar heat has been piped into the swimming pool section where it can also be used during the summer to heat the showers. Honeywell estimates that the collector should save 21,500 gallons of fuel oil a year.

"We're about 50 or 60 percent efficient in retaining heat and using what comes in," Kopecky said, "At this latitude, there is a big gain to be made in winter heating of schools with the sun. A state law says that twenty-five percent of the air in a school must be brought in from outside and heated. I guess it's to keep the kids awake. So even if it's zero outside and the collector can only heat to 30 degrees, that is still appreciable. We usually do not go all the way with the collector; above 55 degrees, a gas furnace takes over. Today, though, our water from the sun is coming down at one hundred and thirty degrees and we don't need the gas generator at all. We use a mixture of water and ethylene glycol with three thousand gallons of storage."

Throughout Honeywell's solar energy operation, there is a reflection of the widely held view that schools are a good place to start with solar heating. Since they are publicly owned, they are subsidized more readily than private housing. Even so, the price must be brought far below what has been paid for these prototypes.

"Honeywell is looking at low-cost collectors for public buildings," Kopecky went on. "There's nothing magic about building a collector. Some big company will tool up and go after the market."

It might be Honeywell or, in Minneapolis, it could be 3-M. I was told by a source who did not wish to be quoted that 3-M may be heavily involved. They know surfaces, he said, but they are very secretive. They do not even use the mails or the telephone for their communications. "If they've got anything, we won't hear about it until they come out on the market."

We took a run out to Brooklyn Park to see the plant at the North View Junior High School. There was little to be said about it; it bore a great resemblance, at least superficially, to the plant at Warrenton. Nothing looks more like a solar collector than another solar collector, particularly at these NSF-funded schools where no effort could be made to incorporate the solar heating system into the architectural scheme of things. When we went out to Brooklyn Park, a new suburb that had obviously been farmland a few years back, it was another windy, raw day. The Minnesota climate is a great spur for inventors of heating systems. There may not be much sun in winter, but the heating season goes on and on. The Minneapolis–St. Paul Survival Guide put out by Hertz has this to say about the local climate:

"Winter (November, December, January): Cold, snowy.

"Spring (February, March, April, May): Cold, snowy."

We saw this for ourselves on a Sunday in mid-April when we joined a student bike hike that started at the Coffman Memorial Union and snaked along the banks of the Mississippi and around the various lakes within the city limits of Minneapolis. Though it was well after Easter, the temperature was around 50° F. Of the estimated ten thousand bike-owners in the Minnesota student body, only eight showed up for the hike and two were ringers, high school students. They were more than welcome because one had brought a hip flask of cherry vodka along with him and it warmed us all up in lieu of a portable solar heater when our little band sprawled for a break on an empty lot near Lake Hiawatha.

Another segment of the student body relies neither on cherry vodka nor hopes for the mass-marketed solar collector of the future. Their activities were related to us by Dr. Emil Pfender, a professor of mechanical engineering who cannot help getting himself tangled up with human affairs. He had recently been named "metric coordinator" of the new Minnesota Metric Center, the sort of job that seemed to fall to him. His academic fields of heat transfer and high-temperature plasma physics had brought him from Stuttgart to the United States and NASA in the early 1960s. Ten years later, he had become interested in solar heating, but more on the individual than the industrial side. He was working on instrumentation to measure the results of the Ouroboros Project, a solar house twenty miles from campus that aimed at complete self-sufficiency, even to the point of biodegrading all waste down to a powder that could be used for fertilizer. Pfender said that students on the project had decided to use a Thomason-type system with water running down roof panels into two big storage tanks. In Minnesota, this raised some new problems. Not only was antifreeze needed for the water, but condensation occurred on cold days. When the sun rose on a winter morning with the temperature 10 or 20 degrees below zero, the double panes of glass covering the collectors clouded up.

The project was a big house, some two thousand square feet of living space, heavily insulated with earth banked up against three walls and a solar collector facing south on the fourth. The roof is covered with sod, the same method that settlers used on the Great Plains a century ago to protect themselves from the same sort of winter. The project gets its name from Ouroboros, a mythical dragon that survived by consuming its own tail and thereby eliminated the sort of resource problems that we now face. The house was designed by Professor Dennis R. Holloway of the university's architecture department and it was being built by his students, some of whom had never held a saw in their lives. Pfender and one of his own students, John Isle, were the only engineers involved. Electric power was to be provided by a windmill charging storage batteries with the added possibility of heating a tankful of water if the batteries were full up. The solar collectors had been put up at a steep angle, 60 degrees, to discourage snow from sticking to them. They form the top level of the south wall-roof with a greenhouse below.

On a more official basis, Pfender was helping on the plans for a new university bookstore that will be built underground to save energy. "This is a several-million-dollar project. We are trying to optimize the heating system. Little is known how energy dissipates into the soil. We think that we can meet between 50 and 100 percent of the heating load in winter with four thousand square feet of solar panels and a substantial fraction of the summer cooling load. Then there will be an atrium above ground to let light in. Professor T. Bligh from the civil engineering department has been instrumental in getting this project started. Many people were afraid that it would be expensive to excavate. But not on this scale. On a big scale, excavation is inexpensive.

"When you go that deep, the temperature is constant. This reduces heat losses in winter and provides cooling in summer. The temperature stays at fifty degrees Fahrenheit, plus or minus five degrees . . . just like a wine cellar. In the past we went up, so why not go down today?" In 1975, the architects of the bookstore, Myers and Bennett, received an award from the magazine *Progressive Architecture* for their design, an encouraging sign that energy conservation was gaining status.

Of the fifty graduate students in the heat transfer division at Minnesota, 20 percent work in solar energy. This has been a source of satisfaction for Pfender. "Students read, they hear about energy problems and they want to get involved where they can make an active contribution. They like it much more than antipollution research. That's a defensive approach, this is active. Students also want support. The universities are moving out of aerospace; we ourselves lost our main NASA contract. I think thermonuclear fusion will be the ultimate answer to our energy problems, but not before the year two thousand. In the meanwhile, many false hopes are being raised. As for nuclear power plants, they tend to grow unsafe with time as their leaks get worse."

Pfender had hardly finished with thermonuclear fusion before he was telling us of a tool-and-die maker in St. Paul, John Lyon, who had added an extra story to his factory on Sibley Highway and was using a "solar room" to collect energy from the sun. Pfender had served as his consultant for the heat storage part of the job. It was the simplest possible affair, something along the line of what Trombe was doing. Lyon put up a concrete wall fifteen feet high and about fifty feet long on the roof of his plant, setting it far back enough so that he could glass it in at a 45-degree angle. Behind one thousand square feet of glass panes, his concrete wall is painted black. "You know," said Pfender, "he's getting a temperature rise of fifty degrees and he thinks it can supply 40 percent of his heating needs. The concrete wall gives him overnight storage and, if the weather isn't too cold, it will last all day long. According to his figures, he should be able to amortize it in two years." Lyon was using the concrete wall as a source to be tapped by a heat pump.

The more things change, say the French, the more they remain the same. Richard Jordan, head of the mechanical engineering department at Minnesota, was writing about solar energy heat pump systems with one of his colleagues, J. L. Threlkeld, in 1954 in the *Journal of the American Society of Heating, Refrigerating and Air-Conditioning Engineers*. They said at that time:

> Any home in the United States may be heated entirely by a solar energy heat pump system if a sufficiently large collector, heat pump and heat storage facilities are provided. . . . It has also been shown that a solar energy heat pump system may result in heating operating costs substantially less than conventional fuel-fired systems. With the accelerated depletion of the world's petroleum resources it is reasonable to expect the relative operating economy of solar energy heat pump systems to become more favorable in the future.

Jordan remembers that research. A "solar assist heat pump" looked particularly appealing in a cold climate like that of Minnesota because it permitted the use of a smaller and cheaper solar collector. During the 1950s, an office building in Albuquerque,

New Mexico, was fitted with a solar assist heat pump that used a large water tank for storage. "It could operate with the water at 40 or 50 degrees Fahrenheit, just as long as the temperature was above freezing," said Jordan. "It cut the energy requirements by 50 percent."

Yes, the examples are well known. Jordan, the forgotten prophet, has a brand-new audience. He is taking it in his stride. "The National Science Foundation has asked me to chair a panel on solar heat pumps. It is like listening to a broken record." He is a massive man in his mid-sixties whose involvement with solar energy went back almost as far as my own nostalgia for the campuses of the Midwest. He has watched time oscillate as I have, yet he has never stayed in the past. He is the kind of a man adaptable enough to have taken up gliding and to have soloed at the age of sixty. Nor had he tied his career to solar energy — having been singed once, he remained wary. Even with the latest prices on oil, coal, and electricity, he was worried by the cost of collecting the sun:

"Honeywell hopes to bring out a collector for between six dollars and nine dollars per square foot. Today, the price is more like sixteen or twenty dollars per square foot. Even at three or four dollars per square foot, you would be spending five thousand dollars for a solar collector, say fourteen hundred square feet, to heat a house in Iowa or Minnesota. If you figure interest at 8 or 10 percent, you'll never be able to pay back the capital. We must get the price down to two dollars a square foot to become effective."

The economics of marketing solar heating for homes have led Jordan back to another of his old interests, direct power from the sun. At the Phoenix symposium of 1955, he and one of his assistant professors, Warren E. Ibele, presented a keynote paper on "Mechanical Energy from Solar Energy." In its field, it is as useful as Trombe's historical study of high-temperature applications, but the tone is more wistful. It was not merely heat that the early inventors were trying to derive from the sun but power, that synonym for domination, a multiplication of muscular strength. It looked so tantalizingly close in the age of steam; one only had to substitute the heat of the sun for wood fires and coal flames. Their heyday began just about a hundred years ago in the second half of the nineteenth century.

One of the first was Auguste Mouchot in France who built a steam engine at Tours in 1866 and got the support of the French government which decided in the end that the engine was too expensive to be made on a scale "sufficient to the demands of commerce." As always, the process of invention moved forward simultaneously in a number of places. John Ericsson, the father of the warship *Monitor,* started to experiment with solar steam engines in 1868. At first, he heated water, then he went over to the hot-air engine which appealed to several inventors of the time because it eliminated the boiler and the danger of explosion. In 1883, Ericsson put together the biggest solar engine ever built up to that point. With two hundred square feet of collectors, it turned up almost two horsepower. He knew what he was talking about when he soon lamented: "Although the heat is obtained for nothing, so expensive, costly and complex is the concentration apparatus, that solar steam is many times more costly than steam produced by burning coal."

Running down the tabular history compiled by Jordan and Ibele, we find another steam engine built in 1876 in India by W. Adams. With a "hemisphere" forty feet in diameter, he focused the sun onto a boiler that ran a 2½-horsepower steam pump. Again in France, Abel Pifre built a smaller engine, only .65 horsepower, that created a stir in 1880 when it ran a printing press. D. S. Halacy, Jr., the author of *The Coming Age of Solar Energy,* notes that it was used to publish a newspaper called *Le Journal Soleil,* the Daily Sun.

It is so regrettable that no one has ever written the full story of this period when the bubble of solar horsepower had not yet been punctured. One needs no great nose for news to sniff what must have been behind the few lines that Jordan and Ibele devote to the "famous Pasadena Ostrich Farm solar engine" which was built by A. G. Eneas in 1901 and backed by "a band of Boston capitalists." It used a steerable truncated cone — that is, a parabolic dish affair with the bottom cut off — to drive a 4¼-horsepower engine that raised as much as fourteen hundred gallons of water per minute. It raised hopes, too. The *Scientific American* wrote: "Dwellers in the East, where rain falls every few days throughout the year, cannot realize what such a perfect motor means to the West, where arid lands await the flow of water to blossom as the rose." Later, the engine was moved to Arizona and Halacy quotes the impression it made on the *Arizona Republican* in Phoenix:

> The reflector somewhat resembles a huge umbrella, open and inverted at such an angle as to receive the full effect of the sun's rays on 1,788 little mirrors lining its inside surface. The boiler, which is thirteen feet and six inches long, is just where the handle of the umbrella ought to be. This boiler is the focal point where the reflection of the sun is concentrated. If you reach a long pole up to the boiler, it instantly begins to smoke and in a few seconds is aflame. From the boiler, a flexible metal pipe runs to the engine house near at hand. The reflector is thirty-three and a half feet in diameter at the top and fifteen feet at the bottom. On the whole, its appearance is rather stately and graceful, and the glittering mirrors and boiler make it decidedly brilliant.

With the start of the twentieth century, a short age of solar steam power began. In Jordan's table, we find five entries for H. E. Wilsie and John Boyle, Jr. They started in Olney, Illinois, in 1902, using a flat collector, a wooden water tank covered with double-glazing. It did not have to be moved to follow the sun; this cut costs but also temperatures. They tried ammonia, ether, and sulfur dioxide in their engine so that they could raise steam at much lower boiling points. In 1905, Wilsie and Boyle moved to Needles, California, for the sun and built engines of 15 and 20 horsepower. In their most advanced plant, they could even store heat from the collector to be used when they needed it. But their calculations showed that their solar engine cost $164 per horsepower, twice to four times as much as a steam plant.

The next name is the most intriguing of this entire period. In 1907 at Tacony, Pennsylvania, Frank Shuman ran a 3½-horse-power steam engine with 1,200 square feet of flat collectors. He did not stop there; he had the same outlook as today's visionaries who conclude that since sunlight is so thinly spread, it must be collected over a really large surface. Four years later, still in Pennsylvania, he was testing a 100-horsepower engine that he had built to run on low-pressure steam coming from 10,000 square feet of troughs lined with reflectors. He must have felt he was on the threshold of a giant step; in these days, horsepower was not around in herds. In 1913, he and Professor E. C. Boys took the engine to Egypt with the backing of the Eastern Sun Power Company, Limited. Shuman decided to construct a big plant at Meadi, seven miles south of Cairo, to pump irrigation water from the Nile. He used seven collectors, each 205 feet long. He had over 13,000 square feet, a quarter of an acre of these monster troughs. A picture in Jordan's report shows two of them rising out of the rushes near the Nile with a cluster of scraggly palms in the background. The thing worked, too; he did not get 100 horsepower from it but on one five-hour run, it developed between 50 and 60 horsepower. No one has done better to this day.

What happened next is less clear. According to Jordan, the plant was abandoned during the First World War because it could not compete economically with other irrigation systems. According to Halacy, the main competition came from the massed ranks of Egyptian fellaheen, the cheapest form of energy at the time. Halacy also credits Shuman with a proposal to tar four acres of land and cover it with three inches of water and a pane of glass so that he could run 1,000-horsepower steam turbines. If solar energy ever grows to the point where historians of technology will consider it worth their time, perhaps we will learn more of Frank Shuman and his wonderful sunshine machines.

During the oil age, interest in sun power waned and there were not many entries in Jordan's history. One should mention Dr. C. G. Abbot of the Smithsonian Institution, a pioneer who recently died at the age of one hundred and one. He built a ½-horsepower engine that he ran in 1936 at an international power conference in Washington and he worked on flash boilers that produced steam after only five minutes in the sun. By reducing the size of his engine to .2 horsepower, he was riding the wave of the future much more than Shuman. At present, there is a demand for motors in such small sizes to pump water in out-of-the-way places in the desert where sunshine is delivered much more easily than diesel oil.

Twenty years after the Phoenix symposium, Jordan is careful when he talks of the next twenty years. What has changed, particularly as the result of the space effort, is the materials available to collect, absorb, or reflect the sun and our ability to use computers to study sophisticated systems. He spoke of G. I. Schjeldahl, a firm in Northfield, Minnesota, that is trying out a "power tower," catching the sun with the help of plastic reflective surfaces it devised for Skylab and other satellites. Then there is the solar power system on which the University of Minnesota and Honeywell have been collaborating. "Here, we have put our blue chips down. We have put one of the world's outstanding heat transfer groups to work."

This is the main difference between solar power research as it was carried out by Shuman or Ericsson or "the band of Boston capitalists" and the way it is done today. Industry and academics are looking for the answer, not just the solitary inventor. Honeywell and the University of Minnesota are so interlinked here that, at times, it is hard to distinguish between them. Many of Honeywell's people got their vocation from the university when they studied there. Both sides received their initial boost from the space program. It is gratifying to see how much of the work done for NASA and how many of the people who did it are now at the disposal of solar science. This applies not just to photovoltaic cells but also to techniques for testing materials and approaching completely new problems. In the long run, this may be one of the greatest contributions of the space program to science: the groups that it put together and the ways they learned to cope with what had never existed before. When unemployment struck first in space, the people and the techniques became available. For the first time, solar power systems can be put through the same sort of testing that space technology had to face.

At Honeywell a member of the group, Jerry Powell, explained how they reached their present proposal to generate power, which might be called a radical updating of the parabolic troughs that Frank Shuman was using in Egypt. They began with a number of possibilities which they eliminated in computer simulation studies. The first to go was the flat collector; they think it cannot absorb enough heat to produce steam for a modern power plant. The flat collector is at an advantage because it need not follow the sun, but the overall efficiency looks too low. Next, they took on the power tower. Powell commented that while the tower needs no piping between the collector and the boiler, it creates a new difficulty at the boiler itself. There is a danger of local hot spots breaking out on the receiver that rises from a field of solar reflectors. "They would burn a hole right through the receiver if you didn't defocus in five seconds. They can melt stainless steel. Perhaps five years from now, the power tower will be the answer, but the materials problem will have to be solved. People are looking at very exotic ceramics; metals can't do it." Through this elimination, the Minneapolis group arrived at a parabolic trough collector which, they said, "should not be construed as a final design but is only representative of our current thinking. . . . However, it does represent a system which, we feel, is within the current state of technology." No matter how the sun is eventually captured in solar thermal plants replacing an oil furnace or a nuclear reactor, certain aspects of heat transfer and suitability of materials will have to be settled. The good old trough serves to carry out these studies and, at the same time, it can be multiplied *ad infinitum*. Powell talked about forty or fifty thousand small collectors covering a square mile and feeding a central station producing 150 megawatts of electric power. Each individual trough would be only forty feet long, much smaller than the concentrators Shuman spread out on the banks of the Nile.

The Minneapolis modules could be assembled into a solar farm. In an artist's conception issued by Honeywell's PR office, one sees ranks of collectors standing far enough apart to allow cows to graze on the grass sprouting between them. Shuman seems to have been one of the first to realize that sunshine had to be harvested over a huge area if even two-digit horsepower was to be the result. The foremost solar farmers of recent years have been Dr. Aden P. Meinel of the Optical Science Center at the University of Arizona and his collaborators, among them his wife Marjorie. Meinel thought that solar collectors could produce steam to run turbines used in present-day power generation, and conceived of a farm with four quadrants of collectors delivering heat to a plant in the center. Ten thousand square miles of such farms would meet the projected electric demand of the United States in the year 2000, bearing out Meinel's conviction that "solar energy is indeed God's greatest gift of energy sources to mankind." Where Meinel and the Minneapolis group differ is on the type of collector. Meinel had discounted any sort of concentrator, including the parabolic trough, because it does not work at all on cloudy days and loses performance even under haze or thin cirrus clouds. He proposes a flat "planar collector" with a special coating to enable it to produce high temperatures. The researchers in Minneapolis prefer their trough at present because they see a major difficulty in getting such temperatures without focusing the sun.

Tests began in Minneapolis in 1972 under a grant from the National Science Foundation. Dr. James Ramsey at Honeywell said that the process has already gone through its first phase, an unbelievably thorough evaluation of the parabolic trough (unbelievable in that no one had ever thought of doing it before). There were no breakthroughs, Ramsey himself remarked: "I can get a little upset about solar sciences — there's no such animal, it's just applied physics." Results were prosaic but useful. Engineers determined how to set up a trough collector so that it needed the least movement to catch the sun. At 33 degrees north, the latitude of Phoenix, it is best to line up the collector so that it faces north and south but then it should be tipped upward at a 33-degree angle. This sounds elementary, but it is typical of the sort of investigations that must be made to transform solar energy from hortatory oratory into a competitive source of power.

Then, during the second year of the program, a ⅜ scale model of the collector was built and mounted on a trailer so that it could be tested around the country. Ramsey thought that even if the collector did not lead to a large-scale power plant, it might be directly applicable to air-conditioning. That was how he had started in the field, working for NASA on a solar refrigerator system that would operate with a dish collector instead of photovoltaic cells.

As luck would have it, the working model was being tested in Roseville in the outer space of Minneapolis where it had been put together in that ex-warehouse. We were allowed a short glimpse of it as it was being run up for the first time. The engineer in charge, John Borzoni, told us this was to be its first full-dress test. "We've been putting it together for a week. We'll run it for three hours, then we'll give the results to our mathematical modeling people." We had a chance to prowl around the model while Borzoni was talking on a telephone that rose from nowhere in the middle of the warehouse floor. The drive mechanism that kept the trough facing the sun was nothing but mere clockwork. We looked at the collector's surface. We must have been looking too hard; Borzoni came out of the warehouse with a warning: "Better not touch the surface of the collector. It's silicon monoxide. It reacts with body oil, it turns blue."

The surface of the collector was one major contribution by the Minneapolis group. The other was the "heat pipe" that caught the sun reflected by the parabolic trough. Borzoni explained how it worked. One took a stainless steel pipe, closed at both ends, and raised one end. At the lower end, there was a puddle of water inside the pipe. Once the pipe is heated, the water evaporates into steam, rises to the upper end where it condenses back into water and runs down to the bottom to start all over again. This is how steam heats a building: steam goes up and water comes down. It is a way of moving heat without moving parts.

Eckert, the director of the Heat Transfer Laboratory at the University of Minnesota, thinks that the heat pipe was devised by a former teacher of his, Ernest Schmidt, at the Aeronautical Research Institute in Braunschweig in 1939. Schmidt showed that if a copper pipe were filled with fluid and heated from below, it would transfer three hundred times as much heat as a solid copper bar heated in the same manner. It was at the Institute of Technology in Danzig

that Eckert received one of the four degrees he has earned in a lifetime that has also seen him acquire four different nationalities:
Austrian, Czech, German, and American. He was born in Prague
in 1904 and did his first work there at the German Institute of
Technology. He came to the United States in 1945, one of several
hundred scientists taken from postwar Germany by the United
States Air Force in Operation Paper Clip. They came into the
United States "illegally" because there was no time to await immi-
gration formalities. Only later did they enter officially, in some
cases by crossing over to Canada and coming back.

We met Eckert in his office on campus. He was a delightful man
with a well-lined countenance and a warm manner. At seventy, he
had retired, but only on paper. He took advantage of his new status
to spend the Minnesota winter in Hawaii with his wife; spring found
him back in Minneapolis with the first migratory birds. He attracted
people as scientists of his level always do — they have been able to
get some top professionalism into solar power studies. There is
nothing complicated about a parabolic trough, but each component
must come up to expectations. Surfaces are basic. First, there is the
coating of the reflector on the inside of the trough. Samples of
surface materials from six different firms have been tested in the
sun of Phoenix, Minneapolis, and St. Petersburg, Florida. These
materials are mainly aluminized plastic or fiber glass; some have
failed while others have shown no degradation at all. There is
another surface on the heat pipe and this is worrisome for it takes
the full impact of the concentrated sun. Eckert's group had to learn
how the coating would resist both weather and high temperatures.

There have been other problems of materials as there are
throughout solar science. Once the heat had been collected, how
would it be transferred inside the heat pipe? What liquid would be
used? At one point, mercury was considered and then dropped for
several reasons. One was "the fact that a thousand-megawatt plant
with half a million modules would require 80 percent of the world's
yearly production of mercury." It was finally decided to use a much
more plentiful substance, water. Eckert also spoke of heat storage.
The university-Honeywell team looked at hundreds of different salts

until they came back once more to water. They found they could bubble steam from the collectors into a container filled with water. The steam condenses, raising the temperature of the water. Then later, when power is needed, the pressure in the container is lowered and some of the water immediately flashes into steam which can be used in a turbine. Here, one is trying to store solar power for only two or three hours so as to meet the daily peaks of electrical demand. This type of energy accumulator has already been used in Europe.

The analyses conducted at Minnesota do not agree with projections for vast solar farms. Eckert said the optimum size would appear to be one square kilometer (six-tenths of a mile on a side) generating forty megawatts. "Since solar power is fairly pollution-free, you can put it close to a community and use the waste heat. In my own view, solar energy will be used in modest-sized plants near communities. A huge system of pipelines causes problems. Solar energy is available in Arizona, New Mexico, and California. There is no industry there now, but industry moves where power is available. I think solar energy will come, it is only a question of time. If fusion works, that would be the end of all our problems — but we said the same thing about atomic energy."

Eckert and his group have made Minnesota a stop on the international heat transfer circuit. When he invited us to lunch, we went along with several of his colleagues and a visiting Soviet scientist from Novosibirsk who was working in plasma physics. But not in thermonuclear fusion; he said apologetically that he only went up to 10,000° C. All of us followed Eckert into the faculty club on the top floor of the Coffman Memorial Union. Conversation at the table revolved mainly around life in Germany and wine in France while the Soviet guest tried to filter our English. When I stammered the few words of Russian I knew, he responded gratefully. I did not get a single syllable, but it was good to see him come to life. And it was a good lunch, even though a non-nonsense, no-wine, cafeteria-style meal, a far cry from the way that the professors must have dined in Eckert's younger days in the Hochschule of Prague and the Technology Institute of Danzig.

Eckert had taken the provincialism out of solar energy, he had brought it into that boundless world of science where references leap frontiers. He himself did not like to travel, not nearly as much as one of his students who sailed the seas while working on the heat balance of the albatross. But the people around him moved to keep in touch and to accept the honors bestowed on them. One was Dr. Gottfried Wehner whom Eckert took great pleasure in introducing to us as the "World's Leading Sputterer." Wehner was another German physicist brought over by Operation Paper Clip. His main interest has always been plasma and surface physics. First he worked for the United States Air Force, then he went over to General Mills in Minneapolis (where his employers allowed him to do as he pleased 50 percent of his time), and he finally came to the University of Minnesota in 1968. He has won a number of awards for his sputtering which he first studied in Munich before he came to the United States. Electrodes sputter as they disintegrate when they are bombarded with ions. Another example is the process of bombarding materials with electrons. They give themselves away, they identify themselves through their ejection of electrons so that the investigator can determine exactly what composes the surface he is testing.

Here we have another process devised for the space program that has been brought into solar energy. Wehner was enlisted by Eckert to examine the coating that Honeywell developed for the heat pipe in their collector. He found that it would stand up to 300° C. and added that "between three hundred and four hundred degrees, we're optimistic." These surface analysis techniques can be applied to substances used to contain heat storage materials to see how they resist corrosion without waiting months or years for an actual failure.

The process is based on Auger electron spectroscopy. Science *is* a tight little world; in July 1973, the congress on "The Sun in the Service of Mankind" in Paris had been inaugurated by Dr. Pierre Auger, a nuclear physicist who became an international figure, first as head of UNESCO's science department and then as director of the European Space Research Organization. Auger is another of those amazing young men of science who never grow old. He had retired time and again; at seventy-five he was busy popularizing science for French radio listeners and devoting himself to such causes as solar energy. He has never lost his enthusiasm which can be qualified as boyish in the best sense of the word. He opened the solar congress in Paris by playing over the loudspeaker system the "Hymn to the Sun" from *Les Indes Galantes,* the opera written by Jean-Philippe Rameau in the eighteenth century. A friend in the audience told him that when the congress began not with words but music, he was not at all surprised: "*Ça, c'est Auger!*"

In the modest apartment on the periphery of Paris, where he lives, writes his radio programs, and does his sculpture, Auger took it in his stride when I relayed to him what Wehner told me: something like four million dollars a year worth of apparatus using Auger electron spectroscopy was being manufactured and sold in the United States. He knew all about that; he added that the French Atomic Energy Commission was using the process as well . . . with apparatus made in Japan.

He first published on the "Auger effect" in 1923 and it became the subject of his doctoral thesis in 1926 at the University of Paris. At that time, it was already known that an atom would emit electrons under X-ray bombardment. Auger learned that the atom would continue to lose electrons even after the radiation stopped. These electrons identified it. From his desk, Auger plucked a bronze medallion. "If we clean this, what will we have on the surface? Copper and brass? No . . . oxygen, nitrogen, carbon. The first layer of atoms on the surface comes from the atmosphere. Then, through ion bombardment, these layers can be peeled away and the metal appears." The method has been particularly useful in space research. He explained that problems always arise when moving parts come into contact in the environment of outer space. This is why tape recorders fail more often in artificial satellites than in the home. Study of the surfaces of the moving parts show which elements are deteriorating to the point where failure is likely.

Auger himself has never applied his technique to solar energy. During his early career, he worked on a photovoltaic cell but he turned to the study of cosmic rays in the Thirties and discovered the so-called "Extensive Air Showers" effect. He came to the United States, joined the Manhattan Project during World War II, then returned home where he helped set up the French Atomic Energy Commission. At this point he went off to UNESCO, first as one of its founders. He has always been an internationalist; it was under his lead that UNESCO ushered into being CERN, the European Nuclear Research Center in Geneva, and this same concern appeared in his opening words to the solar energy congress in Paris. He remarked on the diffuse and generous distribution of sunshine, then said: "This point takes on particular importance in our era of universal patriolatry with its chosen and its outcasts designated by blind fate." Auger now lends his prestige as an elder scientist and his skill as a popularizer to solar energy. One of his most recent publications is a long chapter in an Italian encyclopedia of science and technology on *Prospettive per l'utilizzazione dell'energia solare.*

If these prospects are ever fulfilled, it will be because of a meeting of basic science and sound engineering, the sort of meeting that has occurred in Minneapolis and is best exemplified in the person of Honeywell's Roger Schmidt. One cannot identify Schmidt all that easily, his electrons have not moved in predictable paths. He grew up on a farm in St. Peter, Minnesota, and, at one time, hesitated between the study of science and a life as the foreman of a road construction crew. At forty, he is not so far removed from those years and he has an appreciation for the average man's concerns that is not always found at his academic level. But his findings tell him there is no easy way to bring the sun within the reach of the average man. He is not a solar demagogue, and does not have much respect for the ones that are around.

Schmidt's renown in solar energy has reached the point where he often must go home to get work done. That was where we tracked him down at his house on Bass Lake, one of eighty-three in Minnesota bearing the name. Schmidt's was eighteen miles north of Minneapolis, only half an hour by Malibu from all the solar activity flaring at Honeywell and the university. It might have been in the north woods. Though this was now the end of April, the ice had broken only a week before and Schmidt's golden retriever was paddling hopelessly after a duck on the lake front. Schmidt had built most of the house himself; he had drawn the plans, bulldozed the foundations, hammered some of the beams. To do it, he had taken his only vacation in years. For three months, Honeywell let him work half time for the company while he finished his home on the lake. The result was worth it. The wooden roof of the house soars over the living room almost like a forest canopy. Double-glazed windows separate the room from the lake, reflecting the flames from a central fireplace so they appear to dance in the branches of the tree just outside. Schmidt is attached to this house the way he is attached to the state of Minnesota; Honeywell made a good investment when it gave him time off to deepen his roots. Otherwise, he might be tempted by the offers that he receives wherever he goes. But then, if money had tempted him, he would have stayed on that road gang.

From high school and a small college, he came to the University of Minnesota where he encountered Eckert and worked five years until he got his master's in heat transfer. Schmidt is an easy man to interview. He does not fit the image of the captive scientist in private industry, spinning money for his employers. Schmidt has freedom within Honeywell; it may not be academic but it is freedom. He is free to work on nights and Sundays, free to forgo vacations except to build his house or fly off to Salt Lake City or Sydney to speak about the sun. He is interested in what is going on elsewhere, not in a spirit of competition but, one would think, because it adds to the sum of solar science and the dimensions of his own work within the company. What it does offer Schmidt is a sense of confidence that he transmits all the more readily because he is a big man, a former high school football player. Schmidt need not seek publicity or grants; he has the money, now he only has to produce. In his group at Honeywell, he has thirty-five men; he can concentrate on trying to bring solar energy to the point where it will be ready for production. Yet he has become the best-known solar scientist in his part of the United States. In the spring of 1973, he appeared on the "Today" show with Paul Rappaport of RCA. For months afterwards, he was getting five letters a day from people building solar houses or thinking about building them. A year later, the mail was still coming in and he was still in demand as a speaker. While we were talking and judging the bouquet of his Seven-Up, his wife, Lee Ann, put her head into the living-room: "Barbara Garbin called. Her sister went to high school with you. Her sixteen-year-old son has to write a paper about solar energy."

Schmidt went to work for Honeywell in 1959 and, a year later, he got into solar science through the space program. In his case it was not a matter of obtaining energy out in the cosmos but of getting rid of it. He explained: "There are three modes of heat transfer. First, there is conduction, from molecule to molecule. This is what happens when you touch a stove or put a poker into a fire. Next there is convection. Hot air rises, cold air falls. Conduction is easy to calculate. Convection is easy to understand but very difficult to calculate, to know exactly how much heat is transferred. Then there is radiation, the movement of heat by electromagnetic waves. In space, you cannot conduct heat to and from a spacecraft. You cannot move heat by convection for there is no air. The only way to get rid of heat is by radiation. And radiation depends solely on the surface involved. You can put a coating on the surface that will change its character radically. With very thin layers, either you can make it reflect or you can make it absorb heat."

The true originator of the selective surface, he said, is Tabor in Israel. Working at Honeywell, Schmidt invented a selective solar absorber coating. It looks black and it absorbs heat but, unlike an ordinary black surface, it does not reradiate the heat that it absorbs. Schmidt used it first for such purposes as the inhibition of radiation so as to fool heat-seeking missiles. Of course, it is this capacity to soak up heat without shedding it that one seeks in a solar power system.

The notion of the selective surface is difficult to understand because it goes beyond our senses and we must accept as an article of faith what we cannot see. Schmidt tried hard to explain it to us and he may have succeeded. It is worth trying to grasp because if solar energy ever moves into our lives, then "selective surface" is going to be a strong selling point. The effect is that of the greenhouse where all the energy penetrates the glass roof which then blocks infrared rays from escaping. This can be felt in the greenhouse; it is much harder to seize with the selective surface. One starts with a shiny surface, perhaps bright nickel. It will reflect visible light, we can see our face in it, but it does not reflect the invisible infrared rays. This is what we want, but first we must get the energy in. To do this, the nickel is covered with a very thin black layer that absorbs visible light and now the surface is black to us. But not to the infrared rays. Ten times longer than the visible rays, they go right through the thin surface layer to the nickel and they are held there. Schmidt put it: "To the infrared rays, the layer is like a little bit of scum on the surface. To the visible rays, it's knee-deep, it stops them. Since we can see only wavelengths in the visible spectrum, the coating looks black to us. But rays in the infrared part of the spectrum are trapped by the bright surface behind it as though they were striking a mirror."

Schmidt developed his AMA coating, so-called because it consists of a thin layer of aluminum oxide, a mixture of molybdenum oxide, and then aluminum oxide again. He first used it in 1970 on a NASA project where, like Morse at the University of Maryland, he and Ramsey tried to devise a refrigerator that would work in space on direct heat from the sun. They came up with a parabolic dish that concentrated sunshine onto an AMA-coated ball at the end of a heat pipe. "We could have built it but no one wanted it. They already had solar cells." So Honeywell brought the coating and the collector back to earth and thought about solar power. "The National Science Foundation couldn't fund us then because we were purely industrial. We needed a university. I had lunch with Eckert, my old professor, and he liked the idea. We couldn't have put together a better team. Jordan was one of the leaders of the old guard in solar energy and Eckert had his heat transfer group that was publishing two hundred and fifty papers a year."

By 1973, the National Science Foundation had given a grant of $900,000 to Honeywell and the university. "It was the biggest solar energy contract in history at the time. That's what gave us credibility in Honeywell." Honeywell's association with the university is limited to power generation. Elsewhere, it has gone ahead on its own, notably with a flat-plate collector, the school heating system, and the mobile laboratory we had seen in Warrenton. Schmidt traced the laboratory's genesis for us. It started as a solar heating and cooling system that was to be delivered to a NASA center. NASA canceled the contract but Honeywell decided to build it independently. Schmidt had the idea of putting wheels under it so that it could be tested everywhere from the dry clear skies of Phoenix to smoggy Los Angeles to Cleveland with little or no sunshine at all. He thought of mounting it on a motor home but it grew and grew to its present size. "I thought it would be a beautiful thing for solar energy if we could have set it up on the White House lawn." When the first oil crisis struck at the end of 1973, there was no longer any need for publicity. The trailers serve a more utilitarian purpose. "As a scientist, I need data. Our company is trying to decide if it is worth putting money into solar energy. You can't ask a company to invest ten million dollars without building hardware and taking data. The president of Honeywell wants to know if there is business in solar energy. At a first analysis, it looks like there is one hell of a business. I'm not saying this as a solar scientist, it is the business analysts in our company who are saying it. They think the market would be huge and I'm pretty sure that General Electric is thinking the same thing."

Schmidt keeps an eye on the business side of the sun and that is why he tries to make his collectors more efficient. To heat a house in mid-December in Minnesota, one must cover the whole south roof with collectors. Much of that area will not be used the rest of the year; it is like the peak capacity of the power company. "The more you put on a roof, the greater the chances are that you will not use all of it all the time." This puts a premium on selective surfaces so that the collector can absorb enough heat to run an air-conditioner in summer and help pay for itself. "We can now get 200 degrees Fahrenheit but the question is, can we get it reliably?" Schmidt has gone into the costs.

There is no question in my mind that we can build a solar heating and cooling system. Even though this is true, we still have a very great challenge. Solar energy is not cheap. . . . If you take a square foot and set it a proper angle at one of the most sunny places in the United States, it will collect approximately half a million BTUs in a year. If we compare that to the cost of our most expensive fuel, oil at $2 for a million BTUs, this means that half a million BTUs are worth a dollar. You reap a benefit from that one square foot of solar collector of $1 each year. If you apply the normal 15 per cent for amortization and maintenance and things of that kind, the square foot must not cost more than $6.70 And you have to install it. In Minneapolis, it must not cost more than $3.30.

At Honeywell, we have a collector design which we think is quite new in some respects, but it is very similar to designs that have been used and proposed by people for many years. This system, or this collector, which we are putting on the school and the portable lab, is expected to be 50 per cent efficient both in winter and summer. . . .

Our production engineers have looked at this collector and they believe it will cost $2.70 a square foot to manufacture and put in a shipping crate. You know that industry has to charge at least two to five times more than direct factory costs to stay in business. If we apply a factor of three to that, that is, $8.10 for a square foot, then it is not competitive today. I don't mean to be pessimistic. I think we can be optimistic because that is very close. There is no doubt in my mind that innovative people are going to come along with better ways of producing the same things. For example, there is a company in California that makes swimming pool heaters. . . . It is a very special application, but he is selling those collectors for $2 a square foot or less, I believe, and he has a hard time keeping the customers away from the door. He cannot actually meet his orders. A very viable business.

I don't believe that within the next year you are going to buy a solar heating-cooling system commercially that is cost-competitive. But you will be able to buy collectors that are competitive for special applications. Let me get up on the soap box: I think it is perfectly proper for the U.S. government to fund a major portion of the research and development in such an endeavor as solar energy where it is in the public interest and where the R & D investment and the business risk are too large for industry to go it alone. When you compare solar energy with fusion and a few other things, I think solar energy warrants a bigger cut of the pie.

Schmidt had made these remarks at the round table on solar energy held in New York. A few months later in his home on Bass Lake, his view of the future had not changed materially. There were still all those question marks hanging over solar power generation. Schmidt thought that the Honeywell trough collector looked like a good bet for a pilot plant. "The trouble with the power tower is that you have to build the whole damn thing before you can test it. The problem is in its heat transfer dynamics. There is no way to test it before building it, you cannot simulate solar efficiency at those temperatures. I could build a trough collector and simulate the boiler and plumbing system tomorrow. The next step will be to build a full-scale trough, then five or ten or more."

He had given some thought to heat storage at high temperature for power generation. Along with hot water and molten salts, he saw possibilities in the use of compressed air or even super fly-wheels to put energy aside for use when it is most needed. "We are also considering the idea that we may not need so much storage. The solar plant could be a load-following plant, it would only meet the peak load. You can afford an expensive power plant to do this and you would need only three hours of storage." At present, he likes thermal power. "In solar cells, they need a breakthrough, we just need cost-reduction engineering."

This is the way that Schmidt and his Honeywell group think about solar energy. They are far removed from those of a generation ago who saw the sun being applied first in the developing countries. It has come home to the United States. "We're the richest country in the world, we're such gluttons. This is where an industry can start. The market is so large." Schmidt likes to act as a bridge between the old and the new generations in solar energy. "The new guard, they're mostly industry people like me. This creates a problem. When I go speaking, I don't make dollars for Honeywell. Promotion doesn't help them." His employer wisely lets him mount the soapbox on his own time, on the only vacations he has ever taken except for the months he spent digging the foundations of his house, a return to the construction work of his youth in St. Peter.

He talks of St. Peter and the farm. Once a year, he goes back there and he has never moved very far away. That, too, is part of his strength. He knows what industry wants, he knows what his neighbors want. He talks unhurriedly, a man who has all the time he needs, Sundays and holidays included. He does not teach students, but his group is permeated with his attitude of relaxed self-confidence. No one feels any need to prove anything by words. Schmidt himself does not talk in multimegawatts but in microns. He does not await the breakthroughs, he is getting through. He is not at all controversial, yet he is probably the most effective solar energy proponent in America. It is hard to delimit him in a few interviews. Schmidt is what he does, not what he says. He is his house, he is his selective surfaces, he is that group of young engineers at the Honeywell plant on Ridgeway Road, individuals all in their Spartan cubicles. Schmidt is not locked into any single concept, he can evolve with the state of the art and the progress of the science. He is relatively well known now; we shall hear more of him.

IX.

~~~~~~~~~~~~~~~~~~~~~~

The Challenge to
Replace
These Bullocks . . .

~~~~~~~~~~~~~~~~~~~~

WE MUST GO BACK, historically and geographically. Our itinerary around and about the sun leads from the future back to one of the first of the true believers, Farrington Daniels, who died in 1973 during the last darkest hours. Daniels had been responsible for the founding of the Solar Energy Laboratory at the University of Wisconsin in Madison and acted as patron saint of the Solar Energy Society. In 1964 he wrote a book, *Direct Use of the Sun's Energy,* that was published in a new edition ten years later, still the most complete review of this subject (in which the information explosion has never been more than a weak *phhhht*). Daniels was gone but he had left his mark in Madison. Work there was continuing under Dr. John A. Duffie, the director of the Solar Energy Laboratory and a past president himself of the Solar Energy Society. In answer to a query we had sent him, Duffie had been helpful but with all the reserve of those who had been sawn off the limb of solar energy years before. His letter said:

> There is a lot going on in solar energy in this country. There
> are now a number of industries in the business of research and

development of solar processes. Some are people getting themselves educated at government expense. Others are people who have some interesting ideas and the capability to do something about them. In one sense, there isn't very much that is new. The processes look very much like those discussed in 1961 at the United Nations meeting and in 1973 in Paris. However, things are changing. The technology available to carry out some of these operations is substantially improved. The costs of alternative sources of energy, i.e., the competition to solar, are rising to astronomical levels (judging by my own fuel bill). They still aren't at the level where solar energy can compete today but the prospects appear to look better all the time.

The best indication is the fact that a number of industries are investing significant amounts of their own money to do development work and explore markets in this field. Some of this industrial development has been going on for some time and if there weren't some reasonable prospects, it would have been dropped a long time ago. On the other hand, the money involved in the solar energy field is still quite small compared to that going into other kinds of energy resources.

I think that the industrial interest is one of the most encouraging signs of the times. Without industrial participation, the whole business is an academic exercise. Until someone is manufacturing and distributing equipment so that it can be installed and used in buildings, it really doesn't mean very much. In that respect, things look much better indeed. . . .

Our own research is concerned most entirely with models and simulations of solar processes. That is, we write mathematical descriptions of components (collectors, pumps, storage units, etc.), program these for the computer and assemble the component programs into system programs. These are operated using climatic data for the location in question to see how such a system would perform if it were actually built. Such simulations are useful in evaluating systems and deciding which ones are worthwhile attacking experimentally. We can "build and run" a system on the computer in a matter of a few weeks with the help of a graduate student. It would take a quarter of a million dollars to build and evaluate such a system over a year or two in real life.

Thus our laboratory consists of a bunch of desks piled high with paper and some competent graduate students to keep us going. There isn't much to see, but we do have some ideas on what kind of things should be done.

To see Duffie and his laboratory, we had to backtrack. From Minneapolis to Madison is two hundred and sixty-five miles as the concrete ribbons of Interstate 94 unroll over the Minnesota border and southeast. With her peasant sense of economy, her gift for mopping up every drop and crumb in her plate, Madeleine estimated that we had enough unused mileage on our rented Malibu to make the round trip to Wisconsin by car before enplaning west again. Like most of her ideas, it was brilliant. Once out of Minneapolis–St. Paul, we saw dairy farms along the road and cheeses for sale in the service areas. The Malibu purred the miles away; it was as big a car as I had driven since my dad's 1941 Cadillac. The V-8 hardly whispered at the fifty-five-mile speed limit even when it was nudged to sixty. Americans beat their breasts and tear their hair in guilt over the gas they guzzle but it has not been all waste. The monsters hiss along the Interstates with scarcely a sound but the the whirr of their tires. When the day of compulsory thirty-mile-per-gallon motoring comes, the easy life of the highwaymen will be gone. As in Europe, every journey will be a rally, drivers of under-powered cars clinging like death to their lanes so as not to lose an r.p.m. before whining and raging up the next hill. The Interstates could then become what the European motorways and autobahns are: a swath of heavy decibels where roadside sleep is impossible within a range of five hundred yards. The Europeans handle the matter by routing main highways through the poor parts of their cities and countryside; the trouble with the United States is that there aren't enough poor people to accommodate all the main roads. I cannot see the voters of American suburbs putting up with the environment of an airport in a state of never-ending takeoffs.

And so we moved by Malibu, perhaps for the last time in our lives. The car was a palace, a *Queen Mary* on wheels compared to Madeleine's 3-horsepower Citroen resting in Brittany. She climbed into the back seat with her cameras to draw a bead on a silo here and a farmhouse there, here a red barn there a dairy herd, everywhere the green, green state of Wisconsin. The Americans may have hogged a third of the world's energy but they spread it around evenly once they got it. One hates to think of a shift to European styles of consumption, the rich in sub-Malibus screaming at ninety miles an hour past everyone else mounted on auxiliary-engined bikes or four-wheeled cardboard boxes that get forty miles to the gallon and untold death and destruction to the mile. The Malibu wasn't ecological and it wasn't solar. We loved it between Minneapolis and Madison, it was so much better than jetting. It was almost like a private railroad car; let us wait and hope for the day when Malibus will run by steam and whistle at the intersections. The Malibu was a good vehicle for visiting the past at the University of Wisconsin, sliding back down the curve of exponential growth.

There is no more continuity in time, the "nostalgia craze" is a recoil at what lies ahead. As I sit writing in Brittany in the autumn of 1974 while scrap lumber from Pierre Germain's sawmill burns in my fireplace, the BBC tells me the story of Ella Fitzgerald and I hear "A-Tisket, A-Tasket" coming out of the radio in the pure clear tones of the youthful Ella and I remember my contemporaries at DeWitt Clinton High School in New York singing it. Now the announcer comes onto the BBC with the news of a United Nations vote on Cyprus, another travel agency going bankrupt in Great Britain, the illness of Richard Nixon, the talks of Kissinger with the Shah of Iran on energy problems. Time is a mosaic that has been put together and then suddenly broken up, jumbled, the pieces thrown into the air and strewn on the ground where one must try to pick them up. The unpredicted becomes the everyday, the rising curves break and dip, the futurologists are given the lie before they can get their forecasts into print.

Farrington Daniels was the past of solar energy; he may have been the future. I don't know, I went to see Duffie to find out. He squeezed a few hours for us out of the sort of schedule typical of anyone with a name in solar energy these days. He reserved a room for us at the Ivy Inn on the edge of the Wisconsin campus. We did not have much time in Madison, but it was enough for Madeleine:

"Ivy Inn, 2355 University Avenue, Madison, Wisconsin. Dinner, for $2.95, I had the right to a whole variety of salads. I could take all I wanted because one served oneself: potato salad, macaroni salad, grated carrots with pineapple, lettuce, beets, radishes, raw carrots, green Jello-O with plums, cream cheese, five different kinds of seasoning (I chose 'French sauce' which turned out to be a sugary tomato dressing), a selection of salted crackers, a crock of butter, then three big pieces of breaded fried chicken, a big baked potato (*Translator's note:* the French expression is better, a potato in a country dress), a small pitcher of thick cream, a big pitcher of lemon tea. I keep writing all the time about what we eat because people never stop saying, here as in France, that food is no good in America.

"Before dinner, I patted the snouts of five pigs, scratched the head of an exquisite little brown and white lamb and tickled the noses of four ponies and a strange gray horse with a shaven mane and a haughty air. This was the model farm of the University of Wisconsin, a piece of country in town, tall wooden silos covered with metal domes that gleamed like the church spires of Moscow. The university was vast, carpeted with lawns bordered by bushy-headed trees trembling in the cold wind, washed and cradled by the waves of the big lake that hems the campus, dominated here and there by ghastly modern towers of brick and concrete, but mostly a park, green and sunny, where students of all colors laugh and jostle each other on sports fields, leaping high all at once to sink a basketball, then going home by bicycle, wrapped up in parkas and scarves as if they had to travel kilometers while in fact their houses, delightful as usual with pillars and balconies and frame sides of any and all colors, sit side by side among the trees on the hill just across University Avenue that cuts the campus in two.

"We walked home past these doll's houses carved so intricately out of wood, lighted so beautifully inside. We could see their inhabitants dining quietly by candlelight as we looked through windows framed by two ribbons of pink and mauve curtains, lacy and pleated. Next to the dining-room table, a pale-blue bookcase, very neat, a window above it. I felt as if I wanted to open the front of that house. I would not have been surprised if there had been a hinge on the left and a brass knob on the right so that one could open the front of the house like a cupboard door, to get a better look at all the furniture and tables so well arranged, all the inhabitants sitting in their chairs. Little girls used to be able to do that with their doll's houses.

"That afternoon, we had walked through the campus. Up on top of a hill, we saw a stone bearing a plaque. It was here that Hawk or Running Hawk, the Indian chief of the Sauk tribe, had come to take refuge and die in 1832 on this hill overlooking the lake where he had been driven by the militia and the American troops. Here, almost on the same spot, the first building of the University of Wisconsin went up in 1851. Another plaque said that, during the Civil War, the hot-air heating system was shut off. Each student had a wood stove in his room and it was up to him to get his firewood from the neighboring forest. Life was very hard, said the plaque, the toilets were outside.

"I watched the rich fat students in the fraternity houses who seemed to spend all their time speedboating on the lake or hazing their pledges and I began to dream of the life of the students of 1865 when they had to get their firewood from the forests that covered the surrounding plains and hills, now covered with houses unfortunately not all built of wood and colonnades.

"After we left the fraternity houses, we met a friendly red-headed girl who was practicing with a rope, trying to lasso a stone post. I took her picture and she offered to let me try the lasso. I did, but it was hard because you have to turn it with your wrist, not your whole arm, and I missed the post. Later on, I took a picture of two boys and a girl sitting on the curb. We sat down next to one of them who was hiccuping with every word he said. He asked us if my family in France couldn't send them a little hash."

It was in one of those ghastly towers, the Engineering Experiment Station, that we found Jack Duffie, professor of chemical engineering and associate dean of the Graduate School of the University of Wisconsin. He had got his Ph.D. there in 1951, worked for DuPont for a few months, spent two years with the Office of Naval Research, and then decided he wanted to stay in the Midwest. In 1953 he joined Farrington Daniels at the university in Madison. "Daniels was a fascinating man, a renowned physical chemist. He made a career out of getting into things, making a real contribution and then going on to other things. He pioneered in four or five different fields in this manner. He had this burning desire to do something for poor hungry people. He had been involved in the Manhattan Project, he had been director of the Metallurgical Laboratory in Chicago. He saw the first bomb go off. We talked about it; he told me how they all hoped it wouldn't work. It gave him this dream to do something for people. He was an early advocate of peaceful power and he became an elder statesman. He kept the Solar Energy Society alive, he put his own money into it to pay the interest on its debts. He felt it was something he ought to do. His main contribution was the way he kept a modicum of interest alive in this field through what he did as a chemist. He worked on new refrigeration systems and selective surfaces.

"We worked closely together for ten years, starting with an original grant from the Rockefeller Foundation. Daniels was a dreamer. We made a fundamental error, I think, when we directed our thrust entirely towards the developing countries. Solar energy is capital-intensive, and if there is one thing that the developing countries lack, it's money. Yet to the day he died, Farrington was interested in them."

Perhaps where Daniels erred was in his assessment of the developing countries themselves. When they invested, they did not concentrate on solar cookers and pumps for subsistence farmers. They bought the most modern technology, even nuclear reactors.

They displayed as little interest in sun-powered devices twenty years ago as a Londoner or a New Yorker would have shown. Yet Daniels was not all that far wrong. His writings stand and it is not quite fair to judge them by his book, *Direct Use of the Sun's Energy*. Like many books by academics, it started as a series of lectures and grew into a small volume. The lectures were aimed at an audience both well versed in science and eager to put solar energy to work in the hungry countries of the world. Today, industry talks of bringing the sun into the American home; Daniels was a long way from running central heating systems with solar collectors or electric carving knives with photovoltaic cells. His book was first published in 1964, but he was talking to us:

> We might imagine a conference in the distant future when the shortages of coal, oil and gas are felt. Our descendants might well blame us for our wasteful extravagance in the use of these most important natural resources. They might say: "You consumed gasoline in 200-horsepower engines simply to carry one person to his office and back. You talked about the folly of burning for fuel cow dung which should be used for fertilizer, yet you burned the limited supply of organic fossil fuel that should have been used as raw material for making petrochemicals."

That does not sound very original today but only because so many people have said it since.

Daniels was not worried unduly by automation and labor-saving. He wrote:

> It is possible to build a clockwork mechanism operated by a heavy weight and controlled by a pendulum as in the old grandfather clocks. It would be a good project to build and operate a mechanism to rotate a solar collector, using a structure with several gear wheels, controlled by a pendulum or an escapement wheel similar to those used in spring-wound clocks or watches. A 50-pound sack of sand lifted three feet each day could provide the power necessary to rotate a fair-sized collector. . . . Often, it will be more economical to track the sun by hand operation than by automatic devices.

He literally went back to the start of tool-using. He estimated the operating and overhead costs of various ways to lift water for irrigation, starting with two men and a bucket and going on through various devices operated by bullocks, camels, and motor pumps. He figured that it cost three cents an hour to feed the bullock (a man's labor was estimated at four cents an hour). It was at this level that he wanted to introduce solar power.

In 1955 Daniels was one of the main contributors to the Phoenix symposium on applied solar energy. There he told how he had taken a study tour of India's arid regions the previous year and found the living basis of his calculations.

We saw, among other things, four bullocks and two men working diligently and skilfully for long hours irrigating farm land. Every minute, two bullocks pull up by rope 250 pounds of water from a depth of 50 feet. One of the men collapses the water bag made from hide and the water flows over the land. The second operator drives the bullocks and a second team of bullocks pulls up the next 250 pounds of water. This work done by the four bullocks and two men calculates out to be the equivalent of one third of a horsepower. And the four bullocks cost $600 and have a life of only about six years. Moreover, they have to eat and they must consume a considerable portion of the crops that they help to grow. The project seemed to be barely self-supporting — but the water has to be raised.

If the bullocks' walk-way were covered with a solar heat collector which operated a crude engine operating at only 1 percent efficiency, a one-horsepower pump could be operated. One percent efficiency is not much to ask for. *The challenge to replace these bullocks with solar engines seems more important to me now than some of the theoretical researches in which I have been engaged* [my italics]. . . . We also saw a camel walking around a well and pulling up water by a rotating mechanism. We saw a man getting water with the help of a counterweight and a woman walking back and forth on a balanced beam to pull up pails of water. . . .

Solar energy utilization holds out special hope for improving the standard of living in areas which have not yet become in-

dustrialized. . . . . Solar heating and cooking are definite possibilities if the units are cheap enough. I have a letter from Mexico
telling me that the women of the village spend much of their
time walking to the hills six miles away to collect firewood for
their cooking and heating. The letter goes on to say, "We have
plenty of sunshine all around. Can't you do something to help
us?" Maybe we can. Let's try!

Daniels put all of his considerable ingenuity to the task of cutting
the costs of solar energy. He devotes four pages in his book to the
construction of a cheap collector for a solar cooker which should be
made "with plastic on a hoop laid over a parabolic concrete
mound." Every step of the job is explained, no detail is omitted.
"It is difficult to get a perfectly smooth edge with the hand saw.
Accordingly, a 3/8-inch copper tube is attached along the edge of
the plywood with a series of nails. To the top of the plywood shap-
ing tool are attached two protruding strips of wood, one on each
side, for straddling the wooden pipe. A wooden plug is inserted between the ends of the two strips to give a square hole."

One may smile at the missionary spirit of the man, but Daniels
was no alternative-technology zealot. He had been chairman of the
board of governors of the Argonne National Laboratory from 1946
to 1948; in 1953 he was president of the American Chemical
Society. His status did not prevent him two years later from getting
down to the problem of cheaper sunshine collectors for solar
engines when he was addressing the Phoenix symposium. He badly
wanted to see those engines work. The chairman of the conference
he addressed had been a chairman of the board of the Mutual Life
Insurance Company of New York and there was also a formal
banquet with the president of General Dynamics as its speaker.
That did not stop Daniels from getting his message across:

Two attempts at cheap reflectors are interesting. A piece of
thin transparent plastic was loosely stretched across a long rectangular frame above the ground and filled with water. The
weight of the water produces a parabolic-shaped water lens

which gives a reasonably sharp focus on a pipe which can be moved along as the sun changes its level in the sky.

Several different types of parabolic mirrors have been constructed using aluminum foil and aluminized plastic. The cheapest one . . . is made by placing 2-mil aluminum foil over a parabolic-cylindrical form, and cementing coarse burlap cloth to it. Metal-lath is then draped over it and then coated thoroughly with a mixture of Portland cement, sand and water. After setting for a couple of days, the form with the inner coating of aluminum foil is lifted off and set on the ground in an inverted position. . . . The cost of material is about ten cents per square foot and two men can make the unit quickly and lift it conveniently. . . .

Four rows each with ten of these reflectors, properly spaced with a steam engine in the center, could give in principle ten horsepower if 10 percent of the solar energy at a million small calories per minute were converted into work. . . . Even a wooden form may cost too much. One of the cheapest ways to produce a parabolic mirror would be to scoop out in adobe or firm soil a long parabolic trough, parallel to the sun's orbit, and line it with aluminized plastic or sheet aluminum after proper treatment of the surface. Such an immovable reflector would be quite limited in its use.

Daniels's thinking went back beyond the Industrial Revolution. Unlike Watt's machine, his would be competing not against windmills and waterwheels but bullocks and womanpower. "Still other types of solar engines should be investigated for possible use when the capital costs must be kept to a bare minimum and where manpower is very cheap. A reflector can be moved so as to focus the radiation first on one boiler and then on a second one. The steam from each boiler forces the piston in opposite directions, and thus it is possible to turn a wheel with the reciprocating piston." This is something like asking the driver of a car to work the spark plugs of his engine by throwing hand switches, but the ultimate users Daniels had in mind did not drive to work.

Take any of the often-heard ideas of today and Daniels was saying it.

I have recently had a survey conducted by Mr. V. Stoikov who found that for every calorie obtained from food in the United States, we put a little more than a calorie from fossil fuel into producing the food. . . . Photosynthesis is a remarkable process which we are just beginning to understand. In the laboratory under special conditions, it can be made efficient. Thirty percent of the light energy absorbed can be stored and released at a later time by combustion. In agriculture, only a small fraction of one per cent of the annual solar energy is stored. . . . We realize, as never before, that our fossil fuels — coal, oil and gas — will not last forever. Several careful studies have been published in the last few years which point out that the depletion of our reserves will come sooner than we think. . . . Moreover, the population is increasing rapidly and the demands for abundant energy are increasing still faster. Any estimates of the life of fuel reserves based on consumption at past rates are utterly unrealistic. . . .

I believe that by a judicious combination of fossil fuels, atomic energy and solar energy the whole world can have within this century, all the mechanical power and material comforts that it wants. This development won't solve all the world's problems, because man does not live by kilowatts alone, but it will help.

Twenty years later, Daniels's associate, Duffie, hardly lifts an eyebrow at all the new daily discoveries of the sun that are being made. He is disabused, as Jordan was, he refuses to get excited when his newspaper tells him the wheel has been invented. What is more meaningful is that here, as in Minneapolis, the students are coming around; the researchers and the professors of the 1980s and the 1990s (will they be the Gay Nineties?) are being trained. Not by the hundred or even the dozen, but it's a start. Duffie said that he had five graduate students in 1974 and expected two more the following year. In 1971 he had none at all. That was the nadir in the history of the Solar Energy Laboratory at the University of Wisconsin. Dr. William Beckman, who has been with Duffie since 1963, remembers it. "Solar energy had its ups and downs after the war. It was on its way down when I got here. A grant from the Rockefeller Foundation was tailing off and there was just no way to get money after that. Jack and I had a number of conversations. We kept asking: should we quit? I stayed in heat transfer, I got interested in the modeling of ecosystems and I did some work on the behavior of lizards in the Mohave Desert. Then in 1972 we received a grant from the National Science Foundation and we started our computer modeling."

Beckman's time was even shorter than Duffie's. On top of his teaching, research, and speaking commitments, he ran. Like Running Hawk, he sped every day along the shore of Lake Mendota with a group of five fellow runners. He did forty miles a week and they kept him fit and slim at thirty-nine, his brain as nimble as his ankles. Often he worked at home at night when computer time was cheap and ideas came readily to him. He could telephone the big computer on the Madison campus from anywhere in the country and this seemed to suit him as a way to work. In their two-man team, he had the background in mathematical analysis while Duffie provided practical solar engineering experience.

The computer tells them they have a long way to go. "If there is anything that represents a hazard to our business, it's some spectacular failure," Duffie said. It is the economics that loom large, the matter of balancing the cost of the solar collector against the energy saved by not running the back-up system. "A conventional auxiliary system is a must," said Beckman, "You can calculate the chances of cloudy weather over a twenty-year period, you can design to store for nineteen days without sun and the next year, you'll get twenty-one. You need one hundred percent reliability; that is why you can't have a one hundred percent solar system."

Their first major simulation was a house in Albuquerque, a site for which solar radiation data were available. They assumed a house with 1,800 square feet of floor space under solar collectors of various sizes from 150 to 850 square feet. The collector was supposed to warm the house in winter, cool it in summer, and supply hot water. Duffie, Beckman, and a graduate student, Larry Butz, were able to come up with the costs of running the house under various assumptions for the price of fuel and solar collectors, the sort of information that private builders will need if they are to try to enter the housing market.

There is more to this than collectors and plumbing. Duffie is among those who see political and tax problems. "If I put a solar heating system on my house, my tax assessment will go up. This is not the case if I put in an oil burner." There are local building codes that do not provide for rooftop collectors, the status of the building industry fragmented into small concerns reluctant to innovate, and the matter of sun rights. But Duffie thinks taxes are the main problem. Solar heating costs are high enough without adding a tax penalty.

The public is not put off by these factors. One woman listened to a talk by Duffie, then asked him for the address of the simulated house in Albuquerque. "People buy plans, then they want more. I get a dozen letters asking 'where can I buy a solar heating system?' I say: you can't. But it can go either way. If the political side is straightened out and if we can get some clever engineering, some industry might convert mathematical models into economically viable hardware."

Duffie and Beckman are teachers who think that work without educational value should not be done in their university. They teach a regular course in solar energy at Wisconsin, one of the few of its kind in the world, and seniors can take it as an elective. Besides classes for students, they have offered several short courses on solar energy thermal processes. The first was held in January 1974, then came a repeat in May of that year. "We gave them five full days," Duffie said; "it was not just a smorgasbord of lectures. We worked them, they weren't out drinking at night." The courses led to a book by Duffie and Beckman, *Solar Energy Thermal Processes,* which has been published by John Wiley and Sons in New York and London. The book is written for engineers, not for amateurs. It has nothing of Daniels's idealism — it is mostly tables and equations — yet Wisconsin has somehow come up with two of the most authoritative works on solar energy, each reflecting the state of the science in its time.

There were fifty-three participants in the first course. They started every day at eight and went on through the end of the afternoon with breaks for coffee and lunch at the Wisconsin Center Cafeteria. The courses began with flat-plate collectors and finished with solar power systems, with the economics of heating and cooling considered on the way. Participants came from a number of universities, industrial firms, and government agencies, notably the Atomic Energy Commission. By not doing much more than jabbing a pin in the list, one garners names like Kennecott Copper, DuPont, Monsanto, General Motors, American Can, General Electric, Corning Glass, Ray-O-Vac . . . not likely to be drawn to a steam engine running on a collector focused by hand or made from scooped-out soil. As an example of the content of the course, Duffie and Beckman cite a problem they gave their students. It starts: "Determine the collector area to supply 75 percent of the hot water requirements of a residence of a family of four in Boulder, Colorado, based on the meteorological data for the week of January 8 through 14 as shown below. Each person requires 75 kilograms of water per day at a temperature of $60°$ C. or above. . . ."

X.

~~~~~~~~~~~~~~~~~~~

The Nightmare
of George Löf

~~~~~~~~~~~~~~~~~~~

STUDENTS HAD A CHANCE to design this water-heating system and run it through the computer to see how, for example, a selective surface or an extra heat exchanger would affect its performance. Without bothering with the wan Wisconsin sun, Duffie and Beckman have been able to test a number of systems. They simulated an industrial water-heating experiment conducted on a large scale in Australia. An Australian scientist, Dr. Peter I. Cooper, spent five months in Madison on this job. Then the computers were used to design a house at Colorado State University in Fort Collins that is being heated and cooled by the sun.

Unlike the place in Albuquerque which exists only on magnetic tape, the Fort Collins house stands in three dimensions. At a cost of a quarter of a million dollars, it is the most complete solar heating and cooling experiment that is being undertaken in the United States with the help of NSF. Political pressure in solar energy concentrates on the home. It is people who get cold, it is people who vote. While there may be a twenty-five-year lead time on any new power system, there is no such gap ahead for solar heating. But it does represent a return to other ways of living. Thomason was right when he related how his big collector can give him all the heat he needs at 68° F. Duffie told us how he had been lowering his thermostat in Madison; over the twenty years, the temperature in his home went down five degrees Fahrenheit, and he felt all the better for it. He did not need computer time to figure his oil bill: up from fourteen cents a gallon in 1972 to thirty-two cents two years later. From time to time, Beckman finds that all his computer analyses are not enough to discourage the man who wants solar heating: "People say: I don't care about the cost, I want to be warm. A person gives up as much energy as a 100-watt light bulb, but some aren't so bright."

At Fort Collins, solar heating is being tested in reality. It's a good place for this sort of thing, a western Newark, Delaware. No one would discover nuclear fission at Fort Collins. There seems to be a two-story height limit on university buildings, all in cream stone and red tile. Even the S. Arthur Johnson Hall where business is taught looks like a Spanish ranch house. For the first time since we had begun our tour of solar America, we were in a sunny climate. I remembered Colorado from a short spell at an Air Force base in Denver during the war. The air was clear and the city had the reputation of being the best soldiers' town in the United States.

I recognized it not at all. From the moment we landed at Stapleton Airport, I had the impression someone was building a spare California so there would still be sun and smog somewhere when the big break-up on the San Andreas Fault drops the original into the Pacific Ocean. Of all our stops, Denver was the only place I had ever seen before and it was the most unfamiliar.

Not so to Madeleine, as we took off in our rented Pinto for Fort Collins, sixty miles to the north. "On the way out of the airport, we saw our first accident: an overturned truck, a flattened car, ambulances and fire engines. And here was Denver, a city of white tents pitched side by side all the way to infinity, all the way to the feet of the snowy mountains. Instead of a cathedral, the smoky towers and black piping of a refinery dominated the city. Then came the countryside. I thought I was back in Biskra, the oasis in the Sahara where I lived as a child. The land was flat and red. The houses were small, so low they looked flattened, too, with immense trees planted next to them. These were the *gourbis,* the native huts of Biskra, none of the neatness and the fresh paint of Minnesota and Wisconsin. And the heat, dry and scorching after the chill of the morning in Minneapolis. The little red Pinto reminded me of a brother-in-law who drove a sport car. He lived twenty-five miles from Paris in a village called St. Vrain. As I told this to Dan, we came to the St. Vrain River and a sign that said Fort St. Vrain.

"Fort Collins was a big open town, small houses and gardens, nothing but lampposts and traffic lights along the streets. Over to one side, the neon signs of the restaurants and the motels. Biskra again or perhaps Iraq but everything wider and brighter. The heat, girls in shorts and even bathing-suits, barefoot, and those tall trees, still leafless, and this flat town at the foot of the blue mountains and the sky suddenly turning orange from one end of the horizon to the other. The sun was setting, the dark orange sky went black. The orange and red and white and purple neon signs wink and shine and flash and stream in small shiny beads against the black and orange sky. All these lights and no houses. We could not see the houses, they were tiny and low, spread out over their lawns behind all the lights.

"A luxurious motel for eighteen dollars, two big beds, a humming air-conditioner, a scent of perfume coming from I don't know where and permeating the room. We had dinner in an Arby's exactly like one where we ate in Minneapolis. I sat there and I looked at the small wall facing me, covered with illuminated photos of Arby's Super Roast Beef $1.15 and Pecan Pie 30 cents, and the same girls in orange behind the counter and the machine cutting roast beef in razor-thin slices and I thought I was in Minneapolis. I am always amazed how easy it is to eat in America when we travel, ten times easier than in France where I get liver trouble after three days.

"We drove up to look at Horsetooth Reservoir, an artificial lake lying between two rows of mountains, five miles of calm blue water. Unfortunately, a horde of little speedboats was running over it like cars on a freeway, all their lights on because night had suddenly fallen on the mountains tufted with trees and dotted with red rocks. A dirt road led up to the reservoir. There was plenty of traffic on it that afternoon and the cars left clouds of yellow dust. A few fishermen were still sitting at the edge of the lake, a group of kids had lit an immense bonfire next to a rocky wall. On the valley side, we could see the lights of all the tiny houses that made up Fort Collins."

In Fort Collins, the frontier was not far. The editor of the local weekly had challenged the mayor not to a duel with horse pistols but a bicycle-versus-car race (the editor choosing the bike) during the Monday morning rush hour, to prove that something had to be done about traffic on South College Avenue. Parallel to South College Avenue, the Burlington Northern Railroad ran through Fort Collins to Madeleine's amazement. "As we walked out of the motel that afternoon, we heard the wailing of a locomotive horn right next to us. We jumped into the car and rushed to the railroad crossing. There was the train, it was running next to us, four green diesel locomotives one after another, the engineer leaning out of his window, striped cap over his eyes, working his horn. I begged Dan to speed up, instead he slowed down: 'I can't, honey.' I wanted so badly to get my hands on the wheel. We had to go through an intersection, then make a turn five hundred yards down the road so that we could beat the train to the next crossing and I would have time to get out of the car with my camera. But Dan lost his head. He saw the red light on the railroad crossing in the distance and he slowed down instead of racing the train. We got to the crossing at the same time as the first locomotive. By the time Dan had stopped the car and I had got out, the other engines were gone, too. Only the freight cars were left, parading heavily down the middle of the street lined with trees and houses. A freight train right in the middle of a street. I had never seen that before."

Once the great locomotive chase had ended, we left South College Avenue and the railroad tracks. It was only ten minutes even at the legal speed limit to the Foothills Campus of Colorado State University, right at the base of Horsetooth Reservoir and protected from instant flooding by the dams plugging the gaps in the mountains like bridgework. At the start of the foothills, we took a wrong turn. Instead of the Solar Energy Applications Laboratory, we ran into an area where a sign said we had no business. We heard dogs barking and, after getting directions, we turned around and headed the opposite way towards the Foothills Campus.

"The solar house stood on a butte," Madeleine wrote. "Arid and endless, the immense plains stretched before it, covered with rocks, scant grass, occasional trees. Again, I had Biskra before my eyes, the desert bounded by gray and blue mountains, the shrubs of an oasis. As we walked towards the solar house, I could see a lizard skittering from one rock to another, suddenly stopping in the shade of the next rock, lifting its head to listen for possible danger. The air was cool despite the sun bearing down. Just as in Biskra, I could hear barking dogs, as if from a neighboring village, squatting in front of the huts, howling to each other from one oasis to the next. But I was told that the dogs here were poor Snoopies yelping because they were being used as guinea pigs by a zoology laboratory. They served to test certain treatments that might have worked on human ailments. It was not good to be a dog in this part of Colorado.

"The solar house was being built of wood like all the other new houses in Fort Collins. I enjoyed watching the way all those boards were cut to fit into each other; it was more like a construction toy than a construction job. It was clean, the smells were good. It was not noisy, not like mixing and pouring concrete. On the contrary, I liked to hear the sound of hammers driving nails. Bud Selley, the foreman, white helmet on his head, two pencils in his ears, kept an eye on the work. He was from Kansas and he was forty-six years old. Until a year and a half ago, he had a 1,600-acre farm and five hundred to six hundred head of cattle. He had lost two fingers from his right hand, one in a grain elevator in winter. It was so cold that he had not noticed it at first. He felt a little pain in his finger and when he finally got around to look at it, it wasn't there any more. He lost the other finger ten years later in a power saw. He smiled and he chewed tobacco, for he had stopped smoking. He had sold his farm because he could earn $18,000 a year as a carpenter, as much as he could make on his farm but without all the worry. When Dan told him of the farmers we knew in Brittany, he said: 'They're luckier than we are, they stayed on the land.' He had one helper who came from Vermont and another, a young Mormon, who spoke French because he had spent several months in the north of France as a missionary. The Mormon was only an occasional carpenter; his dream was to own a ranch and become a veterinarian. His father also had a farm but he was the oldest of seven children and he did not have enough money to buy a ranch."

While Madeleine interviewed the construction crew, I talked to Charles Smith who described himself as a hard-knock engineer. That week, he was supervising the installation of the solar collectors on the roof. As Madeleine noted, it was an ordinary house and yet it wasn't. All the components of the heating and cooling system in the basement could be replaced as soon as a test was done. In its first version, the house would be running with an insulated 1,100-gallon water tank for heat storage, an auxiliary gas boiler and an ARKLA Solaire absorption cooling system. "You could drive a small truck into this basement and take everything out," Smith said, "You can change the storage tanks, the air-conditioning system, the auxiliary gas boiler. That was the first criterion that the National Science Foundation laid down: the house must be completely versatile."

Unlike Karl Böer's trademark in Newark, the Fort Collins solar house blends with its neighbors. It is a two-story, three-bedroom house with 1,500 square feet of space on each floor. It is not over-insulated and it needs 60,000 BTUs of heat an hour when the temperature drops to 10 degrees below zero. The house will provide offices and work space for half a dozen solar researchers, but no one is likely to live in it, except perhaps a reporter on a story. "Even then it won't be quiet living," Smith went on; "people will be taking data day and night. It is being instrumented for a hundred channels recording every five minutes. Somebody said they're measuring everything here except the traffic going by on the road."

Nor did Selley think it resembled the other houses he had worked on. "This place is overbuilt. They're building it like they don't want any damn trouble."

Dr. Dan Ward, the project manager for the solar heating and cooling system study at Fort Collins, estimated that the house is the equivalent of a $60,000 home with $10,000 extra for a solar system working in four different modes, each to be tested. Various factors have raised the cost. The roof had to be designed so that the collector could be changed instead of incorporated into the structure and thereby saving on shingles. The site itself has been a problem. "The soil is horrible. When it gets wet, it expands to two or three times its original volume. You can imagine what that can do to a foundation."

Ward got his doctorate in nuclear physics in 1971 from the University of Texas. He stayed there a semester as an instructor, then went into the building trades himself as a general contractor. He had done some construction work in the Navy and he liked designing houses. He still did physics but only during his lunch hour when he worked on relativity. "I work purely on my own. If Einstein could do it working in a patent office, why not? Still, I'm not Einstein." Like everyone else even remotely associated with solar energy, he has been spending a great deal of time giving talks and answering questions. On any odd day, he might be addressing heating engineers in Denver or a junior high school in Fort Collins. Besides that, visitors come into the laboratory. "They have specific questions, they want to know where you put that nail. They buy plans, then they realize that do-it-yourself plans do not answer all the questions. The way some of those plans work, they couldn't heat a house at high noon on a clear day in June." Reactions are getting more and more favorable in Colorado. "A guy came in off the street to hook up the gas meter for the solar house. He's been interested in solar energy ever since. Before that, he thought it was the biggest Rube Goldberg around." It was this sort of attitude that led Colorado State to build a house that would look good, that would be residentially acceptable. The neighbors have their eye on the house, so do the people working for it. Selley remembered that he used to be able to fill the propane tanks on his Kansas farm for fifteen dollars to sixteen dollars. The fellow who bought the farm has to pay seventy dollars to fill the same tanks. "I think solar energy would go in Kansas. There are more clear days than they have here."

Nevertheless, the house got 80 percent of its heat from the sun during the first winter of its life, 5 percent more than the designers had expected, and the bugs had been chased from its cooling system by the end of the first summer. Now it has two solar neighbors. A year later, a second house was started with a solar air heating system rather than a water storage tank. And a third house went up with an advanced collector from Corning Glass, a tubular system literally operating in a vacuum to reduce convection losses and gain efficiency. Not only that, but a fourth house has been planned to incorporate an array of energy conservation concepts with solar heating and cooling. Colorado State University is a claimant to the title of world leader in solar heating and cooling development.

It was not just climate that lay behind the initial decision of the National Science Foundation to put $250,000 into building, testing, and running a solar heating and cooling project at Fort Collins. Since the Second World War, Colorado has been one of the world's heliocenters thanks to the presence of a native son, Dr. George O. G. Löf. We have already met him in the accounts of the early experiments at MIT in house heating when he was Hoyt Hottel's assistant at Cambridge. He has also surrounded himself with an aura of controversy because he has never lost a chance to take a crack at proponents of solar heating who seem not to have heard of engineering. Nor does he show much patience with various gadgeteers "who never bother to read about what has already been done and who reinvent the wheel . . . often far from round." His opinions are shared by virtually all scientists in the field, but Löf is one of the few willing to be kicked, bitten, gouged, and scratched by voicing them in public. He may have been hardened by his childhood. Now starting his sixties, he was born the son of a country doctor in Aspen when it was a mining town that had yet to see its first skier.

I ran into Löf the first time at the solar energy congress in Paris where he delivered a paper that he had written with Dr. R. A. Tybout on the costs of heating with sunshine compared to electricity and fossil fuels. He did gain the attention of his listeners; instead of writing equations, he talked about money. He and Tybout had stated:

> Solar heating is cheaper than electric resistance heating in nearly all situations [in the United States] and in a climate such as represented by Santa Maria, California, it is competitive even with natural gas. Solar cooling and combined solar use are virtually competitive with oil fuel in Miami. . . . A doubling of natural gas prices within ten years has been frequently forecasted. A change of this magnitude, possibly accompanied by economies in manufacture of solar energy systems, will make solar heating and cooling competitive with fuel in almost all U.S. locations.

The Löf who delivered data on heating costs from the rostrum in Paris was not at all like the Löf in his laboratory at Colorado State. He was lean and dry, but with a constant twinkle of humor that had nothing in common with his rostrum manner nor, for that matter, with the way he writes his scientific papers. He is a no-nonsense engineer and this comes through in public. He will not turn somersaults to gain attention.

When we first met him, it was only for a quick lunch. The sun, strong as over Madeleine's oasis at Biskra, was coming in through his office windows. He picked up a piece of plastic that concentrated the sun like a burning glass. It burnt a somewhat elliptical hole in a sheet of his notepaper and he gave it to us as a memento: "From the desk of George O. G. Löf." He had used a Fresnel lens in plastic, another term that crops up in the solar scheme of things. Augustin Jean Fresnel was a French physicist who worked at the start of the nineteenth century on increasing the range of lighthouses. He replaced the mirrors previously used with a compound lens. To the eye, it appears as concentric rims on a flat surface; it is as if one took a thin layer off the rounded surface of the lens and flattened it into adjoining curved segments. Fresnel lenses can be made in plastic for $1 a square foot and some speak of them as a feasible way in the future to concentrate sunshine.

Löf uses his merely as a conversation piece, for he has always been more interested in low-temperature use of the sun. Over sandwiches in Al's Lunchroom, he traced the start of his own career. After he finished at MIT where he was mainly involved with heat transfer, he went back to the University of Colorado at Boulder in the early 1940s. There, he turned to solar heating as a way to save fuel for the War Production Board. "I suppose one might say we are closing the circle today. The government was interested in solar heating at another critical time in our history."

By 1943, Löf had come up with a solar air-heating system. To his credit, he used himself and not a student as a test subject when he installed the system in his house in Boulder in 1945. It was a modest bungalow; a photograph in the proceedings of the Phoenix symposium shows a small house like the ones that can be seen in the more rundown neighborhoods of Fort Collins. It was not at all to the taste of the early Seventies. Instead of the 3,000 square feet of the "typical" solar house that was being built at Fort Collins, Löf had only 1,000 square feet of heated space in his five-room bungalow. It was a retrofit, too, although the jargon term did not exist in that happier day. On the south-sloping roof of his own home, Löf added a solar collector, actually another layer of roof, 463 square feet of glass. Since he wanted a hot-air heating system, he did not use water to store warmth but gravel, 8.3 tons in the crawl space under the floor. Hot air from the roof collector was forced into the gravel bed by a blower and then returned to the house when needed. When the stones were not hot enough, the house's gas furnace took over. Löf had his troubles and he did not hide them. He had to wrap the solar collector around a gable and a chimney, long ducts had to be run from the roof down to the gravel bed, glass in the collectors expanded and broke at first, air leaked from the ducts. During the first winter that he ran the system, he got 25.6 percent of his heat but, by April, he had improved it to the point where it was supplying 55 percent. He made no extravagant claims; he saved only $14.14 on his gas bill, but the results were encouraging enough to lead him to build a new home designed specifically to get part of its heat from the sun.

The house had a history even before it went up. At first, it was intended for Los Angeles and then Dallas before it settled in Denver. On the way, the architect dropped his plans for a solar house and set about to give Löf a home in a then-contemporary style: low and flat, with Shoji sliding doors borrowed from Japan to change the dimensions of its inner rooms whenever desired. Instead of covering the south slope of the roof, the collectors were set up in two parallel rows on top of a flat roof.

Löf promised we would have a chance to see the house in southeast Denver, but first he spoke of its problems and solutions. He built the place with his own money while the American Window Glass Company, which had been supporting his work at the University of Denver, paid for the collectors. The system was designed to carry only one-third of the heating and hot water load. It was never intended to do more. "You learn just as much heating a third," Löf said. He certainly is not on the side of the enthusiasts who preach 100 percent solar heating with no auxiliary system at all. "That's as costly and wasteful as buying a big bus and parking it month after month in your driveway so that, once a year, you can take a lot of friends out on a picnic."

The family moved into the house in 1957 and Löf started taking data that winter. "It is the only solar-heated building that has been up for any length of time with sufficient instrumentation to learn where all the BTUs were coming and going. We have records for four complete years and, since 1957, the solar system has continued to supply its heat to the house." Löf said that the National Science Foundation had given Colorado State University funds to measure the performance of the house in 1974–75 to see the effects of the passing years. Subsequently, he reported that nearly as much heat was supplied to the building in 1974–75 as it had received in the winter of 1959–60. He suspected a moderate drop in the efficiency with which it collected the sun but, even after fifteen years, the system was performing satisfactorily. "The collector has had no maintenance for seventeen years. I just didn't have the time. It does need some work; a few internal panes of glass should be replaced and we could clean the others." He engineered it for utmost simplicity. The heat storage system consists of eleven tons of gravel stored in two great columns, sixteen feet tall and three feet in diameter. These stacks, made of laminated paper, came right off the shelf of the building industry; they are the forms used to pour heavy concrete columns. Hot air from the roof collectors is brought into the gravel bed by the same blower that runs the house's gas heating system.

Löf spoke of costs as an engineer. "Gas bills don't prove anything. In Colorado, you can't compete with natural gas; anyone who says you can is crazy." Not everyone in Colorado is on the natural gas network and this is where Löf thinks solar heat might get its foot in the door. At present costs, heating with oil, bottled propane, or electricity runs from five dollars to eight dollars per million BTUs. "Even with today's custom-made collectors, we can heat for nine dollars per million BTUs. We could cut this in half with cheaper collectors. I think that, within twenty years, we will have solar heat at four dollars per million BTUs in 1975 dollars. And I'm a conservative; next to Hoyt Hottel at MIT, I'm the most conservative man in this business. Gas is cheap because the price is regulated. One of three things is likely to happen. We will leave gas regulated and there will be little incentive to bring in new gas fields. They will be found somewhat accidentally by companies drilling for oil. And gas will be rationed. Or we will have deregulation. The price of gas will go way up. Then demand will come down and people will go towards oil, solar heat, and insulation if their gas costs them five dollars per thousand cubic feet — that's a million BTUs — instead of seventy-five cents. Or we will have a combination of both with some deregulation and some price incentives for new wells. Solar fits in as soon as the price of heating goes up to four or five dollars per million BTUs."

His conservatism keeps him far from all-solar systems. "Some people think about putting in crude solar heating systems with no back-up. They may get some heat out of them but you can freeze just as dead at thirty above as at ten below." He is much more prone to look at new developments by manufacturers. He spoke of Olin Industries, an Illinois firm that produces absorber plates for solar collectors at a price of sixty-five cents a square foot. "It's a fantastic bargain. It is a printed pattern in metal, the same thing one uses in a refrigerator. That is all there is to it. The solar collector, in fact, is like a refrigerator. They both absorb heat but at different temperatures."

Like so many others, Löf has returned to solar energy as a first love that had once played him false. When sunshine went out of style, he went into general engineering, working particularly on water resources and environmental problems. His solar activities were extremely modest; there was not enough money for anything else. Then, in 1970, they started to revive. At first, there was the pollution issue, but Löf saw it only as a stimulator of interest. Obviously, it was cheaper to clean up fossil fuels than to go over to the sun. Then when the price of fuel started to rise, he decided that solar would go.

But not for electric power. Never one to dodge a fight, Löf has plunged headfirst into this one. Again, he reasons like an engineer. Sunshine reaches the earth's surface at an intensity of three hundred BTUs per square foot; in a modern steam boiler, heat is concentrated to half a million BTUs per square foot. "One needs at least a thousand times as much heat transfer surface to generate steam with the sun. It would be very different if cheap surfaces were available but high temperatures need concentration and that raises costs. Solar power is not in the cards for many years to come. Since it runs so much higher than conventional sources, it needs a combination of technical advances and large increases in fuel prices before it can compete."

Colorado State University and Westinghouse are engaged in a study of solar power but not as direct competition to conventional ways. "Our study zeroes in on intermittence. Why not sell daytime power to the electric utility? Then you will be replacing fuel that can be used somewhere else." The main direction along which Löf himself works is still heating. He liked the idea of four solar houses side by side, but with different systems. "It is the only way to get comparative data. Calculations and computer simulation can be used when designing, but then you must try things out. The computer cannot tell you if a solar house will leak or freeze or explode."

Löf has always proselytized. That afternoon, he had an appointment in Denver with a group of architects and officials concerned with building codes. Perhaps municipal authorities would be willing to grant tax relief. "It's hard to be against solar energy. It's like motherhood — or the way motherhood used to be." His speech-making and lecturing are legendary. Löf considers that he has four offices: one in Fort Collins, one in Denver, his home in Denver, and whatever airplane he happens to be sitting in. Everyone in the field knows the story of the nightmare of George Löf: he dreamed he was walking down an endless corridor in the Los Angeles airport when he met George Löf walking the other way.

Sunday at his house in Denver was devoted to the sun. When we arrived there from Fort Collins, he was talking to a visitor who had come much farther. John Brooks was an Australian with the New Zealand Wire Rope Company in Auckland and he was on an itinerary somewhat like our own. "I'm a wire-drawer by profession," said Brooks as we crossed paths and photographed each other, "I've come to see his heliotrope." Löf himself was about to take a plane for Austin, but one would never have known it from his manner. He made sure that we got some lunch while he talked to Brooks about business matters. Löf was associated with a company, the Solaron Corporation, which installed nearly a dozen systems during the first year of its life. It uses the hot air principle in these systems to avoid such problems as freezing, corrosion, and leakage.

That day, the sun was heating the big house in southeast Denver. "We've got an overcast sky today," said Löf, "but I can tell it's working by that buzzing sound. There's a duct touching something and it gives us vibration from the blower." He took us up on the roof to see the collectors. "The ladder looks risky but it's quite safe." As he had said in Fort Collins, they had not been maintained in seventeen years and they looked it. But they had not become an eyesore. Each of the two rows was only six feet high and they could hardly be seen from the street, protected as they were by a wind-screen. If anything, Löf's house has been most successful in shooting down aesthetic objections to solar collecting. In the section where he lives, the neighbors were not likely to accept anything that would mar the view. Löf now thinks that people might be only too glad to look at solar collectors.

The long life of the collectors enabled Löf to get rid of defects that could only show up in use. One was elementary. His collector is made of overlapping glass panes fitted into a frame. At first, the panes rested on a notched aluminum support. As they expanded and contracted, they chipped their edges against the support until the panes themselves cracked and snapped. Löf then fitted the panes into a sheet metal support that lets them slide freely. Costs went down and the breakage stopped.

The hot air from the collectors was coming off the roof at 140° F. and then sucked into the storage bed, the paper cylinders filled with gravel. They were no aesthetic problem. Painted red, they rose up the staircase of the house like the stacks of an old ocean liner. "The system is now storing heat," said Löf. Air was being drawn through the gravel stacks into the basement. He closed a relay. The speed of the blower dropped by half, a damper moved and now hot air was going into the rooms of the house and not the storage bin. Normally, the system is controlled by two thermostats, one in the living room and the other on the collector. Whenever the sun is not shining, the damper shuts the flow of air from the collector and the house is heated by storage or, if necessary, by the gas furnace. There is also an eighty-gallon water tank which gives the family hot water in summer when the collectors are not needed for heating.

There is no solar air-conditioning in Löf's home and he does not deem it necessary. "We really don't need cooling here in Denver except in public buildings and offices where people generate heat." The collector keeps on heating, a perpetual system requiring only a new fan belt on the blower from time to time. "The house has been lived in. It's had four kids tearing it apart." We could see traces of the children, long since grown, in the driveway of the home filled with seven cars, including a 1937 Plymouth that one of Löf's sons was restoring.

Löf glanced at his watch and announced it was time for him to leave for Austin. "I always allow five minutes to pack. Then, five minutes before I leave, I allow five minutes to change clothes. Then I allow an extra three minutes on the way to the airport. I've never missed a plane yet." His wife, Davadell, added: "That's possible only because your wife is your chauffeur."

For thirty years, Löf has worked in solar energy. He has not always been the engineer, he has had his spells of dreaming, too. From his cellar, he brought up a strange umbrella, a solar barbecue that had been taken apart long ago and stored in dusty cartons. "I developed it about fifteen years ago. It would broil a steak. It gave you the same taste of carbonized protein that you got from charcoal. But it was too expensive, it cost twenty dollars to make." Löf folded his solar umbrella and started his packing. He was as good as his word; five minutes later he reappeared in the living room with a small pile of clothes and an enormous heap of documents. Within a few seconds, he sorted out the papers he needed and slipped them into a black briefcase.

Then he was off to another airport and another speaking engagement. Before he left, he told us of another activity of his. Convinced though he was that solar heating rather than solar power generation lay in the immediate future, he wanted to give both sides their say. He advised us to look up his collaborator at Colorado State, Dr. Susumu Karaki, who was managing the joint study with the Westinghouse Electric Corporation on power generation.

XI.

~~~~~~~~~~~~~~~~

At the Head of
the Energy Cascade

~~~~~~~~~~~~~~~~

BACK WE DROVE to Fort Collins and, the next morning, we found Karaki's office in the Solar Energy Applications Laboratory. Karaki is a big man and he fascinated Madeleine who, like most Europeans, has trouble sorting out the racial mixtures of North America. He introduced himself as Skook Karaki, a name that had stuck with him since the age of three. Born in Canada some forty years ago of Japanese parents who had a farm in southern Alberta, he is a naturalized United States citizen. Madeleine asked him if he ever felt Japanese, if he did not feel any sort of emotion at the sight of a kimono. And Karaki answered: "Yes, if there's a pretty girl inside it." He had begun his studies at Utah State University and finished them in Colorado State and Stanford where he received his doctorate in fluid mechanics. He had been at Colorado State for seventeen years and he had started to work with Löf on solar energy in 1972. Previously, he had worked with him on problems of desalination and evaporation. He made the changeover easily, for the basic principles of radiation and heat transfer could still be applied. Solar heating is not Karaki's entire professional interest, but he could not help speaking of retired people living in mountain communities in the Rockies, far off the natural gas network. "It can cost as much as two hundred dollars per month to heat a small cabin with propane gas. We in a university, which is part of society, have a service to perform for people living up there on fixed incomes."

He lets himself be drawn into faculty councils, committees, and guest speaking engagements. He works seventy to eighty hours per week. He is in his office on Saturdays and Sundays and he was there until one in the morning the night before. He does not necessarily agree with Löf's ideas about the future of solar power, but this does not prevent them from collaborating. "I have to respect Löf's views but I need not climb aboard a bandwagon. Here, I can do anything I want to do as long as I do it right."

He, Löf, and a team from Westinghouse's Georesearch Laboratory at Boulder and Pittsburgh spent the better part of two years on their study of STEPS, Solar Thermal Electric Power Systems, for the National Science Foundation. The purpose was to put some price tags on the speculation that had been so rife over the future of vast solar power farms. Like Duffie and Beckman with their mathematical house, they chose Albuquerque as the setting for a hypothetical ten-megawatt power plant. They looked at various ways to capture the sun such as flat collectors, troughs, or the power tower, and they ran through the pros and cons of heat storage for short or long periods. Out of this came some ideas that ran counter to popular conceptions.

They remark that the economies of scale achieved in modern plants running on oil, coal, or nuclear fuel do not apply to solar power. Instead, a solar plant must necessarily consist of a large number of small modules. Since two acres of collectors cost nearly twice as much as one acre, there is no saving in bigness. Not only that, but heat would have to be moved from the collectors to the turbines over all that distance, a drawback one does not find in the usual plant which is at least a thousand times more compact, as Löf remarks.

He and Karaki also get away from the idea that a solar plant might fit today's systems by supplying steam to turbines already in operation. They think there is not all that much money to be saved. An estimated $2,500 per kilowatt in capital costs would go mostly to the collectors with less than $100 per kilowatt represented by turbines, generators, and accessory gear. Besides that, these turbines run best at temperatures too high to be supplied by solar collectors without extremely precise and costly optical devices.

Karaki went over the work that Colorado State has already done with Westinghouse. "We have tried to get a computer evaluation to identify a system in which R and D money can be invested. We think that the flat-plate collector is out for power generation and the power tower is in. But the approach may be one that has been used by Francia in Italy, an array of parabolic reflectors on a rack which can be focused onto a central absorber." He displayed a drawing of the system. The reflectors were small, no more than eight inches in diameter, and mounted in twelve rows of twelve on a rack. At the University of Genoa, Professor Giovanni Francia has built three solar plants, the largest covering 2,150 square feet, on this principle.

"We shouldn't look at a solar plant as a substitute for present plants but as a supplement. In the United States, people work an eight-to-five day. There are two peaks in power demand, one at around eleven in the morning and the other about seven in the evening. With three hours of storage, we could meet the evening peak with solar energy. Then we could use the other plants to meet the base load. We don't need a thousand-megawatt plant to prove the concept but, in my thinking, we will have to have a pilot plant. Until then, this is all conjecture."

Yet it is on the wings of conjecture that one may journey from the foot of Horsetooth Reservoir at Fort Collins to the banks of the Neckar River in southern Germany without ever leaving the domain of solar thermal power. At Aldingen on the outskirts of Stuttgart, Nikolaus Laing is pursuing one of the wildest visions in the whole world of energy reverie: an "energy cascade." He wants to use the sunshine of North Africa to turn water into steam to be piped under the Mediterranean to Europe where it would be superheated by nuclear reactors. Then it could be used, in descending order of temperature, to run steam cars, industries, and power turbines. After it condensed back into hot water, it could heat homes, melt highway ice, and irrigate year-round crops.

The strangest part of the story is that the vision is not all that wild. Laing already has some of the components. He can store heat, whether from sunshine or any other source, in a "latent energy battery." His selective surfaces that absorb solar heat by day or reject it at night have been tested. In his plant where he brings his ideas to the production stage, material for a "solar wall" is turned out in great rolls. At Aldingen, there are visitors every day; occasionally reporters, more often emissaries from governments, Laing's and others. Habib Bourguiba, Jr., the son of the president of Tunisia, came here to talk about the production of steam beneath sand dunes; a Kuwaiti sheikh called to look at Laing's roofs that keep room interiors cool; officials from Bonn want to discuss his ideas for high-speed intercity transit. With all this, Laing meets the payroll of a staff of a hundred and forty. He is neither an academic nor a civil servant; he earns his overhead with income from his patents, some seventeen hundred around the world. He estimates that the sale of products based on his inventions amounts to $800 million a year, although he does not collect royalties on anything near that figure. In the United States, thirty million devices with a Laing twist are sold every year, half of them under his licenses. They include hoods for kitchen ranges, air curtains for buildings, and the blowers in 90 percent of American hair dryers (he can name the brands but their manufacturers would rather he didn't). The blower is the kind of invention that typifies Laing's attitude, that of the European who has been bumping his head against energy ceilings all his life. In the usual version, blades suck in air at the rim, working against centrifugal force. Laing's tangential blower inhales at the hub and ejects from the rim, a more efficient way to do the job in so small a machine. Income from these unexciting sources allows Laing to cover his expenses, something like $3.7 million a year, and then let his mind run to other matters more worthy of his fancy.

A trip to Laing's think-plant has something of the aspect of a visit to a house of wonders. On the main street of Aldingen (which is one of the few streets of Aldingen), a turn leads to the Neckar. A narrow bridge crosses the river at a lock, then a towpath runs downstream for two hundred yards through an embryonic industrial park. The road goes no farther except for hikers. For cars, there is a parking lot next to the concrete marquee over the entrance to a building that houses all nine of Laing's companies at this address (there are others elsewhere: among them the Illinois Fan Corporation in Chicago and Laing Vortex on Fifth Avenue in New York). Inside, an elderly workman serves as a receptionist. Young men in jeans and checked shirts scurry up and down stairs in a state of New World relaxation and a life preserver hangs near the door in case any visitor lands in the Neckar instead of the Physikalisch-Technisches Entwicklungsinstitut Laing.

Soon Laing appears. He is a slender man, tieless, wearing a corduroy sport coat. He does not look as old as he does in his photos that have appeared in the popular German press and in the publications he puts out. He is fifty-two, well-preserved by a regimen of twice-daily swimming, winter and summer, in the outdoor pool next to his house on the hill overlooking the plant. He took us up to the house, out of the way of the scurrying young men. It is a spacious place; when an architect started to build Laing's home, he and Inge, his physicist wife, were too busy to see what was going on. When they finally did look, they hired a bulldozer to knock it down so they could put up the house they wanted. The basement is somewhat of a Laing Energy Science and Technology Museum which enables investors in his companies to keep track of what he is doing with their money. "I do research and development work. In R and D, the problem is always money. I have five hundred partners; I said 'partners,' they are not stockholders and we are not a public company. They want to make money with money. I always have a rope around my neck. If I'm not successful, they'll kill me."

He started as a student of aeronautical engineering, but that was another profession closed to Germans at the end of the Second World War. He changed to meteorology, a subject that demanded little of his time and effort at the University of Göttingen. A professor had a group working on a research contract; they had been stopped four months by a problem. Coming in with a fresh mind, Laing found the answer in four days. "It taught me that, in R and D, one can compete with large groups." When he was graduated, he started a laboratory under government sponsorship. "I couldn't work with civil servants. They're not aggressive — and I am. So I bought the government out fifteen years ago. In fourteen instalments, I paid back everything they gave me."

His first independent research led to a pump seal. He sent it to several manufacturers and the highest offer that came back was eight thousand marks. It looked as though he would never make a living on research and development. He tried to sell his discovery on a royalty basis. A vacuum cleaner manufacturer explained that he did not like to pay royalties and offered a flat two hundred and fifty thousand marks. "His sales with my invention amounted to two hundred and fifty million marks. That taught me that theoretical R and D is not possible. If you cannot ensure that the product can be manufactured, no one will pay for it."

Pumps and fluid mechanics were interwoven into the early days of his independence and they led to the conservation of energy. Examples are on display in the basement of his home. One of the most convincing is the heat pump, before and after Laing went to work on it. His house of wonders is also a house of horrors; he has several appliances built the old way. The mere sight of his old heat pump makes it easy to understand why few bought it. The machine juts out a yard from either side of a thick brick wall and runs with the clatter of a freight train. "It has a life of five thousand hours," said Laing, almost compassionately. His own heat pump, he added, can run almost indefinitely. It consists of two rotating heat exchangers, one inside a wall and the other outside, along with a compressor to provide the heat. In summer, it is converted into an air-conditioner. He spun a heat exchanger and it rotated without a sound. Bearings are another of his obsessions; to reduce friction, he uses water in some instances and a magnetic field in others. Laing estimates that if his heat pump were introduced into Germany, three times more heat would be supplied for the same consumption of electricity, an annual saving of thirty-eight million tons of fuel oil. Laing would also be competing against Laing for it was he who devised the mineral oxides that are used to collect and store warmth at off-peak rates at night in many electric resistance heaters in Germany.

His effort to wipe out friction led him into the high-speed movement of people and freight where it is the biggest obstacle of all. It can be overcome at great cost by jet airliners and, so experience has taught, at not much less cost by air cushions, magnetic levitation, and other devices on the ground. Laing proposes a system based on his water bearing: a groove filled with water under pressure. It can work for two rotating parts or, he believes, it can become a water cushion for a high-speed cabin hanging from a monorail inside a tube. This is the basis for the Laing Impulse-Bahn. He calculates that it could make a three hundred-mile trip from city center to city center in thirty-six minutes as compared to four hours by train, two by plane, and an hour and a quarter by currently projected high-speed ground vehicles.

Water under pressure would drive the Laing Impulse-Bahn. He finds that a high-speed vehicle weighing three hundred tons would need as much energy to accelerate to six hundred miles an hour as an ocean liner needs to get a hundred thousand tons up to twenty-four knots. But the ground vehicle must dissipate this power when it comes to a stop and its brakes would release enough energy to heat a large building. Instead of getting rid of the heat, Laing proposes to put it to work. To start his impulse-bahn, he would use a linear turbine; that is, curved blades would be fixed in a row beneath the cabin. Over the first seven kilometers (4.2 miles) at the start of a journey, nozzles spaced seventy-five meters apart would throw a jet of high-pressure water against the blades, sending the cabin on its way. It would need no power at all during its journey through the tube on a water bearing; the initial impulse would be enough. Then, twelve kilometers (7.5 miles) from its destination, the process would be reversed. The cabin would be braked by sucking up water into tanks and compressing a gas to two hundred and fifty times atmospheric pressure. This would immediately become a flood of heat to be stored by a "latent energy battery." When the cabin started again, it would use the stored heat in an onboard rocket, supplementing the water nozzles.

It all sounds far out, but not too far for German government officials who have consulted seriously with Laing. Germany is nearing the situation that Japan has met with the high-density, high-speed rail line between Tokyo and Osaka. In the future, even railroads threaten to be swamped by the need to move passengers quickly over hundreds of miles to maintain the connective tissue of economic centers. With the kind of congestion in Japan or the Ruhr Valley, there is no room for Los Angeles. Frankfurt planners estimate that if all commuters drove to work, they would need roads covering two and a half times the city's area.

The impulse-bahn shows how Laing regards energy. He dispenses it in measured doses, never losing a drop to friction, then scoops up the leavings so they can be put to work again and again. Not a calorie is frittered away. He is like the oilman who runs crude through a refinery and extracts everything from it, from jet fuel and plastics to pesticides and heating oil. Throwaway living is for the consumer, not the producer. To mop up spilled heat, Laing has his latent energy battery which, unlike the impulse-bahn, exists off the drawing board. It can be seen in his basement. A radiator heated to the boiling point of water needs only eleven seconds to lose 35° F. when the heat is turned off. The latent energy battery holds the same amount of heat for half an hour.

Laing is circumspect when he explains his heat battery and certain other patented processes. "I can't give away what other people are paying me to provide." It would appear that he is using the principle of a change of phase from solid to liquid to store heat, which we have already seen. His system involves a manmade zeolite, a stone composed of water for all but 3 percent of its weight. If heated to the boiling point, the zeolite becomes as brittle as cigar ash. In Laing's battery, as in the zeolite, only 3 percent of the material is a permanent solid while the rest is a salt melting at a predetermined temperature. He claims that this enables him to eliminate the stratification that has bedeviled others with eutectic salts. Laing states that the structure of his zeolite forms a skeleton that keeps its shape while the salt changes phase. It requires only a light protective envelope instead of a true container, yet it is strong enough to be used for heated floors.

When it comes to moving heat around a house, he has gone the opposite way into a "fluidized" system. Here, Laing explains, minute crystals are introduced into a carrier liquid as a fine powder. Charged with heat, the liquid flows like water. When all the heat is released, it crystallizes to the consistency of condensed milk. Consequently, the heat storage medium can be pumped from a solar collector to an underground tank and then to the radiators of a house. His substance also responds immediately to seeding; heat spreads through it "at one centimeter a second, not one centimeter a day." Depending on changes in its chemical composition, the battery can be used to store heat at temperatures from $-55°$ C. to $800°$ C.

He was working on a "heat-when-you-need-it" wall with a German manufacturer of prefabricated houses. He demonstrated a panel for us: it was a sheet of clear plastic, the protective envelope for the heat-carrying substance stored in a number of channels. It was charged; then he pricked the sheet to seed the change of phase from liquid to solid. We could see the channels change hue as they released their stored heat. Now the whole sheet was hot. "We think we can store heat in a wall like this. You will be able to release it by telephone half an hour before you come home so that your house will be warm when you get there." The demonstration was convincing and so was Laing. The material was being installed in a test building on the grounds of his institute, which cover eleven acres. He had taken the great drawback of the eutectic salts — their refusal to change phase without seeding — and turned it into an asset. His material, with seeding, can change from a solid to a liquid and back again "for a hundred thousand cycles . . . indefinitely."

Laing has gone ahead with plans for a solar house built of roof and wall panels that collect the sun on the outside and radiate heat inside while his storage material flows between the two surfaces. These sandwich panels, developed with Farbwerke Hoechst, were shown at an international sanitation and heating fair in Frankfurt in 1975. They are part of an experiment to heat a house in Central Europe at a latitude of 50 degrees N., farther north than any of the lower forty-eight states in the United States. Meteorological data indicate that in January, the coldest month of the year, the mean duration of sunshine here is only forty-five hours. Under such conditions, a small oil burner would raise the temperature of the heat storage material to the point where a Laing heat pump could take over. During most of the year, Laing states, the solar collector could more than meet the building's demand and supply not only radiators and a domestic hot water tank but also heat a swimming pool and a greenhouse. His model of the house resembles a small chalet, its roof almost reaching the ground. Sandwich panels, which Laing thinks can be mass-made for the same cost as the building elements they replace, make up 80 percent of the outside of the house. He has assumed a building containing 14,000 cubic feet which would require a spherical underground storage tank holding 280 cubic feet of his latent heat material. No standby central heating system would be needed; the heat pump and the small burner would do the job. Laing calculates that his house would save enough in heating and air-conditioning costs to amortize an added investment of about four thousand dollars.

One hesitates to use the word "breakthrough," but it would seem that Laing has gone a long way in heat storage. He need not wait for the solar house to be built or the impulse-bahn to run before applying his latent-energy ideas. An early use has been a "heat-rectifying roof," an air-conditioning system that can serve for summer cooling and winter heating. During the hot months, warmth inside a room is absorbed by a layer of heat storage material just above the ceiling. The cooled air sinks to be warmed again and, by the end of the day, the storage layer has been melted. At night, the roof gets rid of this heat by radiating it out to the sky. The storage material cools back into a solid state and it is ready to start work again the next morning. In winter, a valve stops heat radiation at night and turns the roof into an insulation panel. When Laing calls these roofs heat rectifiers, he means they allow only a one-way flow of heat. An important aspect of the system consists of selective surfaces which have also been developed at Aldingen. In hot dry places like North Africa, the Arabian peninsula or Australia, Laing claims his roof can supply air-conditioning with no power consumption. He states that he has been able to maintain 64° F. inside a room while the roof went up to 187° F.

At far greater temperatures, his latent energy battery is to run a steam car. In terms of heat, it stores thirty times as much energy as the lead-acid batteries used in electric vehicles. It fits right into the battle that Laing, the builder of better bearings, is waging on the internal combustion engine that grates against everything he believes about waste, noise, friction, and heat. In a brochure, he remarks:

> Ninety years ago, cars already had internal combustion engines. Since then, performance and air consumption have risen a hundredfold and the engine speed by ten times, yet the principle remains the same. Engines whose combustion phase lasts only a few milliseconds do not allow a complete conversion of fuel and the result is the emission of harmful gases. The process of taking in air and, above all, emitting exhaust gas at the speed of sound is still today, as it was with the first motor car engines, a source of noise in city centers most difficult to suppress.

Laing wants to suppress it with a steam vehicle. In its first incarnation, it would run on a burner of his own design which, he claims, offers clean combustion (as an old newspaper reporter, I keep hedging with "he claims" or "Laing states," but my impression of the man is that he is honest and he knows what he is talking about, with money behind his words). This will operate the car outside cities while charging up a latent energy battery to supply power in town. The battery can already be seen in action, running a small steam cart on the grounds of Laing's institute.

Many of his devices are incorporated into the car. Rotating heat exchangers serve as condensers for the steam power plant, a rotary engine based on the Wankel. Since energy will be stored in the battery, a large boiler with its attendant dangers of explosion is not needed and, instead, a film evaporator is used. In a later version, the car would use its burner only in emergencies when it ran out of stored steam. Normally, it would charge its battery at steam stations with each refill giving it a range of one hundred and fifty miles. The stations would be one of the steps along the "energy cascade" which could be brought into operation by 1985. By then, a country like Germany could be using only one-third of the primary energy it now consumes but with the same results.

Laing wants to achieve the cascade in easy stages. The first is at hand; his heat pumps are ready to permit more efficient use of the same power while reducing noise. In the second stage, we find his heat-rectifying roofs and solar collectors. These devices have been tested and Laing says they, too, are ready for mass production. He thinks oil consumption could be halved by such climate control systems in the home.

Next is a giant step: the construction of artificial lakes as heat accumulators, a suggestion that Laing has adopted from another German scientist. Waste heat from power plants would not be dumped into rivers but stored in these lakes. Layers would form, hot water on top and cold below. The top layer could be heated to 195° F. by low-pressure steam from power turbines and then piped away to heat buildings. Water from lower layers at about 160° F. could feed hot water to households and industries. Laing writes: "Heat energy is conveyed over great distances and with minimal losses by insulated co-axial pipes [that is, one pipe inside another], the outer carrying water at 160° F. and the inner at 195° F. The supply of energy to customers remote from the heat generator becomes practical and economically sound." Laing is a great believer in moving heat in pipes instead of electricity in cables.

The idea of artificial lakes goes with another inspiration: Laing wants to run hot water through the present system of mains. Homes would make their own cold water as they needed it, using air-cooled tanks with fins to radiate the heat. Now we have waste heat turned into a resource in the form of warm lakes. In Germany, Laing thinks that "the return from various heat consumers could supply

over 250,000 acres of land under cultivation with warm water, producing harvests all the year round as in the tropics." He wants to
cover the lakes with a ten-inch-thick insulating layer of foam material, strong enough to bear the weight of tropical crops. Laing
has already covered the surface of his own swimming pool with
plastic spheres that clack and rustle when he dives in, maintaining
heat inside the pool.

Water going out from the lakes to heat homes could be mixed
with cold water and supplied to the condensers in power stations.
If it is not cold enough, it could be used to heat roads to free them
of ice and snow, even to dry them after a rain. Laing finds that
piping water to only 10 percent of the German road system — just
about the mileage of roads built in any given year — would be
enough to get rid of unwanted low-temperature heat. If necessary,
hot water would make seven or eight round trips between the
householder and various public services until it lost its temperature.
Since water returned to the system must be cool, the customer who
used the most heat would pay the least. Heat for greenhouses,
garden paths, cultivation, cleaning ice and snow from roofs, and
warming swimming pools would be not only free but profitable.

At this point, Laing predicts that heat release to rivers will be
down to 10 percent of what it is now. Fuel consumption and oil
pollution will be reduced by half as the economy reaches the fourth
stage on its way to the energy cascade. Here we find the steam car.
Laing wants to use nuclear reactors to supply superheated steam to
the filling stations where the batteries will be charged. Since the
reactors will not be generating power, they can be located in remote
mountain valleys, far from uneasy neighbors. And since their sole
product will be steam, they will not be running turbines and wasting
heat. Their entire output will go into the network of superheated
steam lines. Laing says that the steam car and network will start to
appear around 1980 when the internal combustion engine will become obsolescent. Steam will come from the reactors at 650° C.,
too hot for power plants but just right for charging latent batteries
and for use by certain industries. During the charging process, the

steam will lose about 100° C., bringing it down to the temperature that the power stations need.

In the fifth and final stage, all the pieces fit together with solar energy at the head of the energy cascade. Now we go from an oil to a hot-water economy. Laing wants to start in the Sahara Desert. Water might be drawn from the Niger River east of Timbuktu and piped to the solar collectors. Even on the way, it would be heated by sunshine beating down on the pipes (with heat-rectifying panels to cut losses at night). Or geothermal energy could be used at a low level. Twenty-six hundred feet down in Tunisia, the temperature of well water is 175° F., an appreciable headstart before it is heated.

The heating plant itself would consist of Fresnel lenses concentrating sunlight onto flat collectors. Heat exchangers would be installed at five levels beneath the desert to let energy build up. Laing says that it should be possible to store heat for twenty days in the Sahara, a safe margin where the longest period without sun is never more than ten days. By spreading the risk over several heating stations, he would have a better chance to get through sunless spells. With the Fresnel lenses and selective surfaces, he believes he can heat water to 380° C. It would then be sent under the sea to Europe in pipes nearly two stories high. The pipeline would resemble a giant thermos, a vacuum between its inner and outer layer.

Once on the European continent, pipelines would feed the reactors where the hot water would be boosted to 650° C. In 1985, it is estimated that Germany will need 233 million tons of oil a year, corresponding to the energy represented by 2,630 million tons of high-pressure steam at 650 °C. This is the very substitution that Laing envisages with Niger water heated in the Sahara or even seawater taken from the Mediterranean, desalinated and heated on the Iberian peninsula. At this point, oil demand drops to 6 percent of where it is now and coal to 25 percent. Power stations no longer poison the air and ruin the rivers; the importing of hot water also alleviates the water shortage in industrialized Germany.

Laing states this could be done with technology now at hand and that more than eight million dollars has been invested in researching the scheme which should cost no more than the oil and coal that will be needed between now and 1985. Then the energy cascade should pay for itself in eighteen years. Even if it never gets beyond its early stages, Laing is not likely to lose. Overwhelming though the idea may be (as overwhelming as a nuclear Europe), it rests on some sound components. They are as real as the income Laing gets from his inventions and puts into his personal institute at Aldingen. He will go on giving shape to the ideas that come to him; he already has published a new concept of a 1,000-megawatt solar plant on a floating circular island that could be kept facing the sun with only the help of a propeller. The cold bottom water pumped up to run the condensers might fertilize a mussel farm and help pay off the capital costs. One can see Laing's thoughts flowing again; at each stage, there is a leftover or a side effect for which a use must be found. His processes may not be so extravagant after all. He does not suggest that we pile up their by-products in the hope that someday someone will be able to throw them away.

XII.

~~~~~~~~~~~~~~~~~~~~~~

Power Now

~~~~~~~~~~~~~~~~~~~~~~

A MORE ACCURATE HEADING might be "power then and now." As soon as ways are sought to meet demand not for the year 1985 but for late today and early tomorrow, the old approaches are dusted off. Coal is mined, wood burns and windmills spin, forgotten schemes to tap the tides or the geothermal heat of hell come to the fore. What they all have in common is that they are within reach. They can be done because they have been done. They were dropped when oil priced them out of the market. Now they are back in new forms; they look better because their costs need only be shaved slightly, not reduced by one or two decimal places.

George Szego and Clinton Kemp have already voiced their proposal for an energy plantation. Others have looked enviously at how nature collects and stores sunshine. At the National Physical Laboratory of Israel, Tabor has calculated that sixteen hundred square miles of plantations could produce the equivalent of Israel's present capacity of 1,500 megawatts. As far as Israel is concerned, he remarks that the land and the water that would be required could probably be put to better use growing food "of which there is an even greater shortage than fuel." In countries where land is not so scarce, the story might be different.

Unlike Israel, Australia is long on acreage, and the Australian Academy of Science has recommended research on ways to produce methane and alcohol as transportation fuel by fermenting plants. Roger Morse, director of solar energy studies for the Commonwealth Scientific and Industrial Research Organization in Melbourne, deems that less than 1 percent of Australia's land area would be "adequate to provide, on a perpetual cycle, the raw material for a process to produce all our present needs for liquid fuel." What is needed is a cheap way to obtain fuel from cellulose, which Morse regards as solar energy stored in trees and plants. The main problem is to make sure that phosphates and other fertilizers are recycled back into the land in the process. The Academy estimates that seventeen million acres, a third of the country's productive forest but only 4 percent of its total forest area, could meet all its present energy demand. Between seven and ten million acres would supply 10 percent of its needs in the year 2000. Australians are worth heeding when they speak of solar energy. Something like 215,000 square feet of flat collectors to heat water are installed in the country and small factories have been turning out 45,000 more square feet a year. Water heaters have been particularly successful in Darwin on the north coast, far from supplies of fuel. Darwin is only twelve degrees south of the equator which means that the amount of daily sunshine varies but little over the year. A solar water heater there, Morse says, pays off its investment at the rate of 30 percent a year.

A fine summary of the possibilities of using solar energy through photosynthesis can be found in an article by Dr. Melvin Calvin, director of the Laboratory of Chemical Biodynamics at the University of California at Berkeley, which appeared in that energy issue of *Science* published in 1974. Calvin, who was awarded the Nobel Prize in Chemistry in 1961 for his work on photosynthesis, examines the plants that use sunshine most efficiently and takes up the leader, sugar cane. It could be burned; but it would be easier to convert it into alcohol. The price was eighty-four cents per gallon at the time he wrote the article, almost competitive with the price of alcohol derived from petroleum. Calvin believes that such a conversion of carbohydrates into hydrocarbons will represent an economic use of photosynthesis.

If the sugar planters of Hawaii would convert about one-third of their molasses directly into feed alcohol, they would not have to purchase the 15 million gallons of petroleum they need to run their agricultural machinery. Another special system seems to be developing in Nebraska which has about 7 million bushels of spoiled grain per year. This should yield more than 20 million gallons of alcohol which, as a 10 per cent additive to gasoline, would give 200 million gallons of "gasohol." This is the name used by the Nebraska legislature to designate a composition which would qualify for a 3 percent state tax credit.

We have always been taught that alcohol and gasoline do not mix at the wheel, but they get along pretty well in an engine. Professor John Heslop-Harrison, director of the Royal Botanical Gardens at Kew, has remarked that Europe used half a million tons a year of fermentation alcohol, derived mostly from potatoes, to run cars before World War II. It could be mixed in proportions from 10 to 30 percent with gasoline, not only saving petroleum but producing a higher-octane fuel. Recently, the state of Maine has considered a plan to convert five million acres of diseased spruce forest into wood alcohol and thereby reduce some of its demand for high-cost oil. Robert Monks, the director of its Office of Energy Resources, has said: "One thing we have in Maine is trees. It's time to use some Yankee ingenuity and use the resources we do have. Obviously, we can't survive up here on a dependence on domestic or foreign oil. We'll consider anything, because the prospects are so bleak. People may think we're smoking dope when we talk about some alternatives to petroleum, but we're serious."

Calvin sees the possibility of a much higher production of hydrocarbons from *Hevea,* the rubber plant now grown in Malaysia and Indonesia. Since the rubber is a hydrocarbon to start with, conversion into ethyl alcohol is a simple matter. While the yield per acre of *Hevea* is half that of sugar cane, it is only 20 percent behind as a source of alcohol. Rubber planters believe they can raise yields from one to three tons per acre (it was only four hundred pounds per acre at the end of the war) and this would make *Hevea* a source of hydrocarbon formed in a single growing season instead of over geological time.

A much less immediate prospect, but a possibility all the same, is the use of photosynthesis to produce hydrogen. Calvin and his collaborators are among those working on these lines. It has been estimated by Dr. R. M. Pearlstein of the Oak Ridge National Laboratory that twenty thousand square miles of land would be enough to produce a million tons of hydrogen a day.

A number of people are getting around the problem of available acreage by going where there is so much of it: they go to sea. The ocean not only offers space but it is a natural collector of the sun's heat. The energy can be harvested in a variety of ways, all well within reach of present-day technology if not economics. One that is being tested by Dr. Wheeler J. North, a biological oceanographer at California Institute of Technology, amounts to a marine energy plantation. He wants to produce fuel by growing *Macrocystis pyrifera,* a variety of giant kelp. Normally, the kelp grows on the bottom in thirty to fifty feet of water, throwing out holdfasts (which landsmen mistake for roots when they find the seaweed washed up on a beach) to fasten onto rocks. This would make harvesting inconvenient. An artificial sea-bottom has been created, a seven-acre rope raft moored forty feet down. Divers from the Naval Undersea Center at San Diego do the sowing on the energy farm by fastening young kelp plants to the raft. The plants make good use of the sunlight striking the sea; they grow as much as two feet a day until they reach adulthood and a length of no less than two hundred feet.

Howard Wilcox of the Undersea Center foresees a thousand-acre energy farm in operation by 1980 and, five years later, a spread of a hundred thousand acres. The harvest would be used to produce methane gas for power, a process already tested in the laboratory. Kelp cropped from the farm would be loaded into barges where bacteria would break it down into methane in the same way they digest sewage. Wilcox has expressed high hopes for the future:

Assuming a 2 per cent conversion efficiency for converting solar radiation into the stored energy of seaweed compounds, 5 per cent conversion efficiency for the production of human food from the seaweed, and 50 per cent conversion efficiency for the production of other products from the seaweed, the marine farm is conservatively projected to yield enough food to feed 3,000 to 5,000 persons per square mile of ocean area which is cultivated, and at the same time to yield enough energy and other products to support more than 300 persons at today's US per capita consumption levels, or more than 1,000 to 2,000 persons at today's world average per capita consumption levels. Since the oceans, conservatively estimated, appear to contain some 80 to 100 million square miles of arable surface water, this means that marine farms could conceivably support a human population ranging from 20 to more than 200 billion persons, depending on the degree of affluence assumed.

Seaweed is a fixed form of marine plant life but the vast preponderance of vegetation in the ocean drifts to the command of the winds and currents. This is the phytoplankton, of which an estimated ten billion tons are produced in the sea every year. A noted marine scientist, Dr. William Von Arx, writing in *Oceanus,* the journal published by the Woods Hole Oceanographic Institution where he works, has speculated about this figure. If the phytoplankton were harvested for power alone and converted to methane, it would supply ten times the energy that the planet now uses. More modestly, Von Arx suggested that "initially at least, plankton gathering might be a practical way to provide fuel for ships at sea, especially fishing vessels which ply the plankton-rich waters where fish also abound." This implies an efficient harvesting system so the ship will not burn more fuel than it collects.

A final quantum jump in solar sea power was achieved by a *New Scientist* columnist, Ariadne, whose suggestion is not so far removed from Farrington Daniels's "piece of thin transparent plastic loosely stretched across a long rectangular frame above the ground and filled with water." If you share my taste for British humor, read on as Ariadne writes:

My aquoptic friend Daedalus, musing on the energy crisis, recalled those glass-bottomed boats for viewing the undersea scene in comfort. He realized that if the glass were replaced by clear plastic film, the water pressure would bulge it upwards. And if the boat were circular, the even pressure on the transparent bottom would deform it into an exactly spherical domed surface — a big lens, in fact. At first, Daedalus thought he had merely invented a neat magnifying viewer; but he soon realized that this elegant principle makes possible perfect lenses of enormous size. . . . Daedalus calculated that a 100-metre lens-boat, like a giant burning glass, would focus almost 10 megawatts of solar heat to a point 300 metres down, boiling the water into steam at 30 atmospheres (the water pressure at that depth). An insulated pipe terminating in a transparent collector around the heated volume could channel the high-pressure solar steam to the surface to drive turbines. As the Sun traversed its daily arc, simple slewing gear would keep the collector at the moving focus. This beautifully simple and cheap scheme could be extended indefinitely, for square kilometres of ocean could be easily covered by huge lens boats and thousands of megawatts of solar steam harnessed. Daedalus is also thinking of applying the principle to anti-submarine warfare, in-situ instant steaming of deep-sea fish (which should rise to the surface ready to eat) and recovery of wrecks by boiling the water within them to buoyant steam.

"I would be very disappointed if anyone took these ideas more seriously than I take them myself." The comment was made by Arsène d'Arsonval in the *Revue Scientifique* published in Paris in 1881. He was referring not to the lens-boats of Daedalus but to an equally improbable idea.

Let us suppose that we place a steam boiler in the water of Grenelle Well whose temperature, I believe, is 30° C. [86° F.]

Then let us place the steam engine's condenser in city water whose average temperature is 15° C. [59° F.] The loss of heat is 15 degrees. Let us fill the boiler with sulfurous acid or any other liquefied gas. At 30° C., the vapor pressure of this gas is only three-and-one-half atmospheres, which allows it to be used in ordinary boilers. In our boiler, we thus obtain a continuous pressure of nearly two atmospheres which *does not cost us anything.*

We can place our condenser in a glacier and immerse our boiler in a river with a temperature of 15° C. Again, we will obtain a loss of heat of 15°. Note that condensing can be done at a great distance from the engine with the help of an inclined tube on the slope of the mountain which would bring the liquefied gas back to the engine.

Ideally, one should place the boiler in the equatorial ocean and the condenser at the poles. But there is no need to make such a long journey. In fact we know that, even at the equator, the temperature at the bottom of the sea is 4° C. [39° F.] One need only place the boiler on the surface and the condenser a thousand meters below to obtain a sufficient difference in temperature.

What d'Arsonval had in mind was a heat engine working on the so-called Carnot cycle. It requires a source of heat and a source of cold. In a big steam turbine, the heat can come from an oil furnace or a nuclear reactor, the cold is derived from a water supply, whether a river, a lake or the sea. The heat is needed to convert the working fluid into a gas, a complicated way of saying what happens when water is boiled into steam. The cold is needed to condense the gas back into a liquid. If condensation does not take place, then the engine will come to a stop, according to the law formulated by Nicolas Carnot, a French physicist of the early nineteenth century, who studied the relationship between heat and mechanical energy and was one of the founders of modern thermodynamics.

In the usual steam plant, the working temperature may be 500° C. and the size of the operation remains within what are considered reasonable limits these days. To get the same amount of power from an engine running with its boiler at only 30° C., a gigantic amount of water is needed to make up for the missing temperature. The sea supplies it readily, as Georges Claude commented nearly fifty years after d'Arsonval. Claude was a French engineering genius who had put neon gas into light bulbs in 1910 and synthesized ammonia during the First World War. He was worried about bleak energy prospects until he and Paul Boucherot found a new limitless source. On the 15th of November 1926, Claude reported their discovery to the French Academy of Sciences. In the tropics, the temperature of the water 3,000 feet down is about 40° F., compared to 80° or 85° F. on the surface. Put a pipe three thousand feet down and the deep water could be brought up. "It is this fact which, no doubt, will be the starting point of a grandiose solution to the problem of the use of solar heat," Claude said. He apparently reached this conclusion without being aware of the speculations of d'Arsonval whom he credited only later on.

He tested his idea, first in the laboratory and later at sea, but with inconclusive results. In 1960, at the age of ninety, Claude died in obscurity. Perhaps he had been born before his time; a decade and more had to elapse before his gloomy energy forecasts were to reappear and, once again, the thermal store of the sea would stir scientific inquiry. His idea (and d'Arsonval's) received posthumous backing when the National Science Foundation selected "ocean thermal energy conversion" as one of the six areas of solar energy in which it would finance research.

It was NSF's decision that led us to Carnegie-Mellon University in Pittsburgh to see Dr. Clarence Zener but first it took us to a colonial motor inn in juxtaposition to the Pittsburgh airport. As befitting a colonial inn, the motel had a Polynesian restaurant that inspired Madeleine:

"Soft sugary Hawaiian music came from the plastic bamboo ceiling. A vine with heavy purple plastic grapes twined among the ceiling slats while a plastic palm tree swayed its head over my head and a plastic bamboo screen half-hid a big window where a plastic setting sun glowed bright orange. Against this sky was silhouetted a hill crowned by a pagoda where three little figures were talking under other palm trees (all in black wood). We were in the Blue Hawaii Restaurant of our motel where we ordered a very late breakfast. The motel was in red brick and white pillars, green shutters and white windows, built next to the runway of the airport separated from us by a thin screen of birches through which we could see fat white fish roll slowly, then rush towards the sky with a racket of engines that shook the windows, the chairs and the beds in the motel, thereby eliminating the need to put a quarter in the Magic Fingers machine that was supposed to quiver for fifteen minutes to put you to sleep (and it works . . . you do go to sleep). The Blue Hawaii nightclub-restaurant had no windows. It was lighted by multicolored lamps whose shades, driven by the heat from their bulbs, rotated slowly on a frame of gilded, embossed, beribboned metal made in Taiwan. Instead of windows, we had the plastic sunset. To welcome us, there was another big plastic-and-wood panel proclaiming ALOHA in red letters against a background of green leaves. It was strange to have breakfast in a dark nightclub when the day outside was so cool and sunny. Here, in this somber place with every chink stopped up, there was not the slightest whisper of an airplane to be heard, even though they were taking off one after another only two hundred yards away. Dan said that it was not true that the bamboo was plastic because bamboo bamboo is cheaper than plastic bamboo. I could see that he was right, but the rest was all plastic.

"We had frosted flakes, English muffins, jam, tea for me, coffee for Dan, all he wanted, of course he took three cups as usual, all this for $1.75. Next to us, four young men were eating a Chinese lunch of soup, rice, chop suey and egg noodles for $1.50 each. There were white tablecloths on the tables and soft music. Our friends in Brittany, the Lorins, would be glad to pay 50 francs apiece to have dinner in such a setting. Outside, amidst the flowerbeds, there was an immense swimming pool covered with a roof of corrugated plastic and filled with green water. Unfortunately, it was open from 1 P.M. to 9 P.M., too late for us, we were leaving. And all this for eighteen dollars plus sixty-eight cents tax.

"We took a walk behind the motel. A meadow sown with violets and buttercups sloped gently towards some woods. I asked what the trees were, nobody knew. It was a dream spot except that, two hundred yards away, there were those damned silver fish. Dan clapped his hands to his ears and screamed that he wanted to get out. But his screams did not make the slightest impression on a red bird, standing on one leg only a yard away from us. I think it was a cardinal.

"Note: to start our new butter-colored Nova, I had to lift my behind; otherwise, there was a short circuit on the ignition interlock and it would not start. I had already learned how to beat time with my backside to the music on the radio by unhooking my safety belt. You just have to bounce up and down more or less in tune and you work the 'Fasten your seat belt' buzzer."

With Madeleine providing the percussion, we headed for Pittsburgh along the Penn Lincoln Parkway. With the exception of a visit to NSF headquarters in Washington, this was the first time that our solar quest took us to an inner city. The magic carpet of the parkway slowed and stopped, we crawled through streets as we made our way to Carnegie-Mellon, once Carnegie Tech, now so much plusher with the synergy of those two names. An inner city brought back inner-city cares. There was no place to park anywhere within sight of the Carnegie-Mellon campus; there was a visitors' parking lot where we were able to sneak our Nova in and leave it, hoping that the campus police force would allow us some overtime while we pursued knowledge. First we pursued and caught hamburgers in the university cafeteria where the decor was not Polynesian but polytechnical, serious-looking students of both sexes, it was the boys who had the beards. Thus restored, we set out in search of Clarence Zener. He was holed up in Doherty Hall; to find him, we ran a maze of elevators and corridors until we reached his office. He was talking to a student; I could distinguish between Zener and the student because Zener was the elder, in the last year of his sixties as he was to tell us. Otherwise, there would have been no clue. He was tall, lean, only slightly bent by his age. Chalk in hand, one wall of his office covered with a blackboard, the other with bathymetric charts of the Caribbean Sea, he was no more concerned than any bright engineering student would have been with the hang of his trousers or the shine of his shoes.

We sat and talked as if he had known us for years, not for a minute and a half. Zener invented the zener diode, a switching device so widely used in electronics that his name is not even capitalized anymore when it is mentioned. He was born in Vincennes, Indiana; he taught physics until he joined Westinghouse Research Laboratories in 1951 and became their director of science before he went back to the academic world in 1966, first as dean of science at Texas A & M, then as professor at Carnegie-Mellon. He has made his mark widely; that is why he can speak disinterestedly of energy and ways of obtaining it.

He first considered the ocean as a resource when the president of Westinghouse International asked him to work out the cost of getting fresh water from the sea. "That question changed my life. It was the first time I really tackled a cost problem. I became thoroughly disgusted with the standard engineering approach of looking for optimum curves and trying to achieve them at a minimum cost." With Richard Duffin, a mathematician at Westinghouse, Zener conceived a way to design an entire system for minimum cost instead of working on each component. It was to come in useful later.

At Westinghouse, it was natural that Zener should take up power generation. He did not know of Claude's work at the time, but he had the well-known atlas of North Atlantic water temperatures and depths compiled by Frederick Fuglister at Woods Hole. He amused himself trying to calculate the work potential stored in the oceans. His answer was similar to Claude's. "It is more than all the coal and oil on earth and, besides, it is renewable."

About this time, the idea of a solar sea power plant was put forward by a father-and-son team from Yorktown, Pennsylvania. James H. Anderson was a consulting engineer and his son, J. Hilbert Anderson, an MIT student at the time. They were interested in energy and they had already examined geothermal heat. They turned to the Claude process, costed it out and concluded in 1966 that a plant could be built for $165 per kilowatt. "It was the same capital cost as that of a commercial power plant using fossil fuel," Zener said. "It could produce electricity for three mills per kilowatt-hour, about half the cost of conventional plants, but it could not compete with the anticipated cost of power from nuclear plants. In the mid-Sixties, the cost of nuclear power had come down steeply. The Atomic Energy Commission was predicting that it would be one and a half mills per kilowatt-hour. I knew I couldn't beat that. In science, you do a lot of things, then you give up."

He left Westinghouse for Texas A & M, then he returned to Pittsburgh when he joined Carnegie-Mellon. "It was a shock to come back. The ecology movement had taken hold. Previously, one only considered economics in building a power plant. But now I took another look. We were supposed to cut down on our energy consumption. Going back to the simple life didn't appeal to me. I like a simple life, but not that much.

"What I did find was that the nuclear people's forecasts were all wrong. The cost of nuclear power at the busbar (that is, when it leaves the plant) had gone up to twelve mills per kilowatt-hour. For two years, I did nothing but propagandize, I spoke of solar sea power, I was laughed at everywhere, including NSF. All the physicists kept telling me that the efficiency in a conventional power plant is 35 percent; in solar sea power, it is 2 percent. That's a very convincing argument. For the same power output, the heat exchanger within the boiler must transfer more than ten times as much heat. They conclude that the heat exchangers must cost at least ten times more than in a conventional fossil-fueled plant." Zener likes to cite this as a case of what happens when one looks at components instead of a whole. "The joke is that the standard boiler must resist a combination of high temperature and high pressure. The high pressure requires thick walls, the high temperature requires still thicker walls or more expensive material. But the solar sea plant operates at only a very low temperature difference. Since it is submerged, the pressure of the working fluid will be largely compensated by the hydrostatic pressure of the sea water.

The heat exchanger will not cost ten times more but only half as much."

Zener and Dr. Abraham Lavi, a professor of electrical engineering at Carnegie-Mellon, have gone over a number of aspects of solar sea power and none appear insoluble. Corrosion by sea water can be halted through the use of aluminum which forms a self-protecting oxide. A small concentration of chlorine gas is enough to prevent fouling by microbial growth without harming marine life. The plant, as Zener sees it, would be a series of modules — boiler, engine, and condenser — fitted together vertically, the warm-water intake pipe near the sunny surface, the cold-water intake at a depth where the temperature is only 40° F. The plant would be built to float vertically two hundred feet down so that surface waves and storms could not reach it. But then how would it reach customers with the power it generated?

Zener and Lavi envision several markets. The first would be tropical countries where the coast is not far from suitable offshore sites for plants. Aluminum production immediately comes to mind. More than half the Western Hemisphere's bauxite reserves are in Jamaica, but ore mined there is shipped to the United States to be refined into alumina and then hauled to sites with cheap power for final reduction. "Since Jamaica could enjoy an abundance of cheap electric power generated by solar sea power plants," Zener and Lavi state, "a more economical process would be the refining of bauxite into alumina and the electrolytic reduction of alumina into metallic aluminum on the island itself."

At a later date, they think that power generated at sea could be distributed to North American and European consumers as hydrogen produced by electrolyzing seawater. Or, Zener says, the hydrogen could be turned directly into ammonia at sea for use as fertilizer which would find a ready market. His subsurface power plant would also be able to produce hydrogen at high pressure (supplied free by the ocean's depths) to be shipped as fuel.

"We still have to obtain some answers," Zener said. "We cannot yet say this is how to build a plant. But our questions are inconsequential compared to the questions that others are facing." He let himself be pinned down to a schedule. He wanted to start by designing several plants of the smallest possible size that would nevertheless supply enough power to run their pumps. He calculated they would be on the order of a megawatt, a thousand kilowatts. It would take a year to build these plants and two or three years to gain experience operating them and decide on an optimum design for the next step. Then one could build a ten-megawatt system to serve as a laboratory for a full-sized version, perhaps eight hundred megawatts, that would be in operation by 1981. "I like that date," said Zener, "It is exactly one hundred years after d'Arsonval published his first paper." In his cost estimates, he still allows three mills per kilowatt-hour, far below requirements for nuclear or fossil fuel plants. The environmental effects would be minimal. He and Lavi state that "if the world population in the year 2000 were to be supplied by energy from solar sea power plants at the present per capita rate of consumption in the U.S., the surface temperature of the tropical oceans would be lowered by less than one degree C." Zener has gone into the effects of such a change. The most obvious would appear to be a drop in the temperature of the atmosphere in the tropics but things are not quite that simple. That one degree has not been "lost" but rather transferred from the surface to the deep layers of the ocean. Since the surface will be one degree cooler and, therefore, less apt to lose heat by evaporation, the result will be a net gain of heat at the tropics. Zener remarks: "This net tropical heat input must be dissipated outside the tropics, presumably by increased convection currents. Most people in the world would probably welcome a somewhat warmer ocean outside the tropics. Climatologists in particular will welcome an increased transfer of heat from the tropics to the temperate zone, for they are worried that the present interglacial period may be coming to an abrupt end, and that such an end may be accompanied by a marked drop in mean temperature over a period as short as 100 years."

At the time we saw Zener, he had begun talks with several large firms to start a joint program with the government. Manufacturers of aluminum saw solar sea power both as a new source of energy and a market for their product; a leading supplier of deep-sea drilling equipment sensed that his skills could be used in taking the plants to sea and keeping them there. One of the names he mentioned was TRW at Redondo Beach, California. Later, this aerospace company received a $391,000 grant from the National Science Foundation to study the process. The *Los Angeles Times* quoted their project manager, Robert Douglass, on one of the forms that a plant might take: a big cylinder three hundred feet in diameter and stretching as far as two thousand feet down with the capacity to supply power to "a city of 100,000 to 500,000 inhabitants." Like Zener, Douglass spoke of the production of hydrogen or ammonia as one way to use the output of the plant. He also suggested that it could feed electricity directly to the coast by cable over a maximum distance of 7.5 miles wherever a big enough market could be found near a tropical or a semitropical sea. He thought that the operation could be land-based in places like Hawaii where the sea floor sloped steeply away from the coast into deep water.

A different approach would be to use the Gulf Stream as a source of warm surface water well away from the tropics or, in other words, to let the ocean haul water up from the equator after it has been heated by the sun. This idea has been taken up with a vengeance by one of the modern landmark figures in solar energy, William E. Heronemus, now a professor of civil engineering at the University of Massachusetts in Amherst, who seldom overlooks a chance to speak his mind about nuclear energy and ways to replace it with solar sea power plants and windmills.

Heronemus got into solar energy by submarine. He was born in Madison to a family who had come to Wisconsin during the brain drain from Germany in 1848. He received his education in a number of places, starting with Wisconsin and running through Annapolis and MIT. In 1941, he began a career in the Navy and stayed there until he retired in 1965, having spent nearly all that time designing, building, and maintaining submarines. "I love submarines," he once remarked; "I used to wonder why they paid me a salary." He worked for United Aircraft for two years, then in 1967 he joined the University of Massachusetts to help start a new program in ocean engineering. It was these various facets of his background, his life on and in the ocean, his experience with the nuclear reactors driving submarines that led him to the massive use of energy.

At a time when most people were trying to get the most BTUs from coal, oil, or uranium, Heronemus was already worrying how to get rid of the BTUs that were being lost generating electricity. In 1968, he began reading documents on a proposal to build eight new power plants on the Connecticut River. His own conclusion was that it would no longer be a river once it had taken the outflow from the condensers of the plants. Later, weighing the effects of a nuclear plant in the area, he asked:

> How much are the states of Vermont, New Hampshire, Massachusetts and Connecticut going to charge for the Connecticut River water that is to be evaporated away all summer long? How much should be paid for the 93° F. temperature created in the mixing pool of the river, a river whose natural temperature never approaches 93° F. at the height of the summer? No one knows yet how many additional days without full sunshine will be lost during the summer; it was once stated that nine might be expected. What is the worth of the loss of one bright sunshine day in a region whose growing season is already short? Will anyone ever answer those questions? I doubt it.

He likes to go back to the Carnot cycle that we have been talking about. Carnot's "awful expression" (to use Heronemus's expression) limits the efficiency of heat engines to the difference between the temperatures on the hot and cold sides, to put it oversimply. Heronemus charges that mechanical engineers have been looking at the hot side while ignoring the temperature of the cold source as something that nature will deal with. "I am suggesting that we are headed for a trap, if we are not already in it. I am suggesting that at least a few of us should stand back and look at our plans for continued doubling of heat engine installations each ten years for at least another thirty years, look at the total groundwater resources available to this country; then ask ourselves, are we really on the right track?"

He has come a long way from the Connecticut River, he has fined down his language and added to the megatonnage of its striking power. In 1974, we find him testifying before the Energy Subcommittee of the Senate Finance Committee:

> If all of mankind persists in their planned growth of energy demand for another 26 years, at that point in time the world's annual energy consumption will be of the order of one quintillion British Thermal Units (one Q) per year. To achieve that rate of energy release by combustion and fission processes, we will be clawing apart the earth to obtain coal, we will be at the point of maximum feasible production of petroleum and gas resources, we will have a pyramid of high-level radioactive wastes still above ground in South Carolina waiting for that solution to their safe storage, a pyramid that will surpass King Tut's wildest dream. We will have very little clean air, very little blue sky and very little clean, cool ground water left, and we may well have created irreversible weather modifications. While this great horde of humans busies itself frantically in all of that to create one solitary Q per year, the sun will be sending 3,600 Q to us during that same year, available to use or waste as we see fit. Is it not time that we humbly accept the bounty of a nature that wants to see mankind live in peace with the rest of all that which lives on earth?

On the other side of Massachusetts on Cape Cod, Von Arx has expressed the same concern in more scientific and less senatorial terms:

> The power consumed in our industrialized and urbanized societies is all finally transformed into heat. If this heat is *added* to that provided by the sun, it too must be radiated away into space. The balance of heating and cooling in nature is maintained by energetic exchanges within the solid earth, oceans, and atmosphere, and in the growth and decay cycles of life and

death among plants and animals. Modern civilization operates at a power level of 10^{13} [10,000 *billion*] watts, with a consequent heat production (mainly from fossil fuels) which disturbs the solar-terrestrial heat balance by only 0.01 percent — seemingly a tolerable level except, possibly, in instances of unusual heat concentration, such as the thermal islands produced by large cities. Were our power production to be increased to 10^{14} (0.1 percent of the heat balance) or 10^{15} watts (one percent) by adding heat to the solar-terrestrial balance, significant alterations of climate would ensue.

Von Arx is talking about a tenfold and a hundredfold increase which are only about thirty-five and sixty-five years away respectively if we keep doubling power output every decade. He regards the present level as the tolerable upper limit to the addition of heat to the earth and maintains that further power should be supplied by "diversionary methods applied to cycles of energy exchanged and maintained naturally by the solar-terrestrial heat balance."
Then he states:

World climates are fragile, especially in the polar regions. The atmosphere is probably the most sensitively balanced part of the world system, one which can show strongly nonlinear responses to energetic change. The latent heat of evaporation and precipitation is particularly important to the regimen of weather. The oceans store heat and serve as a ponderous buffer in the atmospheric heat balance, thus stabilizing its behavior. But the resources of heat in the oceans cannot be overly taxed. If their store of heat is too greatly altered, their circulations will be disturbed and in turn disturb the atmosphere and the regimen of climate. On land, it is the rocky surface, soil color and moisture, vegetation, and even the works of man which enter this complicated scheme of balances—all driven by the solar flux. It can be seen, therefore, how carefully energy diversion must be practised at power levels approaching that of the solar flux.

Von Arx is worth listening to. One of the world's leading physical oceanographers, he usually busies himself with such matters as the shape of the ocean's surface or the study of the sea from outer space. He is wont to speak up only if he is greatly concerned. When people like Von Arx descend into the arena, albeit as cautiously as in these paragraphs, it is time for the rest of us to take heed.

This matter of the global heat balance is a flaw in the millennium implied by fusion power, free from radiation and atmospheric emissions. Heronemus has his ideas about that: "I am all for fusion. I think we should all thank Divine Providence for solving that problem for us, placing us at just the right distance from a magnificent fusion plant whose life is projected for many more millennia. I suggest that we award a special series of Noblesse Laureates each year to those distinguished gentlemen who are certain that fusion can be made a reality here on earth, but who agree to write their results, then spend their remaining time in the arts and humanities instead of ever allowing it to happen. And at some distant date, when the sun begins to turn into a cooling red ball, then those who inhabit the earth can break out the literature left by those great men, whip together some fusion plants and sustain the Solar System for an extra few thousand years. Their names will literally last until the end of time."

Until that day comes, the sun will go on heating the ocean's surface, melting the polar ice to supply cold deep water and sending the Gulf Stream on its way. Heronemus has worked on a project to use it with the help of what he learned building submarines. It was thought out with his collaborators at Massachusetts, aided by United Aircraft Research Laboratories and the Andersons, all supported with a grant from the National Science Foundation. They drew up a "Mark I" plant generating 400 megawatts of electricity and anchored in the Gulf Stream off Miami. It is a big concept, a pair of cylindrical hulls in catamaran configuration, each one hundred feet in diameter and six hundred feet long. Mounted crosswise above the hulls like two gigantic radiators are the evaporators of the system destined to take in warm water from the surface, but not too close to the surface or they will take oil film along with it. The plant's turbines and condensers nestle inside the submarine-like pressure hulls. Heronemus wants to take them down one hundred and fifty or two hundred feet where they would ride in quiet water out of reach of the "Ancient Interface" of air and sea that wrecked Claude's systems. The first site envisaged is located "twenty-five kilometers due east of the Collier Building at the University of Miami" in about twelve hundred feet of water. A temperature difference as high as 30° F. can be achieved there. Warm water is drawn in through the evaporators topside, literally poured in by the Gulf Stream. Cold water is pumped from the depths through an inlet pipe eleven hundred feet long and designed to handle as much as thirty million gallons per minute. Heronemus regards this pipe, too, as an aluminum hull that can be made buoyant if need be. On its bottom end, it is attached to a concrete anchor, another hull three hundred feet long that can be towed to sea and sent diving to the bottom at its destination. Tethered to the anchor, Heronemus's ocean thermal power plant would ride like a semisubmerged ship. Four access trunks would be provided for the crew. "The access trunks offer great possibilities for valuable by-products," Heronemus says. "A sizeable resort hotel could be built atop any one of these plants or an extensive sports fishing or commercial fishing platform could be carried. If the artificial upwelling is as beneficial to the growth of higher trophic level creatures as some suspect, we may have the making of a huge fishing industry here." The possibilities of artificial upwelling are so great in tropical waters that it has been proposed time and again as worth doing even *without* a power plant. Otherwise, the remains of decayed marine life settle to the bottom and stay there, out of reach of the sun and outside the cycle of biological production.

Heronemus has reported on costs. "Our investigation began with the hope that the total installed cost per kilowatt of the plant, including 15 miles of energy umbilical, would be $400. Our current calculations suggest that $650 per kilowatt may be closer to the mark. . . . One of these power plants, brought on line in 1980 at a cost per kilowatt of less than $1,100, will win the Florida market from any foreseeable nuclear or fossil fuel competitor. Each $100 in capital cost below $1,100 means another 200 miles of transmission distance through which this system's products could win that competition. Ocean thermal difference plants installed at $650 per kilowatt off Miami should be able to win the Chicago market away from the nuclear by 1980. The liquid metal fast breeder reactor doesn't have the slightest chance of competing economically against this system. No breakthroughs or new inventions are required here, only the careful application of straightforward technology. With the proper national or private commitment, an operating demonstration plant could be on site within six to eight years." He has emphasized that nothing in the Mark I plant goes beyond the present-day capacity of a large shipyard.

The capital costs are high, but so are those of the competition. By the end of 1974, the year when he made these forecasts, nearly half of the 230-odd nuclear power plants projected for the United States at the time were being delayed because private electric utilities had trouble raising money in the stock and bond markets. More than the environmentalists, the capitalists were slowing the spread of nuclear power. In October 1974, *Forbes* wrote: "One reason solar power is looking so good is that some of the alternatives are looking rather worse. Nuclear power costs, for example, have been going up faster than food prices. According to Westinghouse, two years ago nuclear power stations that could be operational by 1979 cost $350 to $400 per installed kilowatt of generating capacity. Now, plants planned for readiness by 1981 cost from $600 to $750 per kilowatt." That is still cheaper than power from oil and coal at prices likely to prevail in the 1980s, but is a far cry from the forecast of nuclear reactors feeding industrial-agricultural complexes with electricity "too cheap to meter," as Heronemus likes to recall, giving the knife a slow twist while reopening an old wound.

Heronemus has been breaking lances and anything else he could lay his hands on against the atomic energy establishment for a long time. He has often jousted with Dr. Chauncey Starr, president of the Electric Power Research Institute and dean of engineering at the University of California at Los Angeles. In an article in the *Scientific American,* Starr stated that only one thousand megawatts of windpower could be put to work in the United States by the year 2000. To this, Heronemus has replied: "I must cry 'foul!' I do not believe there is any real evidence to support such a downgrading of windpower potential." As for the breeder reactor, he regards it as "a path to nowhere, an unacceptable approach to life." I have said that I would stay in a neutral corner, but I gave myself away long ago. So, give 'em hell, Bill. He did, at a Senate hearing:

> I have become somewhat outspoken against the peaceful uses of atomic energy, primarily because I have become convinced that its advocates constitute a formidable power group who have been able to block even a semblance of competition with proliferation of nuclear power. The demonstrated capacity of that group to stamp out competition of ideas as to future energy resources and practises is identified by me as a dangerous state of affairs in a democracy.

None of this is new. It has all been said before by all the mouthpieces of the counterculture of the 1960s. What makes the statement disturbing, highly disturbing, is that Heronemus is not one of the flower children. He spent his first career in submarines working on the type of nuclear reactors that became the forerunners of the plants used to generate electricity on shore. One can look at all the safety studies made in all good faith, yet it is impossible not to feel a shiver run down one's spine when Heronemus talks as he did in his office at the University of Massachusetts:

"I know the cost of safety in nuclear power. Are the lives of one hundred twenty-nine men at sea more precious than those of their wives and children on shore? Why not spend the same kind of money to protect the wives and children? Why not have the same quality control that Hyman George Rickover insisted on in building nuclear submarines? The power companies have made their estimates. It would cost them from five to ten times more than what they are paying now.

"I'm upset by what scientists are doing to the country. If the government were to announce that in twelve months there would be no more money for nuclear power but a billion dollars per year for windpower development, within the week, you would see a new sign at General Electric in San Jose, California. It would say: 'General Electric Windpower Division.' "

The University of Massachusetts had been our last campus stop in the United States. Amherst was a mixture of urban and rural, an Old New England motel and New New England high-rise dormitories. Madeleine appreciated the university's library where she was able to catch up on her journal before we saw Heronemus: "Students, boys and girls, all around me. They walk in, arms laden with books, sit down, open their books, heads high, eyes staring at the pages. Then, slowly, their backs hunch, their necks bow, their eyes blink and their heads slump heavily onto their arms, driving their noses into their books. Silence, broken only by discreet snoring . . .

"Suddenly, as if stung by a gadfly, the sleeper lifts his head, rubs his somewhat bloodshot eyes, picks up his books and leaves with dignity. Another takes his place, the same procedure all over again. It's funny, people sleep everywhere in this university (in the others, too), sitting on the floor, legs stretched, backs against a wall, or curled in an armchair or lying on a window sill, books acting as pillows, glasses on the ends of their noses. Someone plays a guitar in the room, perhaps as a lullaby for the others. But there are others who work. They read the fine print in their fat books, then they write in spiral notebooks, their backs bent, their noses down on their desks. Very bad for the posture. Besides, they hold their pens in a strange way, fingers broken in two, bent the wrong way. It hurt me to look at them. Behind me, a boy was lying on the rug as if he was on a beach. He was propped on one elbow, reading a newspaper and drinking a container of steaming coffee. Not a bad life . . . I thought of my boarding school at Jalesnes. Here in this atmosphere, so free and easy, yet so steeped in work, I could have gotten myself five degrees, maybe more, and I would have loved it."

A few moments later, her euphoria gave way to despair. In Marston Hall, we had asked directions to Heronemus's office. "Oh! Our celebrity! One flight up, three doors down. But he's a nice guy." Madeleine agreed: "Yes, he was a nice guy. He reminded me of a friend of ours who writes about fishing. It was easy to see him on a fishing boat, rod in hand, cloth cap crammed onto his head.

The walls of his office were hung with drawings of windmills spinning above the ocean and relief maps of the sea bottom with all the water removed. Heronemus would talk in a serious voice with a touch of humor, then suddenly break out into an enormous laugh that threw back his head and bared all his teeth. Every time, I missed a picture of that laugh because I was laughing myself. It's no use, I'll never be a photographer! I looked at Heronemus, I could imagine him roaming the oceans mounted on a seahorse to inspect his astonishing windmills with their arms crossed. I could see him leaping from crest to foamy crest and his big windmills greeting him, gravely bowing their heads to him and gently swaying at the ends of the long cables linking them to land, carrying the electricity they have manufactured while playing with the wind and the clouds without leaving any dirt."

First, Heronemus spoke of ocean thermal power. This project was being funded by the National Science Foundation while windpower was his own child with no fairy godmother to help. Heronemus looks at the world the way oceanographers do. He sees a water planet. "The zone between the two tropics is 90 percent ocean and 10 percent land but its resources of solar energy are unbelievable, at least 50 percent of all that comes to the earth. They are more than the sunshine falling on the deserts, more than we could collect by photovoltaic process." He indicated a map on his wall. "The Gulf Stream flows south to north five hundred and fifty miles from Sombrero Key to the waters off Charleston, North Carolina. It runs in a path measuring fifteen miles wide, west to east. In that path, the Gulf Stream brings us enough to meet half the demands of the United States in BTUs through the middle of the twenty-first century. It is almost embarrassing that we are not able to complicate this to the point where we can get money for it. No great new inventions are needed, all that is required is to get on with the job. If right now I had the right organization and management team — the concept that the United States Navy used to launch forty-one nuclear missile submarines in six years — I could have the first power plant operating on our site off Miami in six years. I'd go for broke, I'd start with a four-hundred-megawatt plant. I know it could happen if we had the right sympathy in the federal government."

He was even more hopeful about windpower. "With the right organization, we could start delivering electricity to New England at the end of the fourth year. The utilities laugh, maybe that's why we didn't go to American electric utilities to put a man on the moon." Heronemus was in a fine mood. "I was just given tenure last Thursday. Now I can sit here until I'm seventy and throw tomatoes at anyone I please. I'm fifty-four years old. There are days I feel like I'm twelve, there are days I feel like I'm a hundred and twelve." He apologized for the appearance of his desk. "Excuse the clutter, I never throw anything away."

He spoke of sailing ships, power masts and propellers for wind generators. He spoke of sailing; he said he had done very little since he left the navy. He had done very little of anything except work since then. It seemed to suit him.

He went back to the start of his interest in power from the wind and the sea. "I never got along with the internal combustion engine, except for the big diesels we used in submarines. When I was a kid, we had a Stirling hot-air engine. I chopped kindling wood for it. When I was in high school, I built two steam engines." It was during his student days at MIT that he first heard of ocean thermal power from Professor Joseph Keenan. "He was one of the finest teachers I've ever heard anywhere. Never a word too many, never a word too few." Another professor, Dr. Sverre Petterssen, one-time director of the meteorology department at MIT, had told him of windpower. Heronemus later added his own naval experience and realized that the wind could be tapped at its maximum at sea in areas like the Roaring Forties where the gales that once drove the tea clippers could be turned into kilowatt-hours.

There is nothing more likely to bring on a smile or a smirk in any discussion of energy than a mention of windpower. I don't know why this is; perhaps it is the eternal trademark of the windmill as the adversary of Don Quixote, perhaps it is all the bad and facile jokes that politicians can make about harnessing the wind on Capitol Hill or any other site of parliamentary democracy. Or perhaps, I am beginning to think, the wind is too plausible. It works, all the R and D was done hundreds of years ago. It even seems to complement direct sunshine; the very clouds and storms that drive the sun from the sky bring in the wind as replacement power. On the farms of Brittany, the horizon is still punctuated by an occasional tower with a small steel windmill pumping up water for cattle. On the shores of the Bay of Mont-Saint-Michel, the stone towers of old windmills stand on the flat land, their arms gone, their hulks turned into weekend homes. Adversaries of wind-power like to talk of the aesthetic pollution of a row of windmills; I don't think there is much chance of making vacation condominiums from old nuclear power stations. I was once in a windmill myself. Madeleine and I were crossing the plains of Beauce on a back road when we saw arms beating away in the distance. We drove closer, we drove up to the mill and we climbed aboard. I remember the miller, a quiet old man who told us how he had ground grain for animal feed until only a few years before. Now he ran his mill for visitors like us and took his payment in the sale of postcards. But the smell was still there, of wooden cogs and shafts and clean flour. I remember how the windmill pitched slightly as the big sails turned. There was no noise except for the creak of the wooden machinery. It was like crossing the plains of Beauce in a galleon.

But enough of poetry, let's get back to power. This kind of writing can set the cause of wind energy back another few decades when it comes to raising money in the bond market. The invaluable Von Arx makes much more sense when he writes: "The kinetic energy of the world's wind systems amounts to a staggering 10^{20} watts. This represents a stored resource, however, which can be drained at no more than its replenishment rate if it is to perform as a renewable resource. Thus, we are concerned with the 'available' kinetic energy of winds, replenished at a power level not exceeding the solar flux at the surface, 10^{16} watts, because the atmospheric circulation is driven mainly by heating from below." When one removes the jet streams blowing in the upper atmosphere, the remainder amounts to 10^{13} watts that can be collected on or near the surface, ten times more than the earth's hydroelectric potential.

In the same issue of *Oceanus,* Heronemus spoke about windpower at greater length:

Every man of the sea knows of the nor'easter, the roaring forties, the doldrums, the horse latitudes. There are at least 57 winds recognized by name. Each wind has its own characteristics: there are the rather moderate but very steady trades, and there are the westerlies, sometimes blustering, sometimes flat calm. They appear random and capricious, yet when one studies the wind from the viewpoint of energy content — taking measurements at fixed locations and heights for periods of a year — one finds the wind to be a remarkably reproducible phenomenon.

When winds at specific sites are examined for energy content, it is soon apparent that the most energetic of them are over the oceans. Winds leaving a land mass, moving out over open water, show a marked ability to intensify in velocity after passing over a relatively short fetch of open water. The large-scale atmospheric processes which produce such features as the generally low-pressure areas over the northern Atlantic and the similar 'pressure-sink' over the northern Pacific occur in regions rather unfriendly to man. But the winds they create manifest themselves on or close to many shores accessible as work sites. . . .

In 1970, the winds over New England and their seaward extension, the offshore westerlies, became the testing ground for a major feasibility study of windpower as a large-scale source of energy. The idea that windpower could — indeed *should* — provide a major portion of the electricity demanded by a modern industrial society was already well-documented. Many competent men and serious-minded organizations in a number of different countries had worked on the technology of windpower conversion for several decades — had shown clearly that the resource was huge and that utilization was technically quite practical. But they had been unable to demonstrate that wind-generated electricity could compete economically with electricity generated by heat engines burning $2.50-per-ton coal or ten-cent-per-gallon diesel fuel or three-cent-per-gallon residual — costs that pertained some fifteen years ago when these fuels were thought to be nearly inexhaustible. The promise of even less expensive electricity from large numbers of nuclear plants caused the abandonment of *all* windpower programs.

We at the University of Massachusetts reopened the issue primarily because we thought that this country was headed toward a more realistic concept of capitalistic economics in which the externalities associated with energy-industry pollution would be costed so that the energy consumer would pay more of the *total* cost of that energy. Given that eventuality, we argued, pollution-free, wind-generated electricity might be shown to be economically competitive. . . .

We knew of the relative freshness of the offshore winds; excellent long-term velocity data taken at three Texas towers corroborated the intensification-over-water hypothesis. So it was decided that a concept for a large windpower electricity generating system, offshore in the westerlies, capable of selling electricity on demand, would be set down for analysis. The resulting Offshore Windpower System (OWPS) was shown to be large enough to provide most of the New England electricity market projected for 1990. Later studies showed that the system could easily be twice as large, selling up to 360 billion kilowatt-hours of electricity per year, transmitted and delivered throughout the six-state New England region.

The study found that the wind could not produce electricity to compete with the 28.3 mills per kilowatt-hour average sale price prevailing in 1972. Not only that, but all the forecasters predicted that the price would keep going down after 1968 as it had done before. Instead, Heronemus remarks with understandable relish, it was up to 36 mills by 1974. His figures also showed that a smaller offshore wind plant, sending electricity by cable as well as producing hydrogen, could bring in power from 26 to 28 mills per kilowatt hour.

He proposed to build the plants as concrete hulls that could be towed horizontally into position, then flooded to float upright. One system would carry three great turbines, each driven by a two-bladed propeller two hundred feet in diameter. A second consists of an array of thirty-four propellers, each sixty feet in diameter. There are many places where the stations could be mounted. He suggested several.

Where, other than on the front porch of New England, could the winds over the oceans provide large amounts of electricity or synthetic fuel (hydrogen, methanol) or fertilizer (ammonia)? There are many oceanic sites sufficiently near shore, and in shallow enough water, to be connected to the consuming market by pipeline or cable umbilicals. A partial list: the shelf off the Middle Atlantic states; the shelf off Nova Scotia; the eastward edges of the North Sea; the near-shore portion of the Baltic off Sweden and Finland; the shelf adjacent to the entire Aleutian Archipelago; the narrow shelf west of and the broader shelf east of South Africa; the shelf south of Australia; and the shelf adjacent to the Kurils. Any of the islands in the stronger trade regimes and those, like the Falklands and the Prince Edwards, in the Roaring Forties are excellent sites.

Extraction of energy from the winds could proceed afloat on the high seas in the most favorable wind regimes, the wind generator platform doubling as a self-propelled tank ship equipped to generate and liquefy electrically-produced hydrogen. A new concept along these lines, which includes wind propulsion, is under investigation by the author. Further, great amounts of energy could be extracted from a region like the Great Bahama Bank from a forest of wind generators, installed in a way that would accomplish negligible harm to the bank or the biota living thereon. What a magnificent export product for that small nation!

Benefits are not necessarily reserved for maritime powers. Heronemus made a preliminary study of windpower for the Lake Ontario region where he would use propellers on towers at least six hundred feet tall to take advantage of wind speeds that go up with altitude. He estimates that as much as 51 billion kilowatt-hours per year could be produced there by the wind, replacing between seven to eleven 1,000-megawatt power plants. Arrays of generators would be anchored in Lake Ontario and mounted on land. "There is no sense pretending that these arrays would not dominate the landscape. If they were accepted as the best of a number of alternatives, they might be viewed with pleasure."

Heronemus is not at all dogmatic about his giant wind towers awaiting a king-sized Don Quixote worth their mettle. In upstate New York around Lake Ontario, he sees the individual home wind furnace. "The most simple 20-kilowatt wind generator on a 100-foot pole or needle-tower in the Oswego suburbs could produce 30,886 kilowatt-hours of direct current during the nine heating months, September through May. Those 30,886 kWh fed into a d.c. heat pump and an associated small hot water storage tank could take over the task of heating even the most poorly insulated house of average to large size in the region. The economics of this have not yet been determined: heat pumps were relatively costly in the past. But we are much more affluent each year, much more prone to invest in thousand-dollar gadgets and toys which we seldom use: we may in another three years be quite willing to invest in wind-driven heat pump heating for all new construction, residential, commercial and public."

Much further inland, our Hero (nemus) has studied windpower for Wisconsin, particularly the eastern 60 percent of the state combined with Upper Michigan. If thermal plants keep heating the groundwater, he thinks that "one can visualize the new butter cartons out of Rhinelander enclosing 'Land-of-Cooling-Ponds' butter." Instead, he estimates that as much as 238 billion kilowatt-hours per year could be taken from the winds over Wisconsin and part of Lake Superior and Michigan. A quarter of this would meet the area's projected electricity growth until 1990, although Heronemus wonders whether it should grow at such a rate. "Exactly where all this additional electricity is to be consumed in an area of which the great majority is rural and near-wilderness is not clearly understood by the man-in-the-street or, perhaps better, the man-in-the-woods. The inhabitants have a revolutionary change in their way of life projected for them by the year 1990: it is perhaps a question worth asking as to how many of them have any understanding of this at all, let alone a *desire* for all this to happen to them."

With some reluctance, one turns from Heronemus's visions to see what actually has been done. It is even harder to keep track of windpower than solar energy because it is so much easier to purchase, install, and use. Bits and pieces crop up everywhere as one keeps an eye open and scissors at the ready: a California architect buys a windmill for four thousand dollars from a Swiss ski resort and sets it up to electrify two homes; still in California, another gentleman converts his ham radio mast into a 200-watt windpower generator that he uses to charge the batteries of an electric motor-cycle, getting twelve to twenty-five miles a day from a night of wind-charging; a Romanian wants to desalt water, using a windmill coupled to a heat pump; a Swedish wind plant supplies power to an automatic telephone exchange far from any electricity net; in the Canadian Arctic, installations as big as 100 kilowatts have been used.

One hundred kilowatts is the size of the first wind generator to be built with the backing of the United States government. NSF and NASA joined forces to put up a two-bladed rotor one hundred and twenty-five feet in diameter on top of a tower one hundred and twenty-five feet high on a site belonging to the NASA Lewis Research Center in Cleveland. NSF contributed $865,000 to the venture, its first major "wind energy conversion system," while NASA met the overhead costs. The generator is to turn up its 100 kilowatts when the wind reaches 18 mph. If the wind blows harder, the pitch of the blades is increased so that they spill wind instead of racing and, at 60 mph, the blades are fully feathered, shutting the generator down for safety's sake. In a press release, NSF remarked that this is the "first large wind energy system constructed in this country in the past 30 years."

It was referring to the most ambitious system of all time up to this time, the 1.25-megawatt wind turbine that was completed in Vermont in 1941. The story of that adventure was told in *Power from the Wind,* a book published in 1948 and written by Palmer C. Putnam, the manager of the project. Putnam started his story:

"In the fall of 1941, something new had been added to the generating system of the Central Vermont Public Service Corporation. Motorists in central Vermont saw, from 25 miles away, a giant windmill, its polished sunlit blades flashing on top of Grandpa's Knob,* 12 miles west of Rutland and overlooking the Champlain Valley." In the footnote, he explained: "This peak had not been distinguished on maps by a separate name. It was bought from a Vermont farmer whose family always referred to it as 'Grandpa's.' Because of this, and its shape, we christened it Grandpa's Knob."

Then Putnam described what the motorists saw. It was the experimental Smith-Putnam Wind-Turbine. He contributed half its name, the other half came from the firm that built it, the S. Morgan Smith Company of York, Pennsylvania, now part of Allis-Chalmers. The turbine was turning out 1,250 kilowatts, "enough to light a town, and was feeding power into the utility company's system, permitting water to be stored behind the dams when the wind blew more than 17 miles per hour." The work was done by a two-bladed propeller one hundred and seventy-five feet in diameter, its hub one hundred and twenty feet off the top of Grandpa's Knob.

In his book, Putnam described the birth of the idea: "In 1934, I had built a house on Cape Cod and had found both the winds and the electric rates surprisingly high. It occurred to me that a windmill to generate alternating current might reduce the power bills, provided the power company would maintain stand-by service when the wind failed and would also permit me to feed back into its system as dump power the excess energy generated by the windmill." From then on, the chain of events became more intricate. Putnam decided he needed not a small generator but a big turbine that could be hooked up with the network of a utility. He got meteorologists and engineers from MIT involved; General Electric expressed interest in the wind because there was little waterpower left to develop in New England; then the S. Morgan Smith Company, a builder of turbines, got together with the electric utility.

Those were the days before computers and systems analysis. In June 1940, Grandpa's Knob was selected as the site. A two-mile road was built to the top that summer and the work went ahead in winter. Putnam related how it was done. "Low temperature and high winds made rough work of handling heavy steel, but we had only one accident. On a particularly bitter, windy sub-zero day, the forty-ton pintle girder with the twenty-four-inch main shaft and its two main bearings in place was being trucked to the summit on a heavy trailer drawn by the truck unit and two Caterpillar tractors in tandem. At the hairpin turn below the summit the girder rolled off the trailer-bed and turned upside-down into a deep snowbank, the forty-eight-inch main-bearing housings having found the only opening between rocks! After recovery, inspection showed no damage to the bearings or shaft, and only minor damage to some of the plates of the girder. Erection was completed in August, 1941, when the blades were put in place."

Putnam was in Washington, working in the Office of Scientific Research and Development, when the turbine was officially inaugurated, with William Bagley of General Electric in charge of proceedings. "Finally, on the night of Sunday, October 19, 1941, in the presence of the top management of the Central Vermont Public Service Corporation and many of the staff of the S. Morgan Smith Company, and with the Smiths listening in by long-distance telephone, Bagley, having completed his adjustments and made his final inspections aloft, phased on the unit to the lines of the utility company, in a gusty 25-mile wind from the northeast."

May I interrupt Putnam here to remind that this was happening in 1941, not 2001. One begins to understand why Heronemus goes as far as he does when he lets himself go. But go on, Putnam. "There was no difficulty. Operation was smooth. Regulation was good. After 20 minutes at 'speed-no-load,' the blade pitch was adjusted until output reached 700 kilowatts. For the first time anywhere, power from the wind was being fed synchronously to the high line of a utility system."

This was a test unit, built to see what was likely to go wrong in a windpower plant. Bearings heated, rivets loosened, troubleshooting had to start. In May 1942, after about 360 hours of operation, an inspection showed cracks in the skin of the propeller blades near their roots. They were welded, the tests continued; then in February

1943, one of the two main bearings had to be replaced. This was wartime; two years were needed to get a new bearing. On March 3, 1945, the turbine started running again. Unknown to its operators, the welding of the blade had led to further cracks which occurred right behind a bulkhead where they could not be seen. The turbine ran for three weeks during an almost windless March, generating 61,780 kilowatt-hours in one hundred and forty-three hours. By March 26, Putnam wrote, more than 90 percent of the cross section of the blades was weakened by hidden cracks. "The tension of centrifugal force, amounting to several hundred thousand pounds, was being withstood by only a few square inches of Cor-Ten steel."

When the first shift came on duty at midnight, the wind was blowing at only five miles per hour. Two and a half hours later, the wind freshened and the turbine was finally started in a steady southwest wind of about twenty-five miles per hour. Then, at 3:10 A.M. "Harold Perry, who had been the erection foreman and was a powerful man, was on duty aloft. Suddenly he found himself on his face on the floor, jammed against one wall of the control room. He got to his knees and was straightening up to start for the control panel, when he was again thrown to the floor. He collected himself, got off the floor, hurled his solid 225 pounds over the rotating 24-inch main shaft, reached the controls, and brought the unit to a full stop by rapidly feathering what was found to be the remaining blade of the turbine. He estimates that it took him about 5 seconds to get to the controls after the first shock. One of the 8-ton blades had let go when in about the 7 o'clock position and had been tossed 750 feet, where it landed on its tip."

From that point on, the project began to run down. Beauchamp Smith, from the company that designed the unit, estimated in 1945 that it would cost $190 per kilowatt to install a block of twenty generators about the size of the one on Grandpa's Knob. At that time, the Central Vermont Public Service Corporation could only afford to pay $125 a kilowatt for windpower. This was closer than the estimates we are now getting for many oil and coal substitutes, but too far away for private enterprise. In 1946, Smith wrote that, even though there was a chance to get the installed price of 2,000-kilowatt or 3,000-kilowatt turbines down to $100 a kilowatt (for example, mass production could halve the cost of blades), his firm had to give up. "We had already spent a million and a quarter dollars, and it would cost several hundred thousand dollars more to find out whether we could actually sell in this market at a profit. Our stockholders were unwilling to make the additional investment and, accordingly, we reluctantly abandoned the project, placing the patents in the public domain."

Nearly thirty years later, Smith was ready to change his mind. "With shortages of power developing all over the world, with the growing realization that the world's fuel reserves are not inexhaustible, and with the knowledge that our present known methods of using our dwindling fuel reserves are damaging our environment, I believe the time has come for another close and hard look at windpower as at least a partial solution to some of these problems." As the retired president of his firm, he was talking to the opening session of a "workshop on wind energy conversion systems" run jointly by NSF and NASA in June 1973 in Washington. Like similar meetings, it yielded a volume of proceedings, a 258-page affair with cover portraits of four big wind generators out of the past. Unlike other publications written mainly in the future or conditional, this one reads as if it had come out of a time capsule buried by a vanished civilization. Flipping through it, one sees illustrations of big windpower plants built in Europe in an earlier day. There is a 100-kilowatt Russian generator that ran at Yalta in 1931 and looked something like a giant aircraft engine planted on a tower. Even heavier and more massive in appearance was the 800-kilowatt windmill that the French operated from 1958 to 1960 at Nogent-le-Roi with a three-bladed propeller one hundred feet in diameter. Or, on slender guyed masts, there are the machines that Ulrich Hutter built in Germany, the biggest turning up 100 kilowatts in 1966. One blinks at a paper by Marcellus L. Jacobs of the Jacobs Wind Electric Company in Fort Myers, Florida. He built thousands of home sets between 1931 and 1957 in two sizes, 2,500 watts and 3,000 watts. The price for a 2,500-watt plant, including tower and batteries, ran to $400 per kilowatt. Repairs averaged less than $5 per year, replacement of batteries about $36 per year. He installed plants in Alaska, Canada, Finland, the United States, and Antarctica. Some have been running since 1937, but Jacobs has not made a wind generator for twenty years. He had sold his plant and machinery but he estimated that he could make the same generator today for about $800 a kilowatt.

The workshop talked about storage. Like the sun, the wind is intermittent and it may not always be around precisely when it is wanted. The use of electricity to manufacture hydrogen came up in this context, so did superflywheels and storage by compressing air underground. And batteries, too, just as in the days of the Jacobs plants that ran with three-bladed wooden propellers. What should be noted, I think, is that storage is no longer a solution sought only by those who use the wind and the sun. The electric utilities are also interested in a way to get on top of their peak costs. *Business Week* quotes William M. Irving, director of research for Boston Edison and chairman of the Electric Power Research Institute's Advanced Systems Task Force, as saying that, by the 1980s, utilities may be able to use big storage batteries to meet peak loads. A battery cannot distinguish between power coming from a nuclear reactor or a wind turbine.

The idea of wind-driven heating was well received by several others at the meeting. Everett Lutzy from the municipal light plant in Hull, Massachusetts, needed a sound economic proposition because he had to get the local citizens' votes to approve any bond issues. He said: "The highest peak demand in our electric system is at a time when the wind blows the hardest and that's generally when the temperature is the lowest. The chill factor is such that tremendous heat loss occurs at the time the wind is blowing the strongest. So the question is: is there any way to economically convert this windpower to a meaningful use? For us in Hull, its major use would be for heating in the winter. If the total use was 20,000 kilowatt-hours a year, practically 60 to 70 percent or more would be used for winter heating." A representative of the Rural Electrification Administration in Washington, Brian R. Jessop, agreed that heat looked like the best application because it was easiest to store: "Rocks are cheap, water is cheap. Put together a pool in the basement with rocks and water in it, and you've got a real good thermal tank."

Many at the workshop were like Lutzy, just one step removed from the consumer. James Wharton from the Public Utilities Department in Tillamook, Oregon, said his town was interested because it had plenty of wind. "I would like to see a workable unit constructed and in use so that the utilities could look at the costs and aesthetic ecology. I would like you to consider a low profile . . . you are not going to get the ecology movement to hold still for a gang of windmills hanging from a balloon or up on a wire. . . . The wind machine will definitely be in a fish bowl, so it should be engineered with an ecology and an economic basis in mind. The energy squeeze is on us, and utilities will want to make commitment decisions very soon. Therefore, I would urge you to use some haste on a prototype." That was what the meeting's "committee on applications" recommended: first, demonstration of windpower for homes, then three years to work on plants of 25 to 100 kilowatts and, finally, a big five-year effort to get wind into the megawatt bracket, self-contained systems with storage and running from 500 to 25,000 kilowatts.

Strange reading, all this, because it happened in 1973. None of it seems to have been widely reported and that is why I have chosen to give it so much space here. At the time of this writing, the windmills have not started to go up except through individual efforts at scattered locations. Why not? I do not know, but I am tempted to listen to a man from Grumman Aerospace, J. Mockovciak, Jr., who told that meeting: "I frankly think there's an overemphasis on the amount of research that has to be done, and too little emphasis on finding ways to make it happen. I personally feel that we have an adequate technical base and that we should start building these machines and looking for people who want to use this energy."

Palmer Putnam's book tells us that power from the wind looked almost ready to a big public utility and a large engineering company thirty years ago. Since then, a number of improvements have been made in materials and in technologies; we still have to talk of the performance of wind machines on the basis of the Putnam-Smith turbine of 1941 but no one works out airline economics with the characteristics of a DC-3 in mind. Not much money has been invested, there is no military funding to be scooped up here, there hasn't been a wind missile since Old Ironsides. Nor it is just money; this is the type of problem that does not lend itself to vast applications of money and manpower under centralized control.

It is too easy to blame the scientists and the engineers. We get the technologies we deserve. As a layman, I am not qualified to judge if the various ways of producing energy that I have described are feasible. There are some good people, people who have made their reputations in science, who think so. With the expenditure of relatively minuscule resources and effort, we should be able to find out quickly if they are right. I do not know myself, but I would like to see the matter settled. It is much too important, far too much is at stake, to continue to leave it to the mercies of a market distorted by factors so remote from the welfare of society and the individual.

XIII.

~~~~~~~~~~~~~~~~~~~~~

Power Later

~~~~~~~~~~~~~~~~~~~~~~~~

SINCE WE WILL NOW BE TALKING about solar cells, a qualifier is again needed: "power a little later." Here, the writer is on the shakiest ground. While there are any number of courts he can go to for a verdict, there is no supreme court. We have already heard advocates of thermal power shooting down solar cells because of cost. It is true that the solar cell is the most expensive way to make electricity that anyone has ever marketed. It is also the branch of solar science in which the largest sums are being spent to bring costs down. We have watched solar cells operate on the roof of Karl Böer's house in Newark, Delaware. They were cadmium sulfide cells, those stepchildren of the space program that had forsaken them for silicon, more efficient and durable although more expensive. No matter what it is made of, the solar cell is almost irresistible: no heat, no steam, no smoke, no working parts, just add sunshine above and see the electricity come out below. It is so convenient that, despite its price, it finds buyers not only for use in space and on the loneliest of ocean buoys. Industrialists have been talking of a pocket calculator running on solar cells; a solar-powered watch has gone on sale, capable of soaking up enough electricity for four months after only four minutes of exposure to an autumn sun. One learns of a solar battery charger for hearing aids to be sold to rich deaf campers.

This sounds like a few light-years away from the kind of power that could be of any possible use to a home, a factory, or a community. But it would be a mistake to dismiss the solar cell that easily, especially after one has heard some of its advocates. There are many throughout the world, we could only see a few. We began in Washington where we made our way to NASA's Goddard Space Flight Center in Greenbelt, Maryland, to meet William Cherry. He had served as executive secretary of the panel that prepared the NSF/NASA report on "Solar Energy as a National Energy Resource." Of our appointment, Madeleine remembered that we had to wear "Visitor" badges when we called on Cherry in his office, "a magnificent office, beautiful books, mahogany furniture." She added: "Mr. Cherry had wonderful expressions, but Dan was in the frame every time I tried to shoot. I was furious."

Cherry was more at ease than Madeleine as he related the history of silicon solar cells. It went back about twenty years when Bell Laboratories tried them as a way to charge batteries for the Army Signal Corps. In 1954, RCA's laboratories studied a broad range of substances, trying to raise efficiency and reduce the price. "The cost was too high for the Signal Corps," Cherry said. "Luckily, the space program came along. Batteries were used in the very first satellite launched in 1958, but they lasted only a few weeks. On March 17, 1958, we sent up Vanguard I, the first solar-powered satellite. We designed it at Fort Monmouth, New Jersey, and we thought it would last a year. It kept transmitting six and a half years. We had to make cutoffs for subsequent satellites so that we wouldn't tie up the whole radio spectrum. The power was a tenth of a watt. That was enough for a big receiver to pick up the data being transmitted, it wasn't enough to revolutionize power on earth." In the early days of space flight, solar cells were arrayed against nuclear power as a method of supplying electricity. In space, solar cells won for most uses. Cherry said they performed as far away from the sun as Mars and as close as Venus. To go to

Jupiter, however, nuclear power is needed on a spacecraft because at that distance from the sun, solar energy amounts to only 3 percent of what it is on earth. Skylab represented the biggest solar-cell power system ever built for use anywhere. Fully deployed, it was to have generated twenty-one kilowatts but about five kilowatts were lost on the first Skylab flight when the array of solar cells was damaged during the launching.

By the end of the 1960s, Cherry had become interested in the possibility of using solar cells as a major source of power on the ground. The cost of electricity in space from the cells was about two hundred dollars a watt — two hundred thousand dollars a kilowatt. That did not bother Cherry. "On the ground, you can get to a solar cell to fix it. A simpler design will do, you can take away the cosmetic requirements." He thought the price could come down to fifty dollars per watt and, in fact, it had dropped to twenty dollars per peak watt by 1974. The concept of the "peak watt" is important: it is a way of stating how much power the solar cell produces at high noon. To get an idea of what this represents over twenty-four hours (the basis that the power industry uses), the figure must be divided by five or six. So the average cost of a solar cell producing peak watts for a capital cost of twenty dollars a watt is actually one hundred dollars a watt or one hundred thousand dollars a kilowatt.

At first, Cherry thought of a solar farm sown with silicon cells. He calculated that a square mile of solar cells could turn out enough power on the East Coast of the United States to meet the summer need of eighteen thousand homes. In winter, this ground array could still handle ten thousand homes and, using batteries for storage, it could run through four cloudy days. One square mile of cells could power seven and a half square miles of consumers in the Washington area but one could do much better in the Southwest. "Enormous land areas in the arid parts of the U.S. are low productive regions. Many thousands of square miles could be made highly productive, harvesting a crop of electric power for sale in the areas where it is vitally needed. Not only would the land become more productive and valuable, but the U.S. would become less dependent upon foreign import of energy resources." This sounds like everybody's editorial page nowadays and it is to Cherry's credit that he said it in March 1971.

He had a fine idea for coping with that power loss on cloudy days. He came out with it in 1970 and he called it the "solar rug." If only one could get those solar cells above the weather, up between fifty thousand and seventy thousand feet, there would be no more brownouts on cloudy days. "By moving farther and farther north until one reaches the North Pole, the sunlight period increases, reaching 100 percent sunshine hours from March 21 through September 21," he wrote. "This, however, requires the transmission of power over many hundreds of miles to reach the populated regions of North America. If the power station is located over a city, it would be analogous to a power dam some 10 to 15 miles from the urban area. While the problem of 24-hour service has not been solved, there would be abundant power available to cover the peak periods, e.g., the hours between noon and 7 P.M. during the summer, when the threat of blackout is greatest."

To hoist the station, he proposed a helium-inflated mattress floating well above the forty-thousand-foot ceiling of jet airliners. The mattress would consist of helium-filled cubes about one hundred feet on a side. If it were a mile square and one hundred feet thick, it could keep ten thousand tons aloft. Cherry thought of a few other aspects.

> The solar rug could be a manned power station as if it were on the ground. Compressors to supply air at near one atmosphere would be provided. The "stationauts" would wear protective clothing capable of providing necessary shielding against ultraviolet radiation and for supporting breathing. The "stationaut" would necessarily have to wear a parachute as a safety measure in case of a fall. He would be safer than a man working on a tall building because of the large amount of time he would have to activate his safety equipment in case of a fall.
>
> The solar power station would be manned to insure proper operation and allow for maintenance and repair. It would be

reached by a balloon gondola with a propulsion system. At 50,000 feet, this should require 40 minutes from the ground.

Getting power down was not that easy. At first, Cherry thought of sending it through a cable but that worried him. "While designs exist for tethers greater than a hundred thousand feet in length, it's quite another matter to design one which can conduct large amounts of electrical power and survive surface storms as well." He took up the idea of beaming the power down by microwave transmission, eliminating any need for wires. That was his solar rug, otherwise known as "the Cherry pie in the sky."

Since then, he has become more hopeful about using solar cells on the ground. At least four companies in the United States are manufacturing them in arrays from a few watts to a hundred watts. Cherry thinks the best way to introduce them in larger amounts would be to start with a commercial building or a community unit with ten to twenty houses so that the load can be spread. He estimates that before solar power can make its contribution, from three to four billion dollars will have to be spent over fifteen years. "We must explore other things besides the atom. We are going to need them about the turn of the century. Right now, solar energy is enjoying a honeymoon. It is not getting too much money, no one is hurt. But when you start to take away dollars, then everyone comes out of the woodwork."

Cherry thought prospects of cutting costs drastically were good. "Look at the manufacture of photographic film. That's a complicated operation; you have to run it in the dark. It just involves sending a substance through a continuous process at one hundred and twenty meters a minute . . . and the final cost is one hundred and twenty dollars a square foot. Why not apply the same technology to make solar arrays in a continuous high-speed operation?"

Chances for bringing solar cells down to fifty cents a peak watt are promising, if one is to judge from the tone of a broad survey carried out by Alfred R. Rosenblatt, the industrial editor of *Electronics* magazine. He quotes H. Richard Blieden, manager of the solar cell program at NSF, as estimating that by 1990 industry in the United States alone will produce solar cells amounting to five thousand peak megawatts, a figure that will double in 1995 and double again by 2000 when it will reach twenty thousand megawatts, between 1 and 2 percent of the country's needs. On the way to this goal, Blieden had a series of stages in mind. By 1977, he hoped to see the solar cell come down to five dollars per peak watt, then to the commercial threshold of fifty cents by 1979. According to *Electronics,* NSF planned to start installing power stations that year, quickly increasing their size to ten megawatts by 1985, then one hundred megawatts by 1990.

The magazine mentioned an intriguing job undertaken by the Mitre Corporation at its Systems Development division in McLean, Virginia. The company is putting one hundred and thirty thousand dollars into a one-kilowatt array of solar cells, the largest ever built for use on this side of outer space. It will use electrolysis to break water down into hydrogen and oxygen which can then supply a fuel cell to generate power at all times. The four major solar cell manufacturers in the United States were involved in supplying components: Solarex Corporation in Rockville, Maryland; Solar Power Corporation, an Exxon offshoot in Braintree, Massachusetts; Centralab Semiconductor Products at El Monte, California; and Spectrolab, a division of Textron, Inc., at Sylmar, California. Richard Greeley at Mitre is quoted as stating that within a couple of years, the firm will be ready to go up to one megawatt.

Joseph Lindmayer, the president of Solarex, was proposing a solar breeder. *Electronics* reported that he had asked that at least one solar cell manufacturer receive a government subsidy to put arrays on its own roof. Power from these cells would be used to make new cells: a solar equivalent of the breeder reactor. Lindmayer wanted to start such a system and bring it up to half a megawatt within five years at a cost of $10 to $15 million. He was saying in 1974 that he could deliver solar cells for thirty dollars a watt, the price he quoted for a forty-watt array to run a transmitter on a mountaintop for the United States Forest Service.

Although the American space program lends a healthy fillip to the solar cell business by taking $5 million worth a year, things are moving ahead almost as fast outside the United States. *Electronics* said that Japan put $8.7 million into its Sunshine project in 1974, of which $3.1 million was for solar energy studies (the project is so called because it hopes to bring more sunshine to Japan by getting rid of smog). The Japanese schedule a one-megawatt power plant running on solar cells for 1980, then a ten-megawatt system in 1986 and a hundred megawatts in 1991. This is all the more bold when one realizes that, through the mid-1970s, Japan had probably produced no more than twenty kilowatts of solar cell power, less than on the single Skylab array. The Japanese plan to spend about $7 billion by the end of the century on their Sunshine project.

The French have done extensive work on solar cells, particularly with cadmium sulfide, and they seem to be the first to use them to run an educational television set. In 1972, they installed a system in a school in Niger and found it cheaper than buying batteries, the usual way of supplying power for educational TV in villages without electricity. Another line has been taken in Australia where Dr. Laurie Lyons and Dr. Jack Garnett are studying ways to use the natural photovoltaic effect that occurs in plants when photosynthesis takes place. The Australian Academy of Science is optimistic: "Recent results have demonstrated 8 percent quantum efficiencies and there is reason to expect that further research may produce energy conversion efficiencies which will be acceptable commercially. Both theoretical and experimental results show promise that organic materials could provide the required efficiency, weight and cost reductions for use in large solar photovoltaic conversion plants." Various organic chemicals are being tried as a light-absorbing material in the form of a layer sandwiched between plates of electrodes.

Solar-cell power in much larger quantities has been proposed for a number of years by Dr. Peter Glaser, vice-president in charge of engineering sciences at Arthur D. Little, Inc., the reservoir-sized think tank near Cambridge, Massachusetts. Glaser is the father of the SSPS, the Satellite Solar Power Station, and everything about it

is big: a satellite measuring two and a half by six miles in synchronous orbit (that is, always remaining over the same spot on earth)
22,300 miles out, generating five thousand megawatts from the
sun shining nearly twenty-four hours a day and beaming them to
earth by microwave transmission, just as Cherry wanted to do from
his helium-borne rug a mere fifty thousand feet up.

The satellite station is probably the most controversial proposal
in the whole solar panoply. It raises the hackles of the ecology-
minded when they think of getting rid of Con Ed on earth only to
see it reappear in space, an all-powerful sun on its own, pouring
energy into a central receiving station owned by Con Ed. The costs
of putting the station into orbit are well in the order of magnitude
of an Apollo program. Much fuss is made over the fact that it will
capture sunlight that ordinarily would not have reached us, thereby
adding to the thermal burden that the earth must bear. I had met
Glaser at the solar congress in Paris and I was eager to see how
he had fared in the intervening time.

Arthur D. Little, Inc. is big enough to rate a private exit off
U.S. 2 outside Cambridge. It is also big enough to stumble over
itself, as we learned while we waited for Glaser in the lobby. Madeleine noted: "The receptionist said Peter Glaser would see us in
five minutes. Tons of people were asking for him, nobody knew
where he was. After fifteen minutes, Dan turned yellow as he always does. He squirmed in his chair, he was furious. I said we
ought to wait, he kept squirming. At two o'clock, a secretary came
out and brought us into Glaser's office. I expected to see a big
man, ill-tempered and haughty. We were welcomed by a small
monsieur, bald but young, an immense smile displaying all his
teeth. He greeted us warmly, he asked Dan if he hadn't caught
cold wearing a big black turtleneck sweater while everyone else
was in shirtsleeves. Dan grunted, then slowly unwound under
Glaser's kindness. The interview started. Glaser had the sharp eyes
of a fox. He leaped from his chair to his blackboard, he ran to get
a book, he cocked an ear to listen to a question, then he answered
enthusiastically. On the walls of his office hung drawings of his
early space power station projects."

Glaser had joined Arthur D. Little in 1955. He had just received his Ph.D. in mechanical engineering from Columbia, topping an education that had started at Charles University in his native Prague and continued at Leeds College of Technology in Britain. The company put him to work on solar energy as a source of heat to test the behavior of materials on reentry from space. "Nowadays, I call it a poor man's laser. I learned what it takes to work with mirrors, what happens when the sun shines and when it does not shine. It's a hard way to do things." This early interest led him to the International Solar Energy Society, of which he is a past president and the present editor of its journal. Arthur D. Little sold thirty of those small solar furnaces, but Glaser was not sold on them as a way to generate electric power. He had come to that conclusion when the United Nations held its conference in 1961 on the uses of solar energy in the developing world. There he reached another conclusion. "I decided that solar energy would not be of interest to developing countries until money could be made from it. I couldn't see the approach that the developed countries were do-gooders. Solar cookers for India? A bunch of bull. You transfer technology at the top, not at the bottom. The advanced countries had to develop it for their own use, then applications could be found for it everywhere." Glaser's discussions at the UN conference only confirmed what he had thought about mirrors all along. "The likelihood of doing anything with them is very small. The photo-voltaic cell is the only device that is not limited by the Carnot cycle in its efficiency. It has no moving parts, it's the real McCoy."

Four years after the UN conference, Glaser produced the idea of a satellite solar power station. "It was in 1965, solar energy was as dead as a doornail. We had to show that solar energy had a potential large-scale application, that it could use highly advanced sophisticated technology and that it had potential for industry. I succeeded. Today, enough work has been done on the satellite to show that, within thirty years, it can become the most widely used method to generate power at the lowest cost. All the others are restricted by where and when the sun shines. The Sahara? Until mankind changes, until a messiah appears, there will always be the matter of ownership and control. What will happen to Japan, to Europe? Today, we know we can build the satellite power station. What is missing is the space transport."

He was asked about the fear of added heat pollution. "Microwaves can transmit power with an efficiency of 85 percent. That means very little thermal pollution at the receiver. We can adjust the receiver so that it can accept or reject solar energy. There is no danger of thermal loading. I can make the earth hotter or colder."

Glaser is a convincing advocate of satellite solar power. In October 1973, he appeared before the Senate Committee on Aeronautical and Space Sciences during hearings on possible uses for a space shuttle. A feasibility study of the project had just been carried out by Arthur D. Little along with three aerospace firms. Glaser said to the senators: "As we investigate all of the potential uses of solar energy, we find that here on Earth we are restricted by the God-given day-and-night cycle and by weather constraints which we are, of course, unable to control. Thus, we should be looking toward a location where the very large capital investment required to utilize solar energy could be put to best use. This location is in orbit around earth." He went on to explain that the station would be a big rectangle consisting of solar cells and mirrors to concentrate light onto them. In the middle of the rectangle, a transmitting antenna would collect the power and send it down to earth. On the ground, next to the receiving antenna, "we can meet the U.S. permissible limit for microwave exposure which is the equivalent of standing in front of a microwave oven with the door closed."

The committee chairman, Senator Frank Moss from Utah, wanted to make sure that no one misunderstood the nature of space power. He asked: "Now, after this has been converted to energy back here, then it could be transmitted over lines just like any other energy we generate?"

"On Earth, the transmission network is supplied with high voltage direct current. As far as the utilities are concerned, they cannot distinguish electricity generated by the satellite or any other energy source."

The senator asked: "In other words, it is just plain electricity when you get into it?"

"Yes, just plain electricity."

Then Moss wondered about the cost of the project, leading to this dialogue in the *Congressional Record:*

DR. GLASER. A 1,000-megawatt prototype station would cost $1 to $2 billion to construct and represents the second phase of the program. This is about the same cost as the prototypes of other energy-production methods.

THE CHAIRMAN. I do not have the figure, but we have already expended vast amounts in the fast breeder reactor development that has not yet come to a prototype.

DR. GLASER. I believe, Senator, that over the long run, there are two major options for power generation. One is fusion power and the other solar energy. Both happen to use the same process. I just happen to prefer my fusion reactor to be 93 million miles away.

THE CHAIRMAN. Well, at least environmentally you ought to be on the better side of the comparison, is that right?

DR. GLASER. That is right, sir. We believe that from the environmental viewpoint, and we have made a rather detailed study of the possible effects, these would be minimal. For example, the heat release at the receiving antenna site would be less than the heat release in the center of Washington, D.C.

THE CHAIRMAN. We get a lot of heat in Washington, D.C. [Laughter.] Senator Goldwater, are you as far up in the clouds as I, listening to this?

SENATOR GOLDWATER. I am.

In his office in Cambridge, Glaser commented on his schedule for bringing all this about. Before the SSPS starts to provide power, progress must be made along several fronts: microwave transmission of power, low-cost solar cells, heavy loads aboard space shuttles. Glaser was confident. "We have started experiments on micro-wave transmission with the Jet Propulsion Laboratory. A dish antenna eighty-six feet in diameter is used to direct the microwave beam to a tower one mile away where elements of the receiving antenna convert the microwaves directly into direct current with 80 percent efficiency. We can now beam one hundred and seventy kilowatts by microwave. A lot is happening here. During the 1970s, we hope to develop and test our technology so that we can get between ten and one hundred kilowatts into orbit by 1981 as a payload for the space shuttle. Boeing has a NASA contract to investigate the feasibility of a space transportation system to place huge payloads into orbit. They also think the SSPS is a feasible approach and they are pursuing their own design studies. Martin Marietta has a contract with NASA to figure out how to assemble large structures in space and they have studied the microwave transmitting antenna to exercise their aerospace skills. With all this, we should have a prototype in operation in the 1990s. During the next ten years, the program will cost between fifty and one hundred million dollars. We're talking about an option that we must protect. When the space shuttle is ready, we believe that the solar cells will be there, too."

Glaser had an answer for those who mutter that he is seeking to corner the sun. "I'm going to Japan shortly, then I will be in Germany. This is not a U.S.-alone project. The solar power satellite cannot happen on a nationalist basis. We will need international assignment of synchronous orbits in space just as we now have international assignment of radio frequencies. We now have INTELSAT, we will have SUNSAT. With one hundred satellites, we will be able to take care of the power needs of the United States. Another two hundred will take care of the world. We can develop this to the point where we can make an offer you can't refuse. As oil and gas go downgrade, we only have solar energy and fusion. But fusion still needs a scientific breakthrough. Its economics are unknown, so are its environmental effects."

It would seem that a few breakthroughs are needed before all 12,500 tons of the satellite power station are placed into orbit, but Glaser does not think so. He maintains that the project is feasible if the cost of carrying cargo into space can be brought down to one hundred dollars a pound from three hundred dollars a pound, the rate in 1973. The ways of the world are such that it will probably achieve a feasible space shuttle long before a workable solar toaster, if one is to judge by where the money goes. The idea of putting large objects into space has led a physicist from Princeton, Dr. Gerald O'Neill, to postulate the creation of an artificial planet, Lagrangea, between the earth and the moon. It would consist of a cylinder 32 kilometers long and 3.2 kilometers in radius with from two hundred thousand to twenty million people living on its inside walls, getting electricity from solar power stations. A *New York Times* article states that the first version could be a cylinder 3,300 feet long where "mirrored panels on the sides would be swung open to reflect sunlight into the cylinder 12 hours a day. . . . they would be opened and closed to simulate sunrise and sunset." Inhabitants of the space colony could construct solar energy stations to serve the earth, but that would not be their main purpose in life. O'Neill has said: "By 2074 more than 90 percent of the human population could be living in space colonies with a virtually unlimited clean source of energy for everyday use, an abundance and variety of food and material goods, freedom to travel and independence from large-scale governments. The Earth could become a worldwide park, slowly recovering by natural means from the near death-blow it received from the industrial revolution: a beautiful place to visit for a vacation."

Lagrangea has been mentioned here not as another solution to the energy problem but only to reassure those who think there is more fiction than science in Glaser's solar satellite power station. As time goes by, it is taken more and more seriously. After a seminar on advanced technology in Sioux Falls in June 1974, Governor Richard Kneip of South Dakota came out for it:

A discussion of a satellite solar power station in South Dakota seems especially appropriate because of our mid-continent location. It is clearly the logical site for satellite reception stations that serve the entire nation. Solar energy is independent of petroleum reserves, waterpower or harbors — it can be harnessed in a state with none of these resources. A solar power plant can be deliberately placed where it is most needed and where it will do the country the most good. I believe that the logical location is in mid-continent America.

The *Rapid City Journal* went almost as far:

South Dakota has the possibility of generating as much electricity as surrounding states without consuming huge amounts of irreplaceable natural resources. . . . We recommend that western South Dakota pursue the possibility of being selected as the initial ground station site, but carefully weigh each plus and minus such a selection would pose.

There will always be an editorial writer.

Glaser heads a project started by Arthur D. Little, Inc. to study prospects of solar energy not 22,300 miles high but at rooftop level. Support was enlisted from eighty-four companies around the world for a plan to develop a "solar climate control industry" and particularly to see how it can serve as a foundation for successful businesses. The company estimated that the market in the United States could reach $1.3 billion by 1985 with solar equipment in a million homes as well as commercial and public buildings. Arthur D. Little itself has designed a solar heating and cooling system for an office building that the Massachusetts Audubon Society has proposed to build in Lincoln, and a solar heating installation already in operation for the New England Telephone Company building in North Chelmsford, Massachusetts.

This was small potatoes, bits and pieces, when the study was completed in 1974 but interest was much wider. "I estimate that forty companies are looking for various product opportunities in solar energy," Glaser said. "They run from the very small to the very large. Every day, I hear of a new company with solar in its name. 'Solar' is the way 'atomic' or 'electronic' used to be. In Western Europe, a nationalized electric utility has written us to ask if they can be reached by the solar power satellite. You must be aware that the costs of electricity and heating have become astonishing. In solar energy, what matters is not where you are but the cost of competitive fuels. In Nevada, electricity costs .65 cents per kilowatt-hour; in New York, it is 3.6 cents. Solar energy is much more attractive a proposition in New York than Nevada." European companies, at first hesitant, have started to reason the same way and quite a few of them have joined the Arthur D. Little project. "At first, they did not believe there was enough sunshine in Europe. But now that the Germans, the French, and the Swiss are jumping onto the solar bandwagon, industry is beginning to think there is money to be made in solar climate control in Europe as well." Later, in 1975, Glaser attended the first annual meeting of the Solar Energy Industry Association with fifty exhibitors, four thousand visitors, and one thousand representatives from industry. He observed: "The enthusiasm displayed there reminded me of a revival meeting getting ready for the Second Coming."

Some of the embryonic forms that the industry had assumed were on display in the basement of Arthur D. Little, Inc. In particular, one could see the showerbath heater that Hitachi makes in Japan, not much more than a plastic bag on the roof, yet hot water for a capital investment of around ten dollars. There is a cultural twist: the Hitachi heater need not incorporate a storage system because the Japanese bathe when they come home at night, not when they get up in the morning. Swimming pool heaters are making their way in the United States; one of them is produced by a firm called Fun & Frolic, Inc. The public relations staff of Arthur D. Little, Inc. puts out releases on this but they have not published the full list of all the companies that sponsored the solar climate control study. Glaser would not let us look at it. "I can't; these firms are so sensitive. One of the world's major oil companies is among them. They told us: 'If people find out we're on that list, we'll have to appoint a vice-president, two administrators, and six secretaries just to answer the mail.'"

Glaser was less reticent about his own position. "Solar heating and cooling of buildings will happen in the United States if the government comes up with the proper incentives in the way of taxes and interest charges. Windpower will happen, perhaps only on a small scale, not the way Heronemus sees it. And photovoltaics will happen. The only question is whether in five, ten or fifteen years. My guess is that silicon will win. The silicon solar cell has a potential efficiency of 15 to 20 percent. That helps to make up for the cost of the device. And the cost will come down." In his Senate committee appearance, Glaser had said that solar cells should be competitive once their price went down to one dollar per peak watt or one thousand dollars a kilowatt.

I had always admired the matter-of-fact way in which solar cell scientists, Glaser and others, talked about costs. They would speak of a reduction by "two orders of magnitude" which does not sound like such an awful lot until you realize that it is the equivalent, say, of cutting the price of gasoline to sixty cents a gallon from sixty dollars. Any number of university researchers and industrial laboratories have been working toward this goal. A consensus directed us to one: Tyco Laboratories in Waltham, Massachusetts.

We had trouble setting up our base camp. The hotels were full along Route 128, the outer rampart of Boston. We got to one and Madeleine didn't like it. "All the guests were terribly rich and unpleasant, one could smell the proximity of a big city. Our room was enormous, neither better nor worse than the others, but there was no bathtub in the bathroom, just a little tile slab to stop the shower water from slopping over. I was furious: $35 a night with stinginess staring out from all sides." That was not quite true. A portrait of George Washington stared out from the wall of the corridor next to our room and the management supplied a complimentary *Wall Street Journal*. We moved on the next night to Lexington where Madeleine found happiness. "The hotel wasn't far from the village green. That was the main *place* of the early American villages, a square covered with grass and surrounded by white-painted wooden houses where people met and played bowls and had the right to graze their cows (fine lawn mowers, no noise, no gasoline). It was at Lexington that the American peasants fought their first battle. They were a sort of militia originally armed to fight the Indians. They had come here to fight the English army. It was the first battle of the American Revolution about the year 1775, the first time in the history of the modern world that a colony had rebelled, Dan said, except that it wasn't the Indians who were rebelling and it wasn't the Indians who won." History had preserved Lexington, saving it from the fate of its industrial neighbors like Waltham, and the Battle Green Inn had been able to hold out downtown. "It was the most *formidable* of all the places where we had stopped. On the ground floor, a patio just as in Spain, trees in the middle, rooms all around it, overlooking it. On the second floor, a wrought iron balcony that formed an arcade around the patio so that you could go to the restaurant without getting wet. We had a big room, navy blue bedspreads, two arm-chairs next to a small table with a lamp, a bathroom (and a bath-tub) and two television sets (neither seemed to work too well) for $17.

"We left our bags and went for a walk. After a day of rain, the sun finally showed itself. I took pictures of three houses among the trees, all exactly alike except that one was sky blue, the second *bonbon* pink and the third pale green. What a paradise this city would be if there were not all those noisy, smelly big cars around. I imagined the same scene, the same houses of painted wood but buggies pulled by whinnying horses in the streets, clattering over the macadam with their light hooves, and bicycles everywhere, and delightful electric trolley cars, all gilded, ringing their bells, each pulling a 'Friendly Ice Cream' trailer, free portions for the season ticket holders."

Lexington put us in good spirits for our journey to Tyco. The postal address was Waltham but it stood in an industrial park on Route 128. It was more of an industrial Park Avenue. We rolled past Bell & Howell, Raytheon, Sylvania until we came to 16 Hickory Drive, the bucolic address of Tyco Laboratories, Inc., and Dr. A. I. Mlavsky who had a new way to make silicon solar cells. Mlavsky was downstairs. He introduced himself as Eddie and swept us into his office. We had hardly sat down before he was standing up. "I can't talk without a blackboard," he told me, then for Madeleine's sake: *"La plume de ma tante est sur le tableau noir."* He juggled with his chalk. He saw that Madeleine missed the shot and purposely juggled again while she was reloading. Mlavsky was an enthusiast, nothing could stop him, he warmed up like the sun over the Arizona desert. He asked us what we knew about solar cells. Nothing, we told him truthfully; it was so easy to forget all that had already been explained to us. Mlavsky was forty-four, a naturalized American born in London. He knew his subject and he loved it, a perfect interviewee; he did all the work, no questions asked, no questions needed.

"One pound of single crystal silicon in the form of solar cells four-thousandths of an inch thick will supply two hundred peak watts. To manufacture it, you buy a pound of raw silicon, you melt it and you grow it as a single crystal. The cost is thirty dollars a pound and five pounds will give you a kilowatt. So the cost of the raw silicon in a solar cell is only one hundred and fifty dollars a kilowatt. Then how do we get to twenty thousand dollars a kilowatt, the present cost of silicon solar cells for terrestrial use?"

Mlavsky asked us another question, "How did you get here? By Chevrolet? It costs two dollars a pound to make a Chevrolet out of steel, glass, rubber, electronics. Then why should it cost one thousand times more to construct a solar cell with no moving parts at all?"

The reason is the way that raw silicon is transformed into single crystal silicon. We must have looked blank because Mlavsky branched off into a supplementary explanation with every good grace. "You know what cheap wallpaper looks like, don't you? You get a cottage, a horse, a tree, a shepherd with a long staff . . . If that wallpaper is well laid, I'll know where every tree is on the wall. Within a single crystal, every atom is in a predictable position."

It was in 1918, he went on, that a Polish scientist named Czochralski developed a way to grow single crystal silicon. "You put a pot in a furnace, a fused quartz pot, then you put in the silicon and you melt it. You take a seed crystal and you dip it into the pot. The seed draws out heat, the liquid silicon turns solid. You pull on it and it grows, if you do it carefully. And you end up with what we call a *boule*. It's a cylinder three to four inches in diameter and eighteen inches long."

Mlavsky flipped his chalk, caught it, and went back to the blackboard. "Now there are five things wrong with this process.

"One: it's slow; you can only grow a crystal at the rate of a few inches per hour.

"Two: it's fussy; you need good temperature control, plus or minus a tenth of a degree C. If you don't watch the temperature, you'll get Marilyn Monroe." (He described it with his hands: bulges, no cylinder).

"Three: it's a batch process; you can only make one at a time per machine. And four: the machines are not cheap.

"Five: the final product is the wrong shape. There's not a bloody thing you can do with that piece of crystal except cut it."

It sounded like a very inefficient way to manufacture anything. Mlavsky said it was not all that bad. "It's good for semiconductors. Take an integrated circuit. They can put a thousand-bit memory on a chip of silicon an eighth of an inch square in an integrated circuit. They get ten bucks for that integrated circuit, they don't care whether the silicon costs them a nickel or zero. No one is going to break his hump to take that nickel out." He sketched a cylinder on his blackboard. "To make a solar cell, first you take the end off the *boule*. Then you slice it like a salami. The trouble is, one pound of salami will give you one pound of sliced salami, but one pound of single crystal silicon will only give you a quarter pound of slices. You have to grind it away with diamond saws. And it's still not ready; you must lap and polish each slice to get it down to four mils, four-thousandths of an inch thickness. It's like making a school bus from a 747."

Size alone does not explain the cost of the silicon cell. The product that Mlavsky was talking about had to undergo further treatment. "Pure silicon is no good to anyone. It's nearly an electrical insulator. To make it do what we want it to do, we must dope it with the correct impurity. This is a very subtle business, we're talking of the right garbage in fractions of a part per million. So you heat the silicon to a thousand degrees C., then you diffuse the proper impurity into it in the form of a gas. Next, you must lay a metallic grid over the slice to collect the electricity emitted when light hits the silicon. It's the opposite of what happens in those calculators. The light-emitting diodes work when they are hit by electricity." Mlavsky kindly oversimplified his simplification and I have oversimplified a rendering of it, just keeping what I understood and leaving the rest to more serious sources.

Then he turned to the efficiency of his silicon solar cell. On earth, the sun sheds one hundred watts per square foot of surface; a square foot of silicon cells will deliver ten watts. "About half of the sunlight never does anything at all to the silicon. It goes right through it. Then some bounces off the surface, so you can add an antireflective layer to reduce this effect. The theoretical efficiency of silicon is 20 percent in space where there is no ozone layer to cut down energy in the ultraviolet spectrum. The actual efficiency on earth is between 13 and 15 percent; I'm calculating only 10 percent."

He stopped sketching on his blackboard. "There are a few more steps to making a solar cell, but let's forget them. Even so, compared to making Chevrolets, it's pretty simple. Solar cells are expensive because there are lots of little slices that must be handled, put into furnaces, arranged on trays, treated with coatings . . . a gourmet cooking process."

Another pause. "Comes the revolution. Why grow a crystal only to slice it? Why not grow it as a ribbon in the desired width and thickness? The alchemist's dream . . . Only the screwballs said it could be done. They flopped. Some say we will, too. I doubt it." He went back to 1965 when Tyco Laboratories developed a new process to grow crystals in a continuous strip. They first applied it to manufacture "sapphire," single crystal aluminum oxide, not as jewelry but as a thin filament to reinforce plastics and, particularly, for use in sapphire arc tubes to operate sodium vapor lamps.

"Let's go back to the pot." And Mlavsky was back at the blackboard. He drew a different sort of crucible, not just a small tub with a seed crystal held over it. Instead, a tube grew from the center of the pot and a heat shield went over the surface of the molten contents. "If you put a tube in the pot, liquid will rise through capillary action." He worked on the sketch. The tube became a die: two plates standing in the pot, squeezing a ribbon out in the space between them. "There you have it, EFG — edge-defined film-fed growth. Corning Glass is using it under license to Tyco to produce sodium vapor lamps. They're twice as efficient as mercury vapor, they give a better color and more light. They last twenty thousand hours — five years, if you prefer. A four-hundred-watt lamp is the equivalent of four thousand watts in tungsten. We've started with silicon, but we're using the technique to work with other materials. Silicon is only the tip of the iceberg."

E . . . F . . . G . . . , he repeated the letters. "This process has five things going for it. One: it's fast; it produces at the rate of inches per minute, not inches per hour. Two: the product is delivered in the right shape; there is no cutting, no waste. Three: it's unfussy; there are few temperature control problems. The process is self-stabilizing. If the pot gets too hot, the adjustment is automatic. With sapphire, we need only control temperature to plus or minus twenty degrees C. With silicon, it's plus or minus two degrees C., but that's still twenty times better than before. Four: it's continuous; we can replenish the melt during the growth of the ribbon. And five: we can multiply the number of dies in our crucible. We have grown twenty-five filaments of sapphire with a single pot. This can be a really large-scale process."

Silicon is not sapphire. "We assume that we can make it work for silicon. The trouble is, you must have precisely the right amount of impurity and we're working with parts per billion of garbage. This is all the harder because molten silicon is damn near the universal solvent. It picks up anything, it can dissolve your die, the wrong kind of impurities appear. Then . . . remember the wallpaper? When I lay out my wallpaper, I will see seams, grain boundaries, they are called. We have to get around this problem.

"If we do, and if we can grow twenty ribbons at the rate of two inches per minute, then the cost of the solar cells will come down to between thirteen and fifteen dollars per pound, plus the price of the raw silicon. At 10 percent efficiency, that will give two hundred peak watts per pound of silicon. So the cost of converting silicon to a ribbon is between sixty-five and seventy-five dollars per kilowatt. The experts tell me that it should cost about fifty dollars per kilowatt to convert the ribbon into solar cells. Add the cost of the silicon, it's now one hundred and fifty dollars per kilowatt, and you get a solar cell for two hundred and seventy-five dollars per kilowatt. To make his living, a manufacturer has to set a price 60 percent over his costs. That gives us a selling price of four hundred and sixty dollars a kilowatt. But wait, the projected cost of raw silicon is fifty dollars a kilowatt. That will give us solar cells producing electricity at a capital cost of two hundred and seventy-five dollars per kilowatt. No fuel costs to be added, no pollution either. In 1970, the capital cost of nuclear power in the United States was five hundred and fifty dollars per kilowatt. And God knows what it will be.

"Now one square centimeter of solar cells gives you the same efficiency as one square meter. We can take solar cells, we can put them onto a house, we can decentralize the whole energy industry. Solar energy is dilute? Do you know what one hundred watts per square foot amount to? One hundred kilowatts over one thousand square feet. At 10 percent efficiency, solar cells will give you ten kilowatts on a surface just twenty-five feet wide and forty feet long. In Arizona, the roof of a garage gets enough sunshine during the day to run an electric car fifty miles at twenty-five miles per hour. One just needs two sets of batteries: one set in the car while the other is recharged. That is the way we'll have to run cars. If the earth were a hollow ball filled with crude oil, we'd still be out of gasoline by the year 2340."

Mlavsky went back to the comparison that Farrington Daniels had made, but with solar cells instead of parabolic reflectors. "Take a pump driven by an ox walking in a circle. The ox must sleep at night, it eats three-quarters of the food the farmer raises. If you cover the ground swept by the ox with solar cells, cadmium sulfide cells at only 1 percent efficiency, you can run a motor to drive the pump. It will pump the same amount of water, but the solar cells won't get sick. They won't eat the family's food, either.

"If we can bring the price of the silicon solar cell down to one hundred and sixty-five dollars per peak watt, then the war's over. It can be done, this process will be moving out of the laboratory. As of May 1, 1974, we were able to grow a ribbon one inch wide and eighteen inches long. We could use it to make solar cells with close to 10 percent efficiency. As of May 2, 1974, we produced the first long ribbon pulled by an endless belt process. We got it up to six feet long, then we ran out of silicon. In early 1975, we grew a one-inch ribbon sixty-five feet long and eight-thousandths of an inch thick. I can see things happening in the future in three steps. First, a single continuous ribbon, four-thousandths of an inch thick, perhaps one hundred feet long . . . even Saranwrap isn't endless. Next step: multiple ribbons coming out of the same crucible. And step three: ribbon solar cells. You can do it with any kind of a ribbon. You dope it and run it on a drum, again in a continuous process." Mlavsky had all the figures, he hardly resorted to his pocket calculator. "Solar cells are now $20,000 a kilowatt. With the ribbon process, they will come down to $5,000 a kilowatt. There will be a new market, more money, an impetus for R and D. At twenty thousand dollars a kilowatt, you can sell them to yachtsmen. You sell a few, make some money and plow it back. It could take fifty million dollars to go the whole route."

Mlavsky makes heady listening, yet he is almost guilty of understatement. Writing in *Science,* Allen Hammond stated that the work done by Mlavsky in collaboration with Bruce Chalmers of Harvard University was the most successful of all efforts to reduce costs of solar cells through new manufacturing techniques. Hammond added: "Centralab, a major solar cell manufacturer located in El Monte, California, has made and tested cells from some of Tyco's ribbon and they confirm the 10 percent efficiency claimed. 'It's good silicon,' said P. Iles of Centralab's research facility."

Tyco's process received rave notices from *Forbes* with the announcement that Mobil Oil "is ponying up $30 million for a five-year joint project with little ($44-million-revenues) Tyco Laboratories." The business journal explained that Mobil and Tyco are trying to get electric power from silicon solar cells at a maximum of six hundred dollars per kilowatt within five to seven years, "about the time that it takes to *license,* let alone build, a nuclear power station." *Forbes* sounded like all the world's solar enthusiasts: "That would bring solar power about in line with nuclear power in terms of capital costs, but solar would not require uranium fuel and it would be far less likely to frighten people in the neighborhood." As the magazine saw it, an energy farm of solar cells would be "acres and acres of gunmetal sheets, tilted slightly so that rain and snow would run off, and pointed toward the south to catch the maximum possible sunlight." Electricity could be stored in batteries or used to produce liquid hydrogen.

Forbes found Dow Corning, the biggest silicon producer in the United States, engaged in a project with the University of Pennsylvania with the same aim as Mlavsky's: to reduce the cost of silicon solar cells. It quoted Jerold Noel, a physicist on the project: "The chances are very good for significantly lowering the price of silicon. In five years we think that pilot plants can be built for the entire process — from taking the silicon out of sand to the fabrication of working solar cell arrays, at or below the cost of nuclear power. There's no need for a theoretical breakthrough. This is an engineering problem." *Forbes* concluded: "What's the catch? None — though 'mere' engineering problems all too often have a disconcerting habit of growing to giant size. The worst problem just now may be the need for time as well as money to work the bugs out of solar energy. As one solar energy researcher puts it, 'You can't make a baby in a month by getting nine women pregnant.' But that same man adds: 'There's very little reason to doubt that solar power can be a multibillion-dollar business in one form or another. The energy crisis may yet yield to human ingenuity.' " The underground press didn't say that, *Forbes* said that.

Mlavsky was saying more. He may have been small compared to Mobil Oil, but he was big in the sunshine game. Two months before, we had watched welders putting together a bath heater in a barn in southern France. It was good to listen to Tyco's money talking. "I'm not result-oriented, I'm end-result oriented. I like to take something from the laboratory and put it into the hands of someone who can use it. No academic will ever develop this process. It's the difference between pornography and sex. My cost estimates may be wrong, but five or six years will be enough to prove that the process is right. The commercial world can do it, the academics can't." Mlavsky had entered the commercial world himself in 1956, two years after he got his Ph.D. from Queen Mary College in London. He started in semiconductors, growing single crystal silicon in the old way he had described. In 1960, he came to Tyco. "We had no building of our own then. We were in downtown Waltham with a cabbage odor in the staircase. We started doing materials research under government contract. By 1964, we went public. In four or five years, we were up from one million dollars a year to forty million." He did not spend much time on the history of Tyco during the 1960s, but he gave me a clipping from the *Wall Street Journal* that related the story of its founder beneath the head-line: "*Sic Transit Gloria* for Arthur Rosenberg, Grand Ambitions Paved the Way to a Downfall — Ex-Tyco Labs Head Dreamed of a Huge Conglomerate, Ignored the 'Dull' Details — A Search for the Holy Grail." Inserted in the article was a box entitled: "If This Is September, Tyco Must Be Changing Chiefs."

By the time we arrived at Tyco, all the excitement was over but there was just as much activity in the laboratory. Mlavsky took us there to see David Jewett who was in charge of the silicon ribbon program. The ribbon was growing out of a crucible heated to 1,400° C. by a furnace that looked very home-made. The ribbon of silicon emerged from the pot, thin and brittle as glass, as its belt pulled it out from the die. It was like looking inside a movie camera where sprockets move the film along. Remembering what Mlavsky had said about the properties of silicon, I asked him what was in the die. "It's an open secret, but I'd rather not talk about it. We use compressed faith." He touched the ribbon, it broke. "We have now shattered a silicon ribbon for the first time, but it doesn't matter. This is the birth of a revolution. This is the fetus, it just started crying. A guy from NASA told us: 'You said you could do it and you did.' It's still one hell of a thrill to do the predictable when it's never been done before." The ribbon kept coming out of the crucible like an uncoiling cobra. "It's the birth of an industry, it's like Fermi getting his first chain reaction. It's not as profound a feat, but it's just as important with its technology."

At this point, Mlavsky decided it was time for lunch. I won't say where we went, but it wasn't Arby's and it wasn't MacDonald's. The service was better, the menu was longer. Let Madeleine, our gastronomic correspondent, relate the rest:

"I found a pretty little pink worm in my fish, perhaps the one that was used to catch the fish. Dan and Eddie laughed. It couldn't have been a worm, it never would have survived the cooking. It was a Chinese plant, Eddie said, the heat of the warm fish made it appear to move. The worm wound and unwound, not very quickly, I grant you, but it still moved more than a plant. I put it in a napkin and held it in the palm of my hand. We decided to take it back with us because there was a microscope at Tyco. As we left, Dan asked me how the worm was doing. 'He's cold.' It was true, he had stopped moving. Then we put him under the microscope. Eddie switched on the light. *Pouah!* an enormous pink worm, a red tail, pimples all over its head, fat as a pig. It revolted me, but when I saw it again on the napkin, tiny and thin, I didn't mind it. He was almost a friend.

"We left it under the microscope for the next man."

XIV.

The Sun in a Cold Climate

ENERGY AND ITS ECONOMICS are so pervasive that it is tempting to try to get away from them, to go back to a time when they did not seem to rule us as they do now. We seized upon such an occasion when we were called to Cambridge (not Massachusetts) in England (not New England). We packed two minibikes for urban transport and rural sport into my large Citroen (larger than Madeleine's, at any rate) and made the journey via French and British highways and cross-Channel packet, breaking the trip with a night in a village near Calais where we rested before embarking for Dover. The night was not quite that restful because the innkeeper seized upon the bikes as a conversation piece and kept us up past midnight, relating how he had cycled the length of France on a pilgrimage to Lourdes.

Late the next day (by now, we had spent more time on the ground between Paris and Cambridge than in the air between Paris and Denver) we sought another place to stop. We did not want a hotel in Cambridge or anywhere remotely urban. Guideless, we left the main highway at Saffron Walden, a promising name if there ever was one, and started crawling from pub to pub in the hope that

one would provide us with bed and breakfast. Our luck ran in at
The Fox, run by Joe King and his bride, in Finchingfield. Despite
the Saturday night trade, there was room for us, the car, and the
bikes. I must have a Finchingfield from time to time; in that, I guess
I am like many Americans. It is my fancy-land, my private Disney-
land, the independent village of sturdy yeomen, the kind of place
my ancestors would have come from if they had come on the *Mayflower*.

Finchingfield is nothing of the sort anymore. The yeomen drive
out to work every weekday morning to factories and offices within
easy reach of their thatched cottages. On weekends, the yeomen
stay home and the visitors drive in, beer-drinkers on Saturday night
at The Fox, tea-drinkers on Sunday afternoon in the cafe-and-
souvenir-shop next door. Trucks come through, too, during the
week. One was so high that when it crossed the humpbacked
bridge over the local brook, a branch of the River Pant, a load of
lumber brushed the flowers from the lowermost bough of the big
pink chestnut tree shading The Fox. Yet Finchingfield reassured
me. If worse came to worst, the yeomen could go back to their fields
and its aspect would not change.

Such were my fancies as we spent Sunday afternoon on a village
green, more operative and less historical than Lexington's.
Madeleine disappeared from sight. When I spotted her again, she
was slithering in the grass behind a white quacking duck, Madeleine
camouflaged by her shock of hair so that she looked like a tuft of
weed trailing a pair of black corduroy jeans. The duck got away,
Madeleine abandoned her hunt to note her surroundings: "Two
P.M. on a Sunday, we were sitting on the green grass in the middle
of the village next to the duckpond. Two American cyclists with
Peugeot bicycles were sleeping on the grass next to us. Behind us,
two wooden tables covered with white paper tablecloths. For fifteen
pence, boys and girls in jeans and summer blouses were selling a
Ploughman's Lunch: a long loaf of bread, cubes of orange and yellow cheese, two pats of butter, pickled onions and a small tomato
in a paper plate. There are fifteen hundred inhabitants in Finchingfield with its small houses, some half-timbered in black beams,

and their roofs of dark tile. The villagers were all around us, the old ones sitting on cement benches, little dogs leashed next to their feet, the young ones sprawled on the grass."

On Monday, before we drove off to our appointment in Cambridge, we were awakened by the caws of crows and the roars of puny overworked motors on the hill just beyond the humpbacked bridge. If the motors and the drivers go, could Finchingfield really live on its own? Is independence possible, could Finchingfield ever man the *Mayflower* again? Our visit there was fortuitous, yet it turned into a prelude for our meeting in Cambridge University with Alexander Pike and his nucleus of young researchers working on "autonomous housing." I first saw Pike the year before at an international meeting near Frankfurt on the problems of urban environments. After listening to a host of solutions, none very bright, I was awakened by Pike who proposed not to solve the problems but to see if they could be eliminated. His thesis was that many people are forced to live in cities against their will because they depend upon water supply, sewage, electricity, gas, all coming from central networks. If their homes were autonomous, almost self-sustaining, then they could live wherever they pleased, saving not only a modicum of money on utilities but a wad on the purchase of land not in the clutches of the subdivider. Pike said this could be done with the energy available from the sun and wind in the climate of England. That made a story the first time I heard him; now I wanted to see how much further he had been able to go. In the interim, too, a United Kingdom branch of the International Solar Energy Society had been started by John Page at Sheffield University. Brighton and Weston-super-Mare were not outrunning Majorca and the Costa del Sol as summer resorts, but the British had started to take their sun seriously.

So seriously that Pike had difficulty getting his work done. We found him at the School of Architecture in Cambridge in quarters much more redolent of Cambridge, Massachusetts. The architects were stuffed into Scroope Terrace, a rundown streak of Victorian row houses far from St. John's or Magdalene. Behind the row houses, they also had a newer building, a disconcerting brick structure, all multilevels and coffee crannies (or, perhaps, tea nooks). Pike's office there was as snug as a ship's cabin, its window not much bigger than a porthole, the sun looking out from a rough-weave tapestry. The room was small enough to be heated by a professor and two students.

Pike has started his fifties. He has longish hair, a mild manner but strong flat hands inherited from his grandfather, a docker. He practiced architecture for twenty years before he came to Cambridge in 1969. Previously, he had worked on mass production of housing and redevelopment of city areas, two foci of interest that got him out of the commercial world when he saw where they were leading. As director of a technical research center at the Cambridge School of Architecture, he has been trying to change directions.

We learned this in a cafe across the street from his office. "I do a lot of my teaching in a cafe. There are so many interruptions in my office. In a cafe, they can't find me." THEY are the media who discovered Pike in 1973 and rediscover him periodically as he works toward his goal of actually building an autonomous house in Cambridge. At the time of our visit, the Italian television service was at his heels while the Science Research Council (the British equivalent of the National Science Foundation) was dragooning him into yet another press conference to explain what he was doing. Pike dreaded the prospect; a previous exposure to the press had yielded eight hundred letters from the world over. This one led to two thousand more.

The publicity had come from the new concern over energy, but he considered this almost incidental. "The problem goes beyond the use of energy. It is the way people live in network systems. The city becomes a confluence of road patterns, a spider's web. When the pressure of growth is applied, it is accretive. Existing networks expand and new developments grow like the free cells of a cancer. They attach themselves to their host and grow because the networks are there. This leads to all the problems of the city: transportation, pollution, crime, neuroses, finances. The bigger the city, the higher the tax load per capita. Many people prefer to live in rural districts but there are no services there. So they stick to the urban networks.

"Our studies of the autonomous house show that the population need not grow in this manner. If houses were freed of networks, limits could be put on the growth of cities. Unserviced and therefore cheaper land could be used; an alternative planning strategy could be brought to bear. That is what we are doing here. We are trying to create the scope for doing something different if we want to do it. This is not for tomorrow, it's for ten or fifteen years ahead. In the meanwhile, we may decide that houses need not be close together. We may grasp the economy of small systems, we may stop smashing up the countryside around the cities. You can put an autonomous house anywhere, so that it will blend with the landscape and not defile it.

"We have looked at the cost of network systems. In this country, nearly 60 percent of the price of electricity must be charged to the costs of distribution and administration of the central system. At present, three kilowatts of generating capacity cost £750. That means we could afford to put in £750 for electricity. The running costs of the house could be lowered, but the savings on land would be even greater. Around Cambridge, building land near the city comes to £45,000 an acre."

Then Pike arrived at British weather and sunshine. He produced a paper in which he had stated:

If we overlook the standard jokes about the British climate and examine the meteorological data we find that the potential solar energy is much greater than is normally imagined. The actual value varies considerably from one part of the country to another, but the annual average solar radiation measured at Kew is approximately 800 kilowatt-hours per square meter, of which up to 67.5 per cent may be in the form of diffuse radiation during the winter months. If we assume a collector operating at about 50 per cent efficiency and an energy requirement for a typical house in this country of about 20,000 kilowatt-hours, then about 50 square meters [540 square feet] of collector would be needed. We have not yet got a collector with an overall efficiency of 50 per cent. Models currently available may have a 25 to 30 per cent overall efficiency and, as part of our research program, we hope to improve this figure.

Even if the sunshine is there, a lot more is needed. "It sounds so romantic, a house powered by the wind, the sun, the air," Pike said. "But anyone who thinks so isn't doing arithmetic. When we look at the known types of solar collectors, we find we do not even know their output in this climate. We have information about Denver or Dallas or the south of France, but people have so much sunshine there that they don't know what to do with it. In this country, you can't cover the whole south side of a house with an opaque black surface. When we have sunshine, we want to see it. We never want to get out of the sun. So you have two alternatives. You can use an opaque collector which is more efficient, or a transparent collector, less efficient but with more surface. That's what interests me. An opaque black surface has an efficiency of 50 percent as a collector and zero as a window. I would prefer, say, 25 percent as a collector and 75 percent as a window. It might be done with two outer layers of glass enclosing a sheet of heat-absorbing glass, with the interspaces filled with water . . . like a thin fish tank."

He started to sketch. The ideal shape of a house when one seeks the least surface for the most floor area — the basic prerequisite for saving heat — is a circle or its more practical equivalent, the octagon. "The trouble with the octagon is, it's a washout. More floor area is needed for circulation space. The rooms are the wrong shape. You must make them bigger so that you can fit furniture into them. I prefer a square house. It can be planned more tightly with efficient circulation space and, in the long run, produce a house with less surface area for any given floor area." He drew a small house inside a glassed-in space like an object in a display case. That was one approach he was considering. James Thring and Gerry Smith, two members of his division, had gone the other way with an octagonal model, not at all to the annoyance of Pike who regards disagreement in the ranks as a healthy sign and maintains that "one cannot conduct research with an imperial attitude." Pike thinks of the consumers. He keeps talking about giving them all the amenities to which they are now accustomed, while his younger associates are more autonomy-minded. It was Thring and Smith who took a computer run at the old dream of soaking up heat in summer for use in winter. The MIT group had tried it in 1939 and given it up as uneconomic. The Cambridge researchers had another go. Pike thought at first they would need a tank holding twenty-five thousand or thirty thousand gallons, perhaps a tank bigger than their house. They tried a five thousand-gallon tank in their simulation and it worked. As the basis for sunshine during a computer year, they used weather data gathered at Kew from 1960 to 1969. The collector was larger than the floor area and the house itself was battened down to a winter configuration in which it consisted of only six hundred square feet. The windows were shuttered at night to provide as much insulation as walls. Printouts showed that, with heat stored in the tank and gleaned from the winter sun, the house could maintain 66° F. during the daytime and 59° F. at night, except for four or five days in January when the temperature dropped to 53° F. Pike thinks the house could be built of relatively light materials with the water tank serving as its thermal store. He does not like thick walls for they conserve cold as well as heat. While heavy insulation is needed, he does not consider it a panacea. As insulation standards go up, ventilation losses account for a higher proportion of the total heat loss and take on much greater significance. A study by the Building Research Establishment in the United Kingdom shows that full use of insulation could save only 1.5 percent of the total losses in generating and distributing electricity, but 10 percent could be saved through the use of heat pumps.

This is where Pike wants a hand from the wind. He proposes to streamline his house so that it could collect wind for a generator mounted on the roof instead of on a tower. Like everyone else, his unit finds that existing wind generators are hard to amortize. The first 1,200 kilowatt-hours they produce in a year cost four times the normal price (although Pike thinks this ratio may go down if the cost of electricity continues to rise even faster than everything else). The Cambridge researchers think the windmills may be too small. They do not make use of surplus wind and they must be feathered and cut out when the wind blows too hard. Pike suggests using an electromagnetic clutch device to slow the racing blades in a high wind without losing all the power. He also wants to gear his rotor directly to a heat pump to charge the storage tank. Once it had completed this task and if it still had some energy left, the rotor could be used to pump up a supply of compressed air to run vacuum cleaners, washing machines and other household appliances. Pike does not go along with Heronemus's array of giant pinwheels. "Why take energy evenly distributed by nature, concentrate it and redistribute it?"

The British have a gift for survival with modest means; it is more in keeping with the age of the sailing ship than the spaceship. Smith and Thring showed me a folder from a small firm that markets a 200-watt wind generator that can light a home. The trick is the use of special thirteen-watt neon lights claimed to correspond to ordinary sixty-five-watt bulbs. The windmill alone was sold for only four hundred and twenty dollars in 1974. That takes care of the lighting bill; as for gas, the house makes its own. The group quotes figures showing that cooking gas can be supplied through a combination of resources: a digester to convert human waste into methane gas, not much of a help until one adds the conversion of waste from the kitchen and the vegetable garden. This last raw material becomes automatically available because the autonomous house is to produce its own food, half an acre of land serving to provide seventy-five hundred calories per day for three people, with protein coming from goats and chickens on another half acre. As this much land will not be available in all cases, gas consumption must be cut by redesigning kitchen appliances which, Pike thinks, are about 90 percent inefficient.

Too much heat is thrown away in the kitchen and in ventilating houses. The British calculate that a normal home wastes 27 percent of its heat by allowing hot water to drain away from sinks, bathtubs, washing machines or dishwashers. Instead, the drainpipes should be run through a heat exchanger that would transfer the outgoing calories into a heat storage tank, allowing the water to run off only after it has been cooled. This is simple and cheap. Opening a window in winter may be simple, but it is not cheap. By the time the fresh air gets in, all the heat has gotten out. Instead, air should be taken in through the attic. Thanks to the cracks around its windows and doors, an old British home is treated to two air changes an hour while a new house gets one an hour. "Ideally, we would like to seal up a building," said Pike. "There is a nutty idea that would give us fresh air: we could use a plant greenhouse to absorb carbon dioxide and provide oxygen. Here is an air-freshening device that does not involve any loss of heat." One might say it was discovered by Joseph Priestley when, as Eugene Rabinowitch relates, he put a sprig of mint on August 17, 1771, into a quantity of air in which a wax candle had burned out and found that on the 27th of the same month, another candle burned perfectly well in the air, an unwitting demonstration of photosynthesis.

Through all this runs a multicolored thread of convictions and technology. "We have been widely criticized for seeking autonomy," said Pike. "We are asked why we do not simply supplement the mains. The truth is that if we do not achieve total autonomy, the experiment will not be a total failure. There will be many side benefits." While these benefits are not too relevant to solar energy *per se,* they are interesting as an indicator of the British temper in these times. I do not spend enough time in the country to pronounce qualified judgment, but I often sense the mood of World War II from which the British emerged as the only undefeated country in Western Europe. When they talk of self-sufficiency, it is not just kids trying to shake the boredom of suburbia. Pike wants the occupants of his future house to operate its machinery by themselves, to run the solar stills that will get the lead and sulfur out of the rainwater falling onto Cambridgeshire, to close and open shutters by hand to save heat by night and acquire it by day. "People today don't understand what they're using. They drive cars, they watch television, they run washing machines, they don't know what it's about. They are not concerned about what is going on outside. But the occupants of this house would have to respond to nature to operate it, to protect it against frost, to feed their digester with waste products. They won't be able to use bleach to clean pans because it would stop the digester process. People have become reliant upon institutions and this is psychologically bad. It is all part of a care process that makes people lose their independence. The Renaissance knew the universal man. One man could understand everything then, but today most people do not even have the knowledge of the universal man of the Renaissance. The medieval peasant could build his own house and that house still stands. If buyers are given a choice today, they'll choose a peasant house. It is sad that the marvelous, sophisticated, educated average man cannot build his own house."

I cannot help quoting another Englishman, Andrew Singer, who published a book called *Methane, Fuel of the Future* in which he said:

Part of the point of being self-sufficient is to get away from experts. Look at the mass of people in cities: they live in houses designed by experts, they go to work in vehicles made by experts along roads laid out by experts, they make things designed by experts and sold by other experts, they eat a meal produced by experts and they are entertained by yet more experts. All they do for themselves is sleep.

The Cambridge unit has published a "survey of indigenous building materials" as one of a series of booklets on autonomous housing. The writing is good here, too:

Cob is found in Devon and Dorset and is made by treading straw into wet mud. When thoroughly mixed on the ground the mud is laid diagonally in courses of 475 mm–600 mm width on a rubble or brick upper foundation wall. The courses are trodden down by foot, and once a course has been completed all around the wall, the wall is pared down by eye after the 2–3 week period necessary for the wall to dry out. The walls must be allowed to dry completely for 1–2 years before applying the protective exterior coating necessary to prevent the wall disintegrating when rain wets the surface.

Clay lump construction is found exclusively in Norfolk and Suffolk. The soil is first mixed with water until it forms a sticky mass capable of being lifted with a fork.

Thring and Smith have drawn up a tentative schedule for the appearance of the autonomous house. It starts to become economically plausible, they think, when a house is about a kilometer — six-tenths of a mile — away from the utility lines. The researchers assume that the cost of oil, gas, electricity, and food will go up 15 percent a year for the next ten years, not a risky assumption in present-day Britain. As energy prices rise, a 10-kilowatt wind generator starts to become attractive. The rising cost of developed land has helped to kite the private house out of reach of the average Briton. This, too, makes it easier to amortize systems that digest sewage and purify rainwater. John Littler, a chemist in the unit, believes that excess nitrate levels in groundwater are enough to create a demand for individual treatment of domestic supplies.

Thring and Smith base their land costs on derelict sites, such as old mining areas that have been left bleak and barren. "People could use that sort of land," Thring said. "These houses would not need expensive services but they do require low density so as not to overshadow each other. About an acre each is enough in terms of self-sufficiency in food production." As for Pike, he is more interested in providing a fully self-serviced house on an average-sized plot of land. Thring and Smith do not expect to see autonomy appear overnight; they may be wrong. Writing in *Esquire,* Alvin Toffler reports:

Today the talk is of some catastrophic breakdown in the world monetary system. And it is not just long-haired solar energy or windmill freaks who seek remote hideaways to shelter them for the duration of the anticipated decline and fall. In London, Jim Slater, the financier, is supposed to have listed what he regards as his hyper-inflation survival kit: tinned sardines, a bicycle, a supply of South African gold coins, and a machine gun.

None of these, except perhaps the bicycle, would be appropriate to the model of an autonomous house that sat in the office of Thring and Smith. It was octagonal, with its living quarters built over a ground floor used to house chickens, store gas and water, and provide play space for the children. Three sliding collectors supplied solar heat, while a windmill sat on top of the roof, a rotor on a vertical shaft that looked more like a cupola than a power source.

Pike kept his own model at home and he invited us to spend a night at his place. He and his wife Nona, a former actress, live in a banker's mansion put up in 1895 and converted into seven apartments in post-Victorian times. It was a patrician life style in the days when financiers weren't thinking about machine guns. The Pikes' ample bedrooms were the kitchens of the banker's establishment. The billiard room, complete with a false beamed ceiling, gave them all the space they needed for a living room. They liked their rented home. Even with all the new tenants, there were only eight inhabitants to the acre on the estate and eight hundred trees still grew there. Pike indicated a tall silver fir. "The banker's daughter told us she used to be able to jump over that tree."

In the billiard room, he explained his model as he had already done on BBC television. It wasn't bad-looking at all, less of a shock, perhaps, to our conservative eyes than the approach taken by Thring and Smith. Outer glass walls sloped at 75 degrees to get the most from the winter sun in Cambridge. Pike sees the living space of his house shrinking in winter with the glassed-in area unused except as a way to get the children out of the house without shooing them all the way out into the cold. This area would serve as a greenhouse porch, piping heat to the storage tank. On fine days it could be opened to let the English have their taste of sun. Interest in the glass house has been expectedly shown by Pilkington while other manufacturers are looking at different aspects. Later, after our visit, the plans and the Plexiglas models were transformed into a grant of about $85,000 to enable a full-sized prototype to be built near Cambridge for occupancy by a family. The money came from the Science Research Council and the Department of the Environment.

Pike hopes to achieve "the same level of amenity as in the average British home." In that kind of a ball park, it's not hard to hit home runs. Pike's own place is heated by coal, kerosene, and electricity, depending on which room one happens to be in. It does not have central heating; the vast majority of houses in Great Britain still do not. This willingness, nay cheerfulness in the face of what any other nation would consider a state of disaster compensates for the sunshine that the British do not get. If they can put up with their present amenities, they will thrive in autonomous houses. The morning that we left the Pikes, I decided to try a shower. The "gas geyser" in the bathroom roared like Vesuvio to dispatch a trickle of water through the delicate rubber tubing attached like the cups of a milking machine to the hot and cold water taps of the Pike tub. If one turned on the cold water too energetically, the rubber tube on the tap popped loose to wriggle down the tub like a runaway snake while the undiluted hot water came out scalding. The effect was that of an instant sauna when the bather scrambled for his life out of the tub and into the ambient temperature of the bathroom, probably the old banker's cold-storage larder. This cleansed me as thoroughly as if I had stepped out of the shower stall in a Holiday-Sheraton-Ramada Inn. Much wider awake, I took the wheel of the Citroen and, with Madeleine navigating, kept to the wrong side of the road over a maze of byways, the Pike shortcut to Highway A-1 that was to take us to Sheffield and the United Kingdom branch of the International Solar Energy Society.

The road ran north. Distances are not great in England, but differences are. We left green and smiling Cambridgeshire for the old coal tips, now greening themselves, of industrial Yorkshire. Madeleine saw "stretches of blinding yellow mustard wedged among bright green wheatfields, meadows filled with little daisies, and bouquets of leafy trees from which emerged, every so often, the tall chimneys of sturdy Victorian farmhouses built of brown /brick polished by time." Her mind photographed "sheep still fat and swollen in their tan wool, lambs with dark-brown feet and muzzles lying next to them; black-and-white cows as clean as if they had just stepped out of a tub, grazing at the feet of immense poplars; from time to time, the enormous cooling towers of electric power plants, standing in groups of five or seven like headless Phoenician statues, white clouds pouring from their necks."

The city of Sheffield rose like the side of a bowl over the immensity of its steel mills. On the western edge of the city stood Sheffield University and its new engineering and science building that emerged from a sea of condemned streets, red brick tenements, red clay chimney pots, boarded-up windows, crooked TV aerials, a sign saying "Scientology," modest shelters for individuals and individualists, probably destined to serve as breathing space and additional parking lots for Slabsides Tech where John Page perched in his eyrie at the very top, nearest the sun. The parking lot attendant advised us to use the high-speed lift. Otherwise we would need four minutes to do the eighteen floors in the paternoster, about the time that it must take a Cambridge undergraduate to cycle from his college to his classes. Once we had divested ourselves of the car (not nearly as much of a problem as at Carnegie-Mellon, the last campus in a steel mill we had seen), we headed for the student union for a second breakfast of coffee, bacon sandwiches, apples and buns, all served by motherly ladies who said: "Thank you, luv," when paid. With our food, we cooked behind the big plate-glass windows of the coffee lounge that caught and trapped an unwanted sun. By the time we were done to a turn, it was time for our appointment.

Page had caught my attention at the solar energy congress in Paris in 1973, when his contribution came through like a ray of sunshine amidst a thud of papers. A professor of building science at Sheffield, Page was bemused by the breakdown in the distinction between town and countryside. Farming in his part of England had become an industrial operation with a rise of half a million in the pig population within three years on an area of six thousand acres, generating sewage disposal problems equivalent to a human population of one and a half million. He wrote:

We are beginning to see animal slums grow up, which remind me personally as a human urbanist of the reported conditions encountered by people in cities in England around 1820–1840.

In the early days of the Industrial Revolution large cities, choked with immigrants from the countryside in poor housing equipped with totally inadequate drainage and sewage disposal facilities, were frequently rife with cholera and typhoid fever. Factory farming has become, in many respects, an extremely noxious chemical engineering operation quite unsuited for its conventional countryside locations.

At the very moment when farmers are constructing "animal towns," Page observed that city people seem to show all signs of wanting to produce "town countries." This is what led him as an engineer into the energy issue. "One scenario concerned with the future of plant growth in towns envisages the extensive use of electrical energy to provide artificial light for plant growth indoors. This scenario implies the botanical factory and the plant city. It should be noted that the scenario is not based on principles of energy conservation as is the case for intensive stock-raising but rather on the concept of wasteful use of fossil fuels and atomic energy resources to replace sunlight. This scenario I shall reject out of hand as being unsuitable for a world facing a long term energy crisis."

Such ideas are part of the *Zeitgeist*. In an interview with *Science*, Karl Hess tells of his Community Technology project in Washington which would enable city neighborhoods to achieve at least partial autonomy by growing vegetables in roof gardens and fish in basement tanks. Half their heat could come from the sun, while sewage and garbage would be converted into gas and methanol. Rainbow trout have already been raised in a Washington basement and, according to *Science,* a solar cooker and water heater have been conceived by the former senior design engineer of the Atlas missile propulsion system. *Science* has also described the efforts of the New Alchemy Institute on Cape Cod to produce food with renewable wind and sunshine instead of fossil fuels. Its ideas have taken shape as an "ark" consisting of three greenhouse-covered ponds, two growing food for fish in a third. It has devised a water-pumping windmill for India that uses cloth sails and a bullock cartwheel. Both on Cape Cod and in Washington, professional scientists are at the roots of a movement to use human science as an antidote to what they consider inhuman technology.

As for Page, up against British population densities, he wants to exploit the resources of hot air and carbon dioxide that a city produces. Air discharged from an office building is four times richer than normal air in carbon dioxide, while the solar energy falling on its roof is entirely wasted.

If, however, we were prepared to consider the roofs of deep buildings as areas suitable for plant growth in higher latitudes, especially horticulture under glass, the waste energy and carbon dioxide from the building could be fed up to the photosynthesis process on the roof and used to foster plant growth.

This approach would enable the growing season to be substantially extended under glass without making any additional energy demands. The relatively high moisture content and carbon dioxide content of the air supply would also be valuable. It would be an economic way of producing early crops.

In this way, town and country meet.

Similar principles could be applied to ventilation discharges from intensive animal houses where the carbon dioxide contents are likely to be even higher than those from ordinary buildings. In the one type of building where people are kept at the same density as animals in intensive farm units, namely lecture theatres, carbon dioxide concentrations in winter may reach eight times normal levels.

With imaginative design of horticultural activities on buildings, the roof space might also be made available for recreational uses as well as economic plant growth. Thus office and factory workers would be able to relax in a desired horticultural environment instead of in the concrete-clad open spaces of contemporary towns.

When we took the high-speed lift to the eighteenth floor, we wondered what sort of hothouse growth was being fostered there. Outside the office of the author of these thoughts hung a sign:

"Professor Page requests students who have NO REASON to be in this department to keep out for our work is seriously disturbed by UNAUTHORISED VISITORS."

Page, tall and slim, reddish-haired, stood behind his desk, if there was a desk under it all. His secretary had told us of his filing system: "If something does not have dust on it, I know it's important."

The business of the United Kingdom section of the International Solar Energy Society was near the top of the heap. It had been one of the first results of the solar congress in Paris. There, Page had met an old acquaintance, Harold Hay, the creator of the Sky-Therm house heated and cooled by roof ponds. Hay introduced Page to Dr. Mary Archer of the Royal Institution in London who was conducting research on ways to turn sunlight into electricity (the Royal Institution is another of those inexplicable British institutions; one might say it has a more applied bias than the Royal Society, which is the closest the British will come to admitting that they have an academy of science). Page and Archer decided to start a solar energy society in Great Britain. They began with a handful, five or six members in the summer of 1973. "I think we must be one of the fastest growing societies in the country," Page said. In April of the following year, they had two hundred and forty-three members and the roll was getting longer. By 1975, they were trying to curb public enthusiasm by issuing a guide to flat-plate solar energy collectors and solar water heating. They alerted consumers to the sad truth that "very few solar heating systems made in the U.K. will have been systematically tested as the industry in this country is a very new one." They also said: "Do not be misled by extravagant claims such as 'solar heat can provide nearly all your domestic hot water requirements.' This could be true *only* if you drastically altered your way of life and had hot baths only during the summer months and were prepared to face the prospect of storing dirty clothes, cups and saucers, etc., for weeks on end during the winter period while you waited for a few sunny days."

Page is another of the old solar hands who have come back for another whack. As a young man, he became interested in energy in the early 1950s when the Building Research Establishment in Great Britain sent him to India and the West Indies to try to improve tropical housing. "When I saw four coolies pushing a big railway truck [freight car] in India, I concluded that the developing world had its energy problems," Page said. He worked with a small group on them until "cheap oil overtook our work and we found alternate occupations." Page, a mechanical engineer, specialized in climate and architecture, but he must have been less in demand at home than abroad. On a sunny day, he said, the engineering school's skyscraper in Sheffield was hotter than Singapore.

Then he was led into the interaction of energy and architecture, not so much to cut fuel consumption but to get more out of what was being consumed. Page started with a botany professor at Sheffield, David Walker, on ways to use the waste products of power generation in horticulture. There was not only heat but the possibility of getting fertilizer from the gases going up the flue. It was conservation on a macro-scale, not at the level of a more efficient teapot.

The solar congress in Paris confronted Page with the entire range of the field and enabled him to pick and choose what would be appropriate for his own country, less well-endowed than most. He leaned toward the collection and storage of sunshine by plants. "The least elegant approach scientifically," Page wrote, "is to take the packets of energy emitted by the sun at 6000° Kelvin [5727° C.] and convert them into low temperature heat just by using them to start molecules moving faster." Plants do it better; "the most logical process scientifically is to use the high energy photons to construct useful stable chemical structures which can store energy," precisely the way that fossil fuel resources have been put aside. This is all the more applicable to the United Kingdom and any other place at such a latitude without great sun-drenched expanses. Page came back from the Paris congress with a measure of hope:

In the country of cricket, we are all liable to say "Rain stopped play" when it comes to solar energy research. However, when one contemplates that the agricultural growth yield per acre is as high here as anywhere in the world, we should perhaps ponder how the plants do it so well and seek a technology that respects the same logic. Otherwise, once our coal and oil have gone, the nuclear power throw will be our only gamble. It could become a world situation of "Lack of energy stopped play."

Since Sheffield is no place for speculation without action, Page and Walker sought to step up the efficiency of photosynthesis. Walker remarks that farmers in the United Kingdom have no trouble getting twelve tons of potatoes per acre, enough to provide a year's intake of calories for ten persons and there are thirteen million acres of arable land available for the purpose. Such plant material could also be turned into alcohol if liquid fuel were the ultimate goal.

Walker admits that the United Kingdom is not likely to grow sugar cane, always cited as the crop that uses sunshine most efficiently, but he maintains its reputation is somewhat fallacious. Yields are impressive in terms of weight, but less so on the basis of calories produced. He and Page have suggested, instead, that studies be made of blue-green algae which have a high protein content and have long been used by man in their wild state. Easily grown and harvested, the algae may yield as much as fifteen tons per acre. Walker and Page first want to try to grow them under glass with the help of carbon dioxide and waste heat from electric power plants. If this looks feasible, then they propose a pilot plant to produce the algae, adding more carbon dioxide and liquid wastes from animal cities. The end product would be protein, whether directly derived or run through livestock as feed.

Walker has a far-off dream of photosynthetic production of hydrogen in which plants would start with water as the raw material for producing molecular hydrogen. A number of people are engaged in the quest for this philospher's stone. "It is possible," Walker said, "but there is a gulf between what one can do as a party trick and in a practical situation."

His colleague has pulled off what must be the most spectacular party trick in the history of solar science. Page likes teaching: "If you teach people, they'll do their own research. If you don't, you have to do it for them." When the opportunity arose to deliver a discourse at the Royal Institution in London, he seized it with alacrity and modesty. "A Nobel Prize winner from overseas had been scheduled to give a discourse, but he fell ill. So they called on me and gave me only six weeks' notice. In a discourse at the Royal Institution, one must perform twenty experiments and finish dead on the hour."

Page chose solar energy and architecture as his topic and prepared it with the help of Roy Webster, his technician. In their laboratory, they were able to run through several of the experiments for us. It was a good show. At the start, Page did an old experiment with light, showing how the atmosphere scatters the beam from the sun to produce a blue sky. He used a glass tank of water into which he poured flakes of a household detergent. As he poured, the blue turned to the muddy gray of smog.

He achieved his greatest tour de force when he showed his audience several different systems for heating houses, using an electric light as an indoor sun simulator. The first was Hay's newest house devised in conjunction with the School of Architecture at California Polytechnic State University in San Luis Obispo. Like his others, it uses roof tanks for heating and cooling. The trouble was that, when Page and Webster made a model of the Hay house, it did not work in such a small size. "If you put water on the roof, you lost too much heat through convection and evaporation," Page said, "so we used stewed apples . . . apple sauce, you call it. It conducts heat 95 percent more slowly than water and it stays hot. We made it with an apple off the tree of knowledge." The light bulb heated the apple sauce to demonstrate what happened by day. Next, Page covered the model with glass and a bucketload of ice cubes to simulate the effect of the cold clear night sky in California. "That needed a bit of courage with five hundred people watching," he admitted, but the roof turned cold and Page was saved.

Then he illustrated in miniature the startling St. George's School at Wallasey near Liverpool which has been *solely* heated by the sun since its completion in 1961 with hardly any publicity at all. Perhaps no one dared brag about it. The idea of a solar-heated building around Liverpool sounds like something the Beatles might have come up with on Merseyside. The school was designed by the late A. E. Morgan with no back-up heating system at all. It gains heat from two sources: the sun through a big double-glazed south wall, and three hundred and twenty pupils whose natural output of BTUs is hoarded by a ventilation system that can be turned down to control temperatures. It keeps heat because it is built of heavy brick and concrete with polystyrene insulation on the *outside*. In his discourse, Page could show this by substituting aluminum for concrete as the heat storage medium. With the insulation outside, the structure can retain all the heat that comes in through the south window from the sun or, in the demonstration, from a tungsten iodide lamp. With the insulation inside, incoming heat is not put away but immediately shed into the room. At the Wallasey school, the building itself stores enough heat to last five sunless days in winter and also absorbs unwanted heat in summer. Studies made by Liverpool University show an average temperature of 66° F. maintained in the school. Page remarks that the designer used heat-absorbing material everywhere — floors, walls, ceiling — not just in the south wall as Trombe does in France. The result is a very heavy structure that cannot be too economical when construction and fuel costs are compared.

In another experiment, Page modeled the Trombe system with visible smoke to trace the flow of heat from the wall collector along the ceiling, the back wall, and the floor of a room. He resorted to a pink fluorescent liquid to illustrate the way that the collector wall stores heat by day at night. The lights went out and "we had a terrible flood at the Royal Institution with a bright pink liquid creeping up on me in the darkness." It was worth all the trouble because Page likes the principle he was explaining. "I think we can use the Trombe wall on existing houses. Here in Sheffield, I intend to try to install one of his collectors on an old house about to be demolished. It should be possible to have a comfortable bedroom with a pane of glass outside and ducts punched in the wall on top and bottom. The great thing about Trombe's apparatus is that it has no moving parts that need to be maintained. Our approach in building is: if you can stop it moving, stop it.''

Page has lost none of the zeal that sent him out to the tropics thirty years ago, but he now does his missionary work at home. "The United Kingdom could become the prototype country of the future," he said. "We have the most adverse population-resources ratio in the world. We can't do strip-mining here, we're all on top of each other." That was the sense of the message that he left his audience at the Royal Institution:

It is time for my final experiment, which I started at the beginning of my discourse. It is an experiment in statistical climatology. When the stop clock shows the hour has lapsed, as the lecture theatre clock strikes ten, the earth will have rotated 15° on its axis since I came into this lecture theatre. Being indoors, you will have seen nothing of the radiative flux falling on the earth. I can only estimate for you, using the experience of meteorology. There is no time to talk about the result, I can only leave it behind for you to read on the overhead projector. I also leave behind two questions on which to ponder.

Solar energy falling on outside of atmosphere during 1 hour on 16th May	1.687×10^{14} kwh
Approximately 45% is absorbed at earth's surface	
Therefore: solar energy available at earth's surface over last hour	7.92×10^{13} kwh
Estimated world population	3.9 billion
Power available per person over past hour	20,300 kilowatt-hours

Question 1: How much of the 7.92×10^{13} kilowatt-hours have we effectively harnessed?

Question 2: Are we going, as a society, to devote the resources to enable science to discover how to harness a greater proportion before it is too late?

ROGER SCHMIDT

The solar collector mounted by Honeywell at the North View Junior High School in Brooklyn Park, a Minneapolis suburb.

E. R. G. ECKERT

RICHARD JORDAN

WILLIAM BECKMAN

JOHN DUFFIE

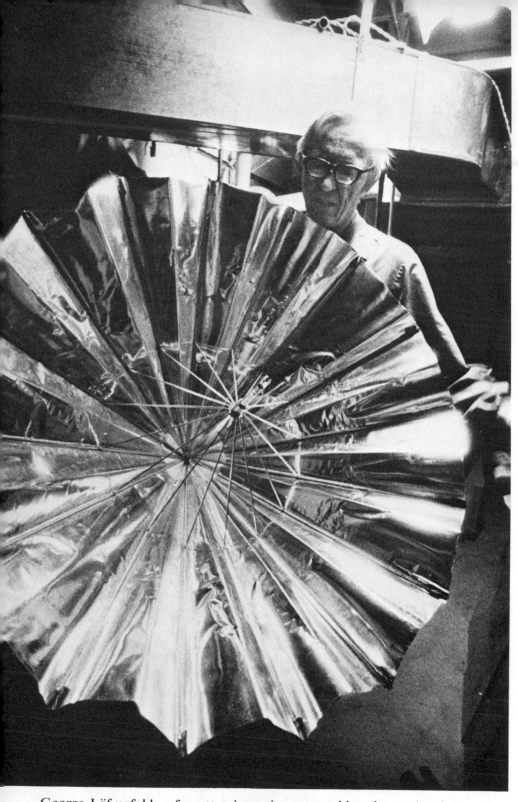

George Löf unfolds a forgotten invention, a portable solar cooker for picnics. It worked, but cost too much to make.

Löf in the staircase of his Denver home next to the two paper stacks filled with six tons of gravel that store solar heat.

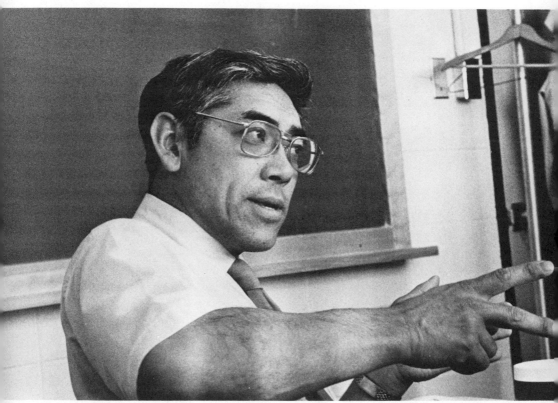

SUSUMU KARAKI

NIKOLAUS LAING

Laing's swimming pool and the clacking plastic balls that keep the heat in.

CLARENCE ZENER

WILLIAM HERONEMUS

PETER GLASER

A. I. Mlavsky of Tyco and the crucible he uses to grow silicon in a continuous ribbon.

Mlavsky shatters a silicon ribbon for the first time.

XV.

~~~~~~~~~~~~~~~~~~

The Sun in a
Warm Climate

~~~~~~~~~~~~~~~~~~

OF THE SUNSHINE that reached the earth during Page's lecture
on a day in May, the kingdom of Saudi Arabia got its full share. It
is as liberally endowed with renewable energy as it is with the
other kind. According to one estimate, the 865,000 square miles of
Saudi Arabia receive in a single year the equivalent of all the oil,
gas, and coal left on earth. If but 10 percent of this heaven-sent
annual income were put to work, it would represent 75 percent of
all the power all of us have used for the past one hundred years.

As usual in solar energy, there is a catch. How can any possible
use be found for the sun in a country that has a quarter of all the
world's petroleum? No one is going to putter here with flat-plate
collectors or mirrors to heat baths or generate a few spavined horsepower. Something really big is needed to compete with oil on an
economic basis.

Dr. M. Ali Kettani, a remarkable man of many roots, thinks that
he has it. Kettani is a professor at the University of Petroleum and
Minerals at Dhahran in Saudi Arabia where he gives shape to his
ideas on energy conversion and conservation in a steady stream of
scientific communications. He started in the footsteps of d'Arsonval

and Claude by regarding the surface of the sea as the cheapest possible way to gather the sun's heat, but he did not follow them very far. As a man of the desert, he looks upon water not as the eternal ocean but as an ephemeral liquid. Over Morocco where he was born and Saudi Arabia where he lives, the sun hoists every loose drop into the sky for windborne delivery into the mountain rivers and lakes of some distant land that uses them for hydroelectric power. In the hot dry lands of the Middle East, there are no mountain lakes, there is only water at sea level. If a way could be found to make water fall below sea level, then the requirements for hydroelectric power would be met. A cycle could begin with the sun evaporating water off the "downstream" side of a dam, returning it to the sky and thence to the sea "upstream." As early as 1933, such a scheme had been proposed for the Qattara Depression in the Libyan Desert of Egypt. The average depth of this dent in the earth is two hundred feet below sea level, its lowest point is four hundred and forty feet down and its area is seventy-five hundred square miles. An engineer proposed to bring seawater into the Qattara Depression from the Mediterranean, using the difference in altitude as the head needed to drive a turbine and produce electric power.

Kettani and a colleague at Dhahran, Professor Lenine M. Gonsalves who was on leave from Southeastern Massachusetts University in North Dartmouth, knew that Saudi Arabia had no such ready-made depressions and thought about making some. One candidate was the Mediterranean itself. They calculated that if a dam were built across the Strait of Gilbraltar, the flow from the Atlantic to the Mediterranean would be cut off. The level of the Mediterranean would drop sixty-four centimeters (twenty-five and a half inches) in just one year. But they did not pursue the thought: "The construction of a dam across the Strait of Gibraltar would be prohibitively expensive, and repercussions on the economies of the Mediterranean countries would be tremendous. All ports of the area would become obsolete and, except for Morocco, Spain, and France, the coastal countries would become landlocked."

Instead, Kettani and Dr. Ronald E. Scott, dean of engineering at Dhahran, took up the possibility of closing off the Red Sea by building a dam across the Strait of Bab El Mandeb. This was not as far-fetched as it may sound. The structure itself would be only seven times the volume of Fort Peck Dam in the United States. In the U.S.S.R., there is a higher dam at Nurek and a longer one at Kiev.

If the Red Sea dam were built, no power would be generated for ten years while the sun started to evaporate water from an enclosed area of some one hundred and seventy thousand square miles. Then, 5,000 megawatts could be produced. Gradually, the level of the sea at the dam would fall and power production would rise. After a hundred years, it would be 40,000 megawatts; after two hundred and forty years, the Red Sea would be down by sixteen hundred and forty feet and 70,000 megawatts would be generated. The two scientists considered the side effects. Behind the dam, the Red Sea would fill with salt as water evaporated and, in thirty-five hundred years, it would be nothing but a big tub of salt. One by-product might be lawsuits from owners of low-lying oceanfront property. Kettani and Scott wondered where the evaporated water would go and calculated that it would raise the level of all the world's oceans by sixteen and a half inches.

Von Arx at Woods Hole has raised another concern. He remarks that the Red Sea is part of the highly seismic East African Rift system. If the weight of all that water were removed, pressure on the sea bottom would be reduced correspondingly and the sea floor might tend to buckle upward. "In that region already under stress, it is conceivable that the reduction of load might have seismic (not to mention political) repercussions which, to say the least, are not very good for dams."

While damming the Red Sea is a spectacular instance of helio-hydroelectric power, Kettani has now put his mind to a project much closer to his home in Saudi Arabia. The western shores of the Arabian Peninsula are washed by the Red Sea; to the east lies the Persian Gulf. A close-up of the gulf shows a long fingerlike bay broadening into a hand. This is Dawhat Salwah; it separates Saudi Arabia from Qatar while the island emirate of Bahrein lies in the middle of the broad upper bay. At his college in Dhahran, Kettani was asked what would happen if a bridge or causeway were built to link Saudi Arabia to Bahrein and Qatar. He thought of the causeway as a dam and wondered what this would imply for the three countries concerned. First, he studied the heliohydroelectric power that could be produced as the sun steamed away the surface of Dawhat Salwah. The idea grew until it became part of a case study of "Energy and Development" made jointly by Kettani's university in Dhahran and MIT to estimate the benefits that could accrue to Saudi Arabia by using natural gas now flared off at wellheads and taking full advantage of solar energy in a "Dawhat Salwah complex."

Kettani wants to tie Saudi Arabia to Bahrein with a dam twelve miles long, then Bahrein to Qatar with another twenty-four miles of dam. Water depths are shallow and no great feat of engineering is needed. As a first payoff, Bahrein would be joined to the mainland and Qatar would have a much shorter highway route to Saudi Arabia. Tolls paid by cars and trucks would maintain the causeway. Once the sun had evaporated forty feet and seven inches of water from Dawhat Salwah, sixteen hundred square miles of new land would appear along the shore. Seven hundred square miles would revert to Bahrein, thereby increasing its area fourfold.

With Dawhat Salwah transformed into a reservoir nearly forty-one feet below the top of the dam at sea level, three hundred million kilowatt-hours per year could be produced. The MIT group thinks it would be done in the way that the French obtain power from the Rance River estuary off the Bay of Mont-Saint-Michel with a tidal plant that also operates with a low head. Turbines designed to work there in salt water could be adapted to Dawhat Salwah, but with much more favorable economics. In the Rance plant, the turbines stand idle while the tide reverses and power production is governed by the tide table and not by consumer demand. Dawhat Salwah, on the contrary, could store the equivalent of weeks of sunshine and release them when needed as any hydroelectric plant does.

Power is not the sole benefit that Kettani sees in the project. Once dammed up, the long narrow bay of Dawhat Salwah will become an evaporation channel eighty-seven miles long and an average of ten miles wide. Water coming over the dam will drive the turbine and then evaporate as it moves toward the far end of the channel, gaining density as the seawater salts are precipitated out. The entire channel would act as a separation factory with specialized salt industries running from one end to the other. Kettani has figured that something like forty-five thousand tons of magnesium a year could be manufactured from magnesium chloride taken from the channel and it would be worth nearly four times the electricity produced. The channel would also yield bromine and table salt but their market is not as promising as that of magnesium.

Finally, 10 percent of the reservoir could be used for aquaculture. The MIT group suggests the culture of brine shrimp, the breeding of fish, and the cultivation of algae. The combination of these uses to serve eight hundred thousand people in the Dawhat Salwah area makes the project viable, but no single one alone would justify damming the bay. To Kettani, another reward is the conservation of oil. He spoke of this on a visit to Paris that gave us a chance to sit and talk in the Tuileries Gardens on a sunny July afternoon. "I identify with Saudi Arabia. I am worried about the future of the country's power and wealth. Oil is false wealth, it is like a chestful of gold. You spend it, then what happens? Real wealth must be based on people and this is why I would like to see Saudi Arabia create a tradition of research in energy. Oil is not the only wealth. The sun is everywhere. It will be a major source of power in twenty or thirty years, but we will have to work for it. Oil prosperity came to Saudi Arabia by chance and history may not repeat itself."

In his writings, Kettani has always insisted that oil must be saved.

Proteins are also a form of solar energy storage. This brings us to a concept that should remain clear to the scientist. Is energy stored in any form good for burning as a cheap fuel? It is obvious that proteins locked in the meat of sheep or fish are better used for food, i.e., fuel for our own life-sustaining natural energy conversion systems, than to be burned as charcoal for the running of a steam engine! Cultivation of high-yield protein-rich algae for the feeding of the hungry and the less hungry millions would, of course, be more meaningful than its use for the production of heat. Isn't it obvious that oil, this rich mixture of hydrocarbons that took millions of years to be formed, could well find a better usage than being burned as a fuel? Should we await an energy crisis to stop wasting such a wealth while such a free source of energy is just waiting to be tapped?

Kettani has the viewpoint of the man in the developing country but he himself is highly developed. He is the author of a textbook on direct energy conversion published in the United States and co-author of a number of works, among them a book on plasma engineering and an article on the control of the water cycle published by *Scientific American*. He spoke to us of the life of M. Ali Kettani. His full name is Mohammed Ali, "like Cassius Clay, but I had the name first." He was born in Fez in 1941, the first of a family of twelve sons and daughters. Though the senior Kettani was a professor of Islamic law at the University of Rabat, he sent his son to a French school because he did not see a future for him in a traditional Islamic education. "French schools were the only decent schools accepted by the French colonial power. It was a culturally castrating experience," said Kettani. "I learned only French history, I memorized only the names of French statesmen. Perhaps because of this experience, I have always been interested in cultural minorities wherever they are. They have become a hobby of mine." The hobby had brought him to Paris on a series of visits to twenty-eight countries to write a book about their Moslem populations. The plight of Moslem workers, Arabs and Turks, in Europe had also caught his attention.

Kettani learned Arabic from a teacher at home, then finished his French education in Syria when his father moved to the University of Damascus. In 1959, he started to study science at L'École Polytechnique of the University of Lausanne in Switzerland, where he was drawn to electrical engineering and power generation. He received his degree there in 1963 and an offer of a job in the Swiss nuclear power industry. He turned it down. "When an engineer has no financial power, he is exploited. I didn't want to be just another engineer."

At this point, Kettani got another stamp in his passport. An offer of a U.S. government scholarship gave him a choice between taking his Ph.D. at Brooklyn Polytechnic Institute and Carnegie Institute of Technology in Pittsburgh. In January of 1965, Kettani landed in New York. "I had gone to French schools; my English was horrible. I went to Brooklyn first and I didn't like it. Coming from Lausanne to Brooklyn is a shock." He left for Carnegie where he was able to work on new sources of energy, particularly in plasma physics. In 1966, with his doctorate in electrical engineering, he moved a few blocks to the University of Pittsburgh where he turned to energy conversion. "The more you study the area, the more you realize that solar energy is the answer. But I was never interested in the little gadgets."

In 1968, he left Pittsburgh for the Middle East. "I wanted to go where the energy was — and to where I belonged." Just as in the United States, his first stop in Saudi Arabia was not the right one for him. He went to Riyadh, the capital. "Thirty years ago, it was a town of twenty or thirty thousand. Now, thanks to oil and the stability of Saudi Arabia, it has half a million people. It's quite an achievement: a modern capital brimming with prosperity in an area that is all desert with only a few oases here and there. The University of Riyadh had started in 1957 and it had a fairly large student body and an international faculty. I met some interesting and bright people in the College of Engineering. But I was full of ambition and energy. I wanted a much more challenging place, a place where my research potential could find an opportunity to be expressed fully. I regard a university as a place of scientific leadership, not just a place to teach. Research and education must go together. The College of Engineering paid a good deal of attention to instruction. It had not yet developed a system to provide full research facilities to those who have the potential for using them. Despite many good experiences there, I was not fully satisfied."

In Riyadh, Kettani met Scott, formerly dean of engineering at

Northwestern, who convinced him to take a look at Dhahran. It was two hundred and seventy miles away and Kettani drove down. That was not much of a trip for Saudi Arabia, where gasoline costs a tenth of what it does in Kettani's native Morocco. "The road between Riyadh and Dhahran is good. The only serious danger is the possibility of hitting a camel if you are not careful when you come over a hill." He liked what he saw at the University of Petroleum and Minerals. "The people there appeared to appreciate my ideas better and they understood the problem of research facilities. The administration seemed much more responsive to the needs of people like myself. If you needed a piece of equipment, you got it; if you needed a technician or a secretary, they would try their best to get you one." The university had been founded as a community college in 1963 by the Ministry of Petroleum of the Saudi government, Kettani recalled. Ahmed Zaki Yamani considered it a top priority project and gave it full support. In 1966, a modern campus was built and the college was on its way to becoming a full-fledged university of technology. Kettani said that enrollment should go up from its present thirteen hundred to three thousand in a few years, and graduate training has begun.

He considers himself a universalist but he likes Dhahran. "Perhaps there is an affinity between the Saudi Arabian and the American temperament. Saudi Arabian society is egalitarian — an egalitarianism characteristic of the Arab of the desert. When you see a Bedouin shake the hand of the king, you know he feels he is the king's equal. The Arab has a strong feeling of justice and equality. He believes no one is better than anyone else. Unfortunately, the Arab is one of the least-known men in the West."

Kettani does not like to see politics impinge on engineering and he thinks this might have happened with the building of the Aswan High Dam. "Perhaps it would have been a better idea to divert part of the Nile to the Qattara Depression and generate power. In Nasser Lake, one-quarter of the Nile discharge is lost to evaporation." In solar energy, he does not limit himself to heliohydroelectric power. He has also considered using the sun to help reach the high temperatures needed to dissociate water into hydrogen and oxygen which could then be recombined in a fuel cell to yield electricity. He shies away from nuclear fission; he thinks it smells of destruction. "All energy that is not controlled — even the energy of the state — must become destructive." While all sources must be put to work, he likes to put his money on the sun. "If we shorten the chain of solar energy, if we can reduce the time needed to accumulate it from millions of years to a few months, then the sun will play a role."

Yet it is not even cast in a walk-on bit in the countries that are short of every other source of power. It is instructive to step out of the megawatt bracket for a moment and look at India. R. L. Datta of the Central Salt and Marine Chemicals Research Institute at Bhavnagar wrote a report for the United States National Academy of Sciences on solar energy utilization in developing countries. India, he says, represents 15 percent of the world's population and 1.5 percent of its energy consumption. And what consumption: "One-third of this energy is in animate form, including human labor; the same amount is accounted for by noncommercial energy in the form of firewood, dung, etc.; the remaining amount is shared by commercial energy sources like coal, oil, hydroelectricity, nuclear energy, etc."

All the barriers against the use of the sun in the industrialized world, whether in terms of cost, maintenance or performance, stand hundreds of times higher here. The first and often the only need for fuel in a subsistence economy in the tropics is to cook food. This requires high temperature, not the room temperature that solar heating devices deliver in cool climates. The solar cookers that have been designed with the best of intentions to save the vanishing forests of India and preserve dung for fertilizer just do not work. There are so many reasons, some technical, others social and economic. In hot climates, people take naps at noon and prefer to eat their main meal in the cool of the evening. Women may or may not want to cook out-of-doors, especially in cultures with a strong sense of hospitality that requires a family to feed anyone passing by. The smell of burning wood whets the appetite perhaps more than the glare of sunshine. All the arguments that we hear against devices concentrating the sun apply to the solar hot plate. Success stories are rare.

One has been related by the Brace Research Institute at McGill University in Quebec. A solar cooker using a collector to boil water was installed by Brace to cook lunches for two hundred and forty students at a school in Haiti run by a religious order. The costs of charcoal there were running to sixty dollars a month, a figure so high that only one-quarter of the pupils could be fed at any single meal. Money was being used to buy fuel instead of food. Brace came up with a cooker in a module big enough to feed eight children at a purchase price of sixty-five dollars. On a clear day, there is boiling water by eight o'clock to cook rice, potatoes, yams, or chicken stew. Even on cloudy days, lunch is usually ready by noon. Yet no one claims that this system could meet the fuel needs of the individual rural Haitian, whose cash flow is such that a cooker is far beyond his means.

In a report that he produced in 1973 for UNESCO on solar energy for developing regions, Harry Tabor of Israel saw a narrow range of places where solar cookers might be used. He limited them to poor communities (but not too poor to buy cookers) that use wood (which they must pay for) and that can be persuaded to eat their big meal at noon and accept solar cookers. The problem, said Tabor, is that no one will go out to sell this sort of household appliance because costs must be kept so low that there is no chance for a manufacturer or his salesmen to make a profit.

It may be hard to cook with solar energy, but it is so easy to get burned again and again. Tabor has been involved in one of the most promising ways of all to harvest the sun in a hot country and yet, so far, the economic yield has been zero. An account of his solar pond exemplifies only too well how promises and disappointments can alternate like sun and shadow. As Tabor has related the story, it began during the First World War when a small boy named Rudolph Bloch went swimming in a lake in Hungary. He noticed that, contrary to what usually happens, the deeper he sank, the warmer the water became. Many years later, as a scientist in Israel in 1948. Bloch thought of his childhood experience and suggested that it might lead to a low-cost solar collector. A body of water like that Hungarian lake would store up heat from the sun if convection could be avoided; that is, if somehow the water warmed at the bottom could be prevented from rising to the surface. He thought this could be done by adding salt to a pond so that water on the bottom would be saltier, denser, so heavy that it would stay below instead of rising when heated. He remembered a report on that lake of his in Hungary which showed that his childhood investigation was right: both the temperature and the density of the water increased with depth down to about six feet.

Ten years later, the National Physical Laboratory of Israel under Tabor decided to try to duplicate the upside-down thermal conditions of the lake. They took an old evaporation pan, sixty-seven hundred square feet in area and three feet deep that had been used by the potash industry on the shores of the Dead Sea and converted it into a solar pond, lining the bottom with black plastic to absorb heat. They added salt and it worked: the water on the bottom went up to over 200° F. Not only that, but they had excellent storage capacity: heat built up in the surrounding earth as well as in the water. Losses on cloudy days were slight. Tabor and his group found a beautiful way to get the heat out without going to the expense of laying a network of pipes under the pond. They realized that the hot water behaved the way that layers of different temperature do in the ocean, where they remain stratified for thousands of miles. One needed only to tap the layer at the sides of the pond to remove "horizontal slices of heat." Calculations showed that a square kilometer (.386 square miles) could produce the equivalent of forty-five thousand tons of fuel oil a year. To convert this heat into work, Tabor's team devised a turbine that would run on the temperatures supplied by the pond or other simple collectors. The turbines sold well; in fact, they developed into a multimillion-dollar business but not one that was sold has ever run on solar energy. "The reason is that solar devices are capital-intensive and simply not available to the people who need them," Tabor has said. "These are the people who do not have capital. If the question of international financing is not taken up, we can forget that program and many similar ones." When Tabor worked out the economics of the solar pond, his employer asked him how much his electricity would cost. At the time in 1966, it came to between one and two cents per kilowatt-hour. The price of electricity generated with cheap oil was only half a cent per kilowatt-hour and that was the end of the solar pond. With the end of cheap oil, Tabor returned to his old project but more as a way to produce fresh water by distillation than as a new source of electric power. He was cautious: "There is some revival of interest in solar energy exploitation in Israel, but it is much slower than in the U.S. because it is really hard to find R. and D. funds in the substantial amounts needed for projects of this kind."

Elsewhere, the solar pond continues to attract interest. The biggest extant would appear to be the one operated by Dr. G. C. Jain at Datta's institute in India. It measures seventy-two by one hundred and eighty feet with a depth of three feet. Like Tabor, Jain wants to use the heat to desalinate brackish water or seawater. The history, too, of the solar pond is repeating itself. Two scientists from the University of Windsor in Canada, P. P. Hudec and P. Sonnenfeld, had an experience similar to Bloch's when they waded in December 1973 into Lago Pueblo, a lagoon about three feet deep on the island of Gran Roque, one of the Venezuelan Antilles lying ninety miles north of Caracas. They found hot brine — that is, seawater denser than normal because of evaporation — on the bottom of the lagoon with a cooler surface layer. The water on the bottom was uncomfortably hot — between 111° and 116° F. — and did not lend itself to investigation by wading researchers. In *Science,* Hudec and Sonnenfeld reported:

We measured the temperature of the water . . . by placing a thermometer at the desired level and then reading the temperature in place by means of a snorkel mask. Temperature readings at depths greater than 75 cm. [29½ inches] were difficult because of (i) the blurring of water caused by thermal mixing, (ii) the extreme discomfort occasioned by the hot brine in the deeper parts and (iii) the buoyancy effect of the dense brine.

They estimate that 90 percent of the energy of the sun can be trapped under such conditions and they report on several other cases that have been found. Perhaps the most unexpected is that of Lake Vanda in Antarctica, where the surface water under twelve feet of ice was just at the freezing point while one hundred and ninety-six feet down, the salty water was at 78° F. Hudec saw the same thing going on in 1958 at an artificial brine reservoir in the state of New York where liquid propane was being stored in salt beds. A scuba-diving geologist found cool water at the surface and then, six feet down near the bottom, dense brine at 122° F., too hot for swimming. The brine had even begun to melt the tarred floor of the reservoir. The two scientists think that the process is worth a second look as a way to collect solar heat cheaply, even in climates as cold as that of New York State and Antarctica. The principle of a pond, but without stratification, has been used by the Lawrence Livermore Laboratory of the Energy Research and Development Administration to supply hot water to the uranium mine and processing plant that the Sohio Petroleum Company operates near Grants, New Mexico. It was described in a report published by *Industrial Research*. The first pond was two hundred feet long, twelve feet wide, and three to four inches deep. It was wrapped in a long flat plastic bag to halt evaporation and placed under a transparent plastic arch to gain a greenhouse effect. The magazine wrote that if this experiment succeeds, six acres of shallow ponds will be laid down to feed five hundred gallons a minute into the processing plant and reduce oil consumption by twenty thousand barrels a year.

Developing countries now seem to adopt the same criteria as petroleum companies: they demand that the sun show a profit. This could be seen at an international course in solar energy that UNESCO organized early in 1974 in southern France for a group of young scientists from Bolivia, Egypt, Indonesia, Mali, and Mexico. They divided their time between classes in theory at the University of Perpignan and practical work in Trombe's laboratory in Odeillo where we had a chance to talk to them. Five Indonesians, all from the Bandung Institute of Technology, formed the largest national delegation and the one farthest from home in every sense. The clear mountain air of the Pyrenees left them with headaches and chapped lips; much of what they were taught involved ways to distill and pump water for dry lands while they were mainly worried about how to get rid of dampness in crops and buildings. Yet to them, Odeillo was a justification of the work which they had been carrying on at home in the isolation that is so often the fate of the solar researcher. Indonesia had a long history in the field. In 1961, Farrington Daniels had lectured there, spreading the word. Aldy Anwar had heard Daniels speak of the possibility of applying solar energy in the tropics. "Before, we only dreamed about solar energy. After hearing Daniels, we understood that it was not a dream. It was practical, it could be done." Anwar and another student started a group with help from the government and Shell Oil. Anwar began to gather data on solar radiation while his colleague developed a solar water heater following a visit by another apostle of the sun, Roger Morse from Australia. Information on sunshine from the meteorological service was not very satisfactory. "The data were unreliable, the instruments had never been calibrated. It was all frustration." Students at Bandung built a small solar still and put together a concentrator to heat a solar furnace. By 1964, the work was slowed by an economic crisis and Anwar left to spend four months in the United States, touching solar bases at MIT, Phoenix, Berkeley, and Gainesville, Florida. Upon his return, the economy had still not recovered and he went into student politics, emerging finally in 1969 when a solar energy project was started at Bandung under the late Professor M. U. Adhiwijogo. "Our philosophy had changed from 1961 when we set out to test and construct solar devices. We had learned the problem was not technology but economic feasibility. We put a priority on measuring solar radiation and using local construction methods that did not need technical skills." Help came from Australia when Morse's organization sent a professor, Robert Dunkel, to teach the basics of heat and mass transfer to Bandung students. All this has left Anwar hopeful. "Is solar energy practical? We use it daily to grow food, to dry crops, to achieve thermal comfort. The problem is to improve the efficiency of natural processes by developing new systems." He saw an immediate application on rice estates where new high-yield varieties produce two crops a year. They would produce three if drying and storage could be carried out during the rainy season. The government has imported oil-burning driers, but Anwar fears they will be hard to fuel and service in outlying regions. One of Anwar's countrymen at Odeillo, Muhammad Taftazani, said that the Indonesian Rice Board was ready to use solar driers if financial aid were forthcoming from the government. Taftazani was all for any use of the sun that would save oil. "Indonesia's main export is oil, we must not consume it ourselves. We are an oil-producing country but we must not take the path of the industrial countries by becoming too dependent on oil and polluting the environment. In our cities, people use air-conditioning and artificial lighting, yet European architectural design is not compatible with Indonesia. A building in a cold climate is a shell against the elements. In the humid tropics, the climate is gentle. People stay outside, they can sleep under a tree."

All the Indonesians at Odeillo saw solar techniques as part of a movement from primitive to intermediate technology in the countryside. One of them, Aryadi Suwono, was working with Dr. Filino Harahap, a Bandung professor, in forcing the trend. Suwono said they aim first at innovations like a foot-operated blower that allows a blacksmith to work with two hands instead of one. In Odeillo, Suwono was interested in the research that Ducarroir had done on refrigerators running off a parabolic collector. He wanted to achieve a small model that could produce one hundred pounds of ice a day so that village fishermen could stop a catch from decaying before they could transport it to market. He weighed the economics of ice-making and concluded it would be best to build a refrigerator to run on the sun by day and on wood at night. It would have to be tilted every half hour to follow the sun; then, every two hours, a valve would have to be opened and closed. "In my country, labor is cheap. We plan to make a manually operated refrigerator. In our development of technology, we want to make processes more labor-intensive."

Nevertheless, this has its limits. Aman Mostavan joined the course to study the use of silicon solar cells, first as components in instruments to measure solar radiation throughout the country, then as power sources. "Silicon cells are expensive, but people don't calculate how much it costs to bring a jerrican of oil to the mountains." This has been worrying the builders of a trans-Sumatran telecommunication system who need power for their repeaters. They chose diesel generators but they still had not decided how to fuel them. Mostavan said they were wondering whether to build a road or a helicopter. "With a helicopter, they would have to train pilots and maintenance personnel — and this is not intermediate technology. The first cost of silicon cells is high but, perhaps, they would pay off over five or ten years." He had hopes of studying photovoltaics at the University of Marseilles after getting over his traumatic introduction to the French language.

The two Mexicans at Odeillo, Everardo Hernandez, a physicist, and Carlos Rondero, a mathematics professor, were much less involved at this level of decison-making because solar energy had not reached a point of applicability at home. It was in the air, however. The Mexican government was interested in buying solar pumps from France, solar stills were being mooted to supply drinking water to islands and, most significant of all, a solar energy program was about to start at their National Autonomous University of Mexico in Mexico City. In Egypt, where so much had been done sixty years before by Shuman, a solar energy laboratory has been maintained in Cairo, where its main interests are distillation, refrigeration, and the drying of crops like apricots and dates, all applications of concern to agriculture. Dr. Sadek Soliman, the senior member of the course, used his stay to absorb what the Odeillo library had to offer. He spoke of getting more science into his own work on distillation which had already led to a small evaporator producing two gallons of water per day. The output was not high, but the apparatus could be built in multiple units if production were to move out of the laboratory stage.

Of all the countries represented at Odeillo, Mali seemed to have gone farthest in making the sun pay. The two Malians in the course, Amadou Doumbia and Mamadou Touré, were attached to the National Solar Energy Laboratory in Bamako that was started in 1965 by Dr. Abdou Moumouni from neighboring Niger and is now headed by Cheikna Traoré. Moumouni is Africa's major figure in this field. In 1969, he became director of the Office of Solar Energy at Niamey, the capital of Niger, and he has worked on using the sun to meet the all too evident needs of the Sahel. By his estimates, the benefits are considerable. The introduction of solar cookers, even if used on a half-time basis, would save twenty-five to thirty million tons a year of wood in these countries. In some areas, he once told me, the nearest firewood sources are twenty miles away. I asked what families did for firewood and he said: "They walk twenty miles out and twenty miles back." Solar water heaters and stills would save an equivalent amount of wood along with two billion kilowatt-hours that could be employed more usefully. Both cookers and water heaters have been developed at Niamey and a solar engine is being tested there. Moumouni might tend to agree with Glaser that "simple" devices made to measure for developing countries will not fill the bill. On the contrary, Moumouni protests that these countries can no longer be kept in a "technological backwater."

In Mali, where Moumouni had started things going, the domestic market for solar appliances has been surveyed to see where the biggest payoffs are to be found. The laboratory learned that Bamako was buying imported distilled water for use by hospitals and garages (to fill batteries) at a price of more than twenty-five cents a liter (there are 3.8 liters to a gallon). In Gao, a country town five hundred and fifty miles to the northeast, distilled water is worth two dollars a liter after it has been brought in by plane or truck.

The laboratory estimated that it could be done with the sun for no more than five or ten cents a liter. Doumbia has helped design a small still to produce fifteen gallons of water per day with one hundred and ten square feet of collector. The laboratory believes that there is an immediate market for one hundred thousand square feet of solar stills in the country. Most of them would be used to purify brackish water that comes up from wells in areas that must rely on army trucks to bring in their drinking water. "There is plenty of sun in our country," said Doumbia. "In May, it crushes you. I decided to work in solar energy so that the sun can do people some good instead of crushing them."

It is not profit that motivates him. "We are trying to get some business so that our laboratory can buy what it needs to work." Touré explained that the laboratory had to break out of a vicious circle in which lack of interest led to lack of money. As a start, it planned to build about thirty-three hundred square feet of stills and distribute them in small units throughout the dry north of Mali. In 1971, the laboratory caught the public eye by producing fifty-gallon solar water heaters and selling them to hotels, public buildings, and private customers. Radio Mali and the Bamako paper, *L'Essor,* were enlisted in the campaign and production began only when firm orders had been received for one hundred heaters. They were built in the laboratory's shop where the staff worked day and night. "When we were short of time, the director used to come down and we would tell him what to do," Touré said. "If we had just stuck to basic or applied research, we would still be where we were ten years ago." The first heaters were shown at a fair in the Kayes district, where the solar scientists were the guests of the local governor. "From then on, the heaters snowballed — or perhaps, one should say they spread like wildfire. We're now at a turning point. Before we made the heaters, we were housed by the army and we didn't even have our own offices."

The laboratory has looked hard for areas where the sun can do its bit. Touré visited a brickworks at Magnabougou where he learned that clay had to be dried for two or three weeks before it could be fired. "We told them that, with solar energy, they could cut their drying time." The sun need not do all the work; it can be used as a booster to save fuel. Certain industries in Bamako burn wood to raise steam, but the laboratory thinks that the sun could be used to preheat their water from 70° to 150° F. to save wood. At lower temperatures, solar driers have been used to preserve mangoes and onions. Touré said they should be able to do the same for catches of lake fishermen. Under the traditional method, fish are split open and set out to dry in the sun, accumulating dust and losing value. To preserve vaccines or milk, Touré wants to get low temperatures with the technique Trombe has developed to use outgoing radiation under a clear sky at night.

Touré kept busy in Odeillo. "We are here trying to collect techniques and information not only for ourselves but for those who were not lucky enough to be able to come here. For me, it has been a pause, a chance to let me look at what I have been doing and where I want to go."

There is plenty of sun to heat and distill water in Mali; it is the water that is not around when and where it is needed. Mali is one of six West African countries — the others are Chad, Mauritania, Niger, Senegal, and Upper Volta — lying in the Sahel, the southern hem of the Sahara, the largest and probably the hungriest stretch of aridity we know. Yet water is there, often not more than a few yards down. In the megawatts that pour onto the Sahel in lieu of rain, there is the energy needed to raise it. This is the way Farrington Daniels reasoned when he sought ways to let the sun confer more benefits upon humanity. The same idea occurred to Henri Masson who was dean of the University of Dakar in Senegal until his death in 1972. Twelve years before, a young Frenchman, Jean-Pierre Girardier, was teaching physics at the university during his period of military service. "I didn't have much to do," Girardier remembers, "just a few hours a week of teaching. I asked Masson if he had a job for me. He was in his office. He had his feet on his desk and he was looking out the window. He told me: 'There's the sun out there, it's a source of heat. There's water under the ground, that's a source of cold. According to the second law of thermodynamics, you should be able to run a heat engine on that.' Masson was an extraordinary man; he always went straight to applications. He inculcated me; after I left Senegal, we wrote each other almost every week for ten years."

It was a happy case of cooperation between industry and a university even if the two parties involved were not exactly neighbors. Girardier returned home after his service in Africa and soon took over the small factory that his family owns at Amilly, a large village near the town of Montargis, seventy-five miles southeast of Paris. It is called Établissements Pierre Mengin and a photograph of its late founder, as bushily and blackly bearded as a Smith Brother, hangs eternally inside the entrance to the offices. The company makes sewage and irrigation pumps, doing pretty well at it with one hundred and twenty workers and an annual turnover of four and a half million dollars, 10 percent represented by sales to the United States. Girardier, the president and general manager, has a delightful wife, six fine children, and a comfortable home in a converted farmhouse ten minutes from the plant. The trouble is that sewage pumps do not fascinate him. Nor does making more money. What he wants to do is bring water to the people of the Sahel whom he saw for the first time when he was in Senegal. He is a religious man and this, too, is a motive.

He and Masson got their first working version of a solar engine into operation in 1963. It was hardly more than a demonstration model. With one hundred and thirty square feet of collector, it pumped three hundred gallons of water an hour. It was set up at the Institute of Meteorological Physics in Dakar where, two years later, a much bigger engine with a collector seven times the size went into operation and ran nonstop for a year. It could pump fifteen hundred gallons an hour from a depth of eighty feet and yet it went almost unnoticed, something of a dead branch on the evolutionary tree of technology. It was expensive and inefficient; it could convert no more than 1 percent of the energy falling onto the collector into actual working horsepower. Such criticisms have never bothered Girardier. The regions whose needs he had in mind were not worried by efficiency. They had all the sunlight they wanted, but they had no source of power.

It took faith to believe in the solar pump. The first time I saw it by chance in 1972, I could hardly credit my eyes. I was visiting a friend in Amilly who ran a lathe in Girardier's plant. He told me about a solar machine that his boss had devised and offered to show it to me. It was a Sunday morning when we cycled out to the factory, but Girardier was there, a thin, almost bony man, perched ten feet off the ground on scaffolding that surrounded what looked to be . . . it was hard to say, I had never seen anything quite like it. It might have been an engine for an early steamboat or a new way to make applejack from cider, a local cottage industry. Girardier paid no attention to us until we asked for a look at his machine. He invited us up and provided an explanation. Like most people wrapped up in what they are doing, he could not imagine anyone wrapped up in anything else and I hardly understood a word of what he said. Later, at my leisure, his lecture sank in. The principle of his solar engine is not very complicated. It runs in somewhat the same manner as the Claude process at a temperature far below the boiling point of water. In Africa, Girardier used flat solar collectors to heat water to $170°$ F. At his factory, he achieved the same result with an oil burner when he was making tests. The hot water was brought into a boiler, a beautiful piece of craftsmanship with more than four hundred small copper tubes, much more closely packed than in an ordinary steam boiler. Here, the water transferred its heat to liquid butane before returning to the collector to acquire more heat. Then the butane vaporized to provide the "steam" needed to drive the engine and pump water. The water itself was more than an end product; it had a job of its own to perform. Coming from below ground, it cooled the butane gas, condensing it back into a liquid so that the cycle could start all over again. Pumping for irrigation or a drinking supply is an ideal function for a solar engine since water not only runs the condenser but can be stored in a reservoir to free the user from variations in the solar flux. This was the background that Girardier gave us that Sunday morning in the loneliness of his empty factory.

The next time I saw him, he had attracted more of a crowd. At the Paris solar energy congress in 1973, he had set up a full-sized engine. Every correspondent in Paris must have climbed up onto his machine during the congress, and he had to struggle to explain it in English as well as French. Three more solar pumps were in operation in Africa, one at Ouagadougou in Upper Volta, another under the aegis of Moumouni's Office of Solar Energy in Niger, and the biggest of all at Chinguetti, an oasis in Mauritania.

At Chinguetti, Girardier had teamed up with two architects, Mr. and Mrs. Georges Alexandroff, to put into three dimensions an idea he had long cherished: a solar pump not as an alien machine but integrated into a community. The Alexandroffs had designed a collector surface consisting of what they called "canaletas," channels down which water flows much as on a corrugated roof. In the collector, water is warmed in the channels and gives up its heat to the butane before returning to the top of the roof. The collector is big but cheap, making it a good roofing material. At Chinguetti, the Alexandroffs designed it to shelter a school. At one end of the school stands a square tower that might have been a desert fort but, in this case, acts as the home of a solar pump bringing drinking water to the two thousand inhabitants of Chinguetti. In 1973, it also offered a cooling glass to visitors who had come to Mauritania to observe a solar eclipse under the clear desert sky. This led to a modicum of renown for Girardier and his pump which had been built with financial aid from the Iron Ore Company of Mauritania.

In 1974, the great drought brought the Sahel to the depths of hopelessness. A French magazine, *La Vie Catholique,* started a campaign to raise $50,000 to install a solar pump and build a dispensary at Dioïla, a small town in Mali. It is easy to stand off at a distance and snipe at such undertakings. They reek of do-gooding, their effect on an area of two and a half million square miles with a population of twenty-two million is on the same order of magnitude as a spit in the ocean. Yet they are heartening, perhaps not because of what they offer but because they give people a chance to do something, not just sit in front of a TV set and deplore. The call for a solar pump was heard around France. Instead of collecting fifty thousand dollars, the magazine was swamped with five hundred thousand dollars. To raise money, kids in a Paris suburb washed cars and ran a flea market; in an Alsatian village, a club sold potted plants and airplane models; in the Midi, teenagers picked up three tons of wastepaper in their spare time. Dioïla got its pump and solar dispensary in March of 1975. Then *La Vie Catholique* turned from Mali to Upper Volta and decided to put in two more pumps there as a start on using the surplus it had collected.

One of the pumps was intended for the small town of Djibo while the other went into a new building on the grounds of a hospital at Koupela.

The hospital is one of a multitude of examples of what happens in the Sahel to aid rendered with more good intentions than wisdom. It was the kind of a place where a patient could come in with a broken leg and leave with an intestinal infection. To supply one hundred patients, it had a well that hardly yielded one hundred gallons a day. The builders of the hospital had installed sinks and toilets, but no one had budgeted for water. The pipes from the sinks were connected to thin air and the flush toilets were used as storerooms. A diesel pump was out of the question; there was no money to pay for oil or maintenance. The hospital could not afford gasoline for its ambulance after the twentieth of each month.

This is the sort of thing that drives Girardier into a state of rage. "You can't imagine the scandal of the Sahel. Wells are dug, reservoirs are built, but there is no money for recurring costs. I know a big ranch opened with foreign aid in the north of Upper Volta. It's a dead city. In the rainy season, trucks can't bring in diesel oil to run the pumps. A hand pump was introduced in the area. It was invented by people who had never been there. A hand pump usually runs a few hours a day. In Africa, it runs twenty-four hours a day. At the end of a week, it won't work. It's finished. If someone needs a piece of iron, he takes the handle. You can try anything you please in the Sahel. If it breaks down, that's all. If you cannot buy oil for a diesel motor, the pump stops and people take the parts. Of six thousand diesel motors in one African country in 1974, four thousand had broken down and there were no parts to fix them. My machines do not break down."

His pump has a watchman who operates it. When the sun comes out, the needle on a pressure gauge rises until it reaches a mark. The watchman opens a valve to start the pump. That is all that he has to do, but his is a respected function and the pump is respected. There is no greasing or maintenance to be done on the solar pump. It is driven by a big two-cylinder engine with bronze pistons, slow-turning and long-lived. It can do as many as one thousand revolutions per minute and as few as one hundred as the sun sets on the collector. It is no throwback to nineteenth-century technology; on the contrary, it is at the forefront of research on materials and lubrication.

I once saw Girardier operate an engine under test in his plant, which ran on compressed air. It started with a hearty chug, then ticked over with no sound but the clicking of valves. "It works without oil," said Girardier. "We know how to do this, but we can't tell you." He did say that it took the firm three years to develop a lubrication system. "We had breakdowns at first, but we worked ten years on these engines without any reporters around. At Chinguetti, the pump ran a year without maintenance. We had only two incidents during that time, neither involving the solar engine."

In 1973, Girardier had to sell an engine rated at one kilowatt for fifty thousand dollars. Two years later, he had halved the price and he thought he could cut it to fifteen thousand dollars once he got into a production run. Even at the 1973 price, the pump seemed a good investment to a place like Chinguetti, four hundred miles away from the nearest supply depot from which diesel oil could be shipped by truck over desert tracks. Girardier said that the cost of maintaining a diesel under such conditions can run to four or five thousand dollars a year. In the Sahel, his solar pump is competitive. "I want to bring water to African villages, not to Montargis."

The pump can lift enough water to take care of five hundred to a thousand head of cattle but no more. What Girardier is trying to avoid was described by Claire Sterling in the *Atlantic:*

By now, the Sahel is crisscrossed with thousands of deep boreholes, drilled at a cost of between $20,000 and $200,000 apiece. Whether by pump or natural artesian pressure, water gushes up in such prodigal abundance that ten thousand head of cattle at a time can drink their fill. The trouble is that wherever the Sahel has suddenly produced more than enough for the cattle to drink, they have ended up with nothing to eat. Few sights were more appalling at the height of the drought than the thousands upon thousands of dead and dying cows clustered around Sahelian boreholes. Indescribably emaciated, the dying would stagger away from the water with bloated bellies and struggle to fight free of the churned mud at the water's edge until they keeled over. As far as the horizon and beyond, the earth was as bare and bleak as a bad dream. Drought alone didn't do that: they did. What 20 million or more cows, sheep, goats, donkeys and camels have mostly died of since this grim drought set in is hunger, not thirst."

Girardier wants a different approach taken. "During the rainy season, the cattle in the Sahel are in good shape when they are brought to Niamey or Ouagadougou. But during the dry season, they can't make it. The wells along the cattle trails are too dry. What the herdsmen need are small wells along the trails and then irrigation projects in river valleys where the water table is near the surface so that forage can be grown. Instead, millet for feed has been brought in by airplane. You have to see this to believe it. Experts are sent in, too, and this is bad. You don't want an expert for this job, you want to see the chief of the village. He's the smartest man in the village. He knows that you will change people's lives if you bring equipment in. That's why I want a school to go with a pump. If you give a kid free time without a chance to go to school, you're not doing him a favor. He's better off fetching water. I discuss things with the governments, then I send people in to see the local authorities and the chief before I put in a pump. If I don't, the chief will sacrifice a black goat and the well will be cursed." Girardier has a fierce faith in the intelligence of country people. A French TV reporter, interviewing him at the controls of his solar engine, once told him: "Tell me how this works and make it simple so my peasants in Auvergne can understand it." Girardier replied: "In the first place, they're not your peasants. In the second place, they're a lot smarter than you are. In the third place, you'd better take pictures of Brigitte Bardot's backside. Get off this machine and stay off."

His approach to public relations notwithstanding, Girardier is successful. By June of 1975, the plant at Amilly had delivered its thirtieth solar engine with ten pumping stations already in operation and new ones going up at the rate of one a month. Two million dollars worth of engines had been sold and the product had been redesigned down to a quarter of the size of the monster I had first seen in 1972 so that it could be shipped by air. The firm was no longer operating solely in Africa. Two pumps were at work in Mexico, at Caborca in the state of Sonora and Ceballos in Durango, while a third was serving as a portable demonstration model. The pump was the hero of a government-produced documentary film, *Mi Amigo el Sol*. Orders for more pumps had come in from Brazil, Cameroon, the Cape Verde Islands, Chad, Iran, Niger, and Upper Volta.

In Mexico, the solar engine was approaching the heroic dimensions of the projects that were attempted before the First World War. The Ministry of Public Health had ordered a 25-kilowatt solar turbine to supply drinking water to a new town being built at San Luis de la Paz one hundred and twenty miles north of Mexico City. To operate the turbine, 18,300 square feet of collectors were being installed, using the same "canaleta" system introduced in Chinguetti. The collectors were to spread out from a central pump and reservoir. Under the canaletas, the people of San Luis de la Paz would be able to build their homes as they always had. Instead of sheltering a single school or dispensary, the solar pump had become the nucleus for a community. With a twenty-five-kilowatt turbine spinning at 7,200 r.p.m., it was almost twenty times as powerful as the pumps that Girardier had installed in Africa.

The solar turbine was derived from a compressor that had been built to run ten years without a break in a highly radioactive zone at the Pierrelatte factory where the French separate U-235 from uranium. By borrowing from the nuclear industry, Girardier brought the French Atomic Energy Commission into solar energy. His corporate structure underwent some changes. Initially, the pumps had been made by something called Sofretes, an acronym composed of the initials in French of the "French Company for Thermal and

Solar Energy Studies" with, as its trademark, a toothed sun driving a gear wheel. Fifty-one percent of Sofretes had been owned by Établissements Pierre Mengin with the rest belonging to French government bodies. It was this latter block of stock that changed hands when it was bought up by the Atomic Energy Commission, the French Petroleum Company, and Gazocean, a firm that transports liquid natural gas. Then Renault bought 34 percent of Mengin, not to develop a solar GT to run at Le Mans but to use its worldwide sales network to find a market for solar pumps.

This was all to the relief of Girardier who disclaims any trace of a head for business. He could now concentrate on important matters. At Amilly, he was setting up a test facility, an electrically heated 1.5 megawatt boiler that could simulate down to the nearest degree solar collectors in any size from 10 square feet to 50,000 square feet, enough to run a 100-kilowatt turbine. On the south wall of the test facility, he intended to put up 3,700 square feet of collector to provide heat and hot water for his plant. The first solar engine that he put to the test was the turbine for San Luis de la Paz, producing what he described as the world's first kilowatt-hour generated at a temperature of less than 160° F. Instead of driving a pump directly, the turbine supplied power to run two electric motors. It represented a step along the path that Girardier has taken towards solar power. If used to pump water, the solar turbine is more expensive to buy and install than a diesel, but cheaper to run in Africa. Girardier has calculated that a solar turbine can deliver electricity under conditions in the Sahel at a cost of between twelve and eighteen cents a kilowatt-hour at the power station. Power produced by diesels in remote areas in Senegal now costs twenty cents a kilowatt-hour at the plant. At present, a solar plant could feed electricity to small industries for six or seven hours a day while running refrigerators that could store cold overnight. "In the near future," Girardier remarked, "we will be able to supply electricity twenty-four hours a day." He refused to speculate how; he has a refreshing habit of not talking about something until he has done it. With those turbines running forever on air-cushion bearings, all sorts of possibilities spring up. Girardier once said that they run at no more than the temperature of hot desert sand.

A process has been started. The small pump at Chinguetti led to a school; at Koupela, the solar engine was incorporated into a new building for a hospital. Yet these were hardly more than first-aid operations. A twenty-five-kilowatt turbine is something else again. If hundreds of acres can be irrigated, then communities develop with needs that go beyond water. They want refrigeration and electricity. It is almost as if Girardier's solar machines have grown organically to keep up with such new communities. Mexico has been thinking about one thousand small pumps for remote villages; a French project known as New Energy for the Sahel talks in terms of three or four turbines of fifty kilowatts each. In Senegal, there are plans for a pumping station next to the Senegal River that would irrigate five hundred acres with engines running off 32,000 square feet of collectors. Girardier does not look for business but it seems to come his way. He needed orders to pay for his research and now he has them. There is not so much talk about low thermo-dynamic efficiency. "For ten years, people had been saying that my machines don't work, but they do work." Not only do they work, but they represent more mechanical horsepower than anyone else has yet to produce with the sun.

XVI.

~~~~~~~~~~

Where the Sun
Is Red

~~~~~~~~~~

WE BROUGHT BACK a helter-skelter of impressions from our solar
cruise across continental Europe by train from Paris to Moscow,
then by air to Ashkhabad on the Iranian border, where the apricot
trees bloomed while snow still clung to the pines and birches in the
forests of the north. The impressions amount to much more than a
few hours of meetings and interviews in the Soviet Union. The train
journey itself might have been in some Wellsian machine, eighteen
hundred miles in forty-four hours at an easy overland jog conserv-
ing energy in all forms, psychic and nervous included. A mile out of
Gare du Nord in Paris, we were sitting in our cabin in a Soviet
sleeping car, drinking Russian tea from the samovar down the corridor while the northern suburbs of Paris slid past the window.
Through that magic window, full color, 3-D, never a rerun, we
watched the many ways of Europe, west and east: the green
hedged-in garden of agricultural France; the almost-English aspect
of the aging brick towns of Belgium; then West Germany at twilight, rebuilt and cleaned up five minutes before just for us, the
Rhine and the spire of Cologne cathedral; the emptiness of stations
glowing orange at night and peeked at through parted compartment

curtains; Berlin in the cold light of morning, Berlin Zoo, Berlin Friedrichstrasse, our West Express a subway local; through the outskirts of Berlin behind a steam engine, the forgotten smell of yellow coal smoke streaming by the cars, the cantering motion of the train accelerating behind its iron horse; Frankfurt-on-the-Oder and Poland, a day in a bygone day, horses hitched in teams to plows breaking the black soil, horses in the shafts of long boatlike wagons stepping out at a trot over the dirt roads, freights sidetracked for us behind simmering coalburners, a roomy diner all the way to Warsaw, full of beer and good cheer. Then the countryside waned and quieted at the end of our second afternoon on board; we crossed into the U.S.S.R. at Brest-Litovsk. The sensation was almost that of a voyage by ship: across the border, we moved our watches ahead two hours, the sleeping cars were jockeyed into a large shed that could have been the hold of a ferry. There, they were jacked up and their trucks changed so that they could continue on Russian broad-gauge rails. The movement was imperceptible, it did not stop us from sleeping. Everything looked broad-gauge in Russia the next morning: marshes and forests that ran with the train forever the way they do in the United States; land, lots of land, no one fenced it in; villages by the trackside, small wooden houses, women and children in boots, walking through the empty lanes; trucks and rough-and-ready cars waiting at the grade crossings, even a horse-drawn sled in the snow at one crossing, but only one that day; then the thickening of suburbs and Belorussky Station in Moscow.

Intourist always get their man; they got us when we stepped off the sleeper, our home for two days. We were sorry to leave it with our two conductors and their bottomless store of tea and melba toast, their coal furnace that kept us warm and the wash water hot through six countries. In Moscow, which I had not seen for nine years, there was the same eerie feeling of inner-city emptiness as Intourist's Zis limousine, no doubt a veteran of ministerial service in its younger days, whisked us to the Hotel Metropole, a stone's throw (but don't try it) from the Kremlin where it has sat for lo these seventy years, riding out history with hardly more than a few frayed edges on its carpeting to show for it all. I do not know what the nine years had done to me, but Moscow looked well. The grayness was gone from the crowds; the lines were short in the shops and the shelves well-stocked with necessities, if not the superfluous. As they had nine years before, Muscovites wore their hearts on their sleeves and gave them away. In no time at all, Madeleine had a collection of Pioneer buttons, gifts of children with whom she had struck up conversations in the street, I don't know in what language.

All this went on at the pace of Moscow. I will not call it a pre-automobile pace because I do not know what is supposed to come after it; perhaps it was the postautomobile pace that we will all know when the last drops of oil are being squeezed from the wells and the tar sands and the shales. Beneath the windows of our third-floor room at the Metropole the trolleybuses moved off towards the Bolshoi Theater and Marx Prospekt. They were not silent but they permitted sleep without earplugs; even the taxis run with tamed motors that sounded more like New York than Paris or Rome. It was an exercise in futurology, an analog model of a modern city in an advanced industrial society without a debauchery of oil and gasoline. There was far less freedom for me, the hurried traveler, to move about on private wheels to take care of my affairs; there was more freedom for stay-at-homes to move about in public transport, buses and subways, or on foot or even under water in a gigantic outdoor heated pool where steam boils off swimmers and lifeguards patrol in polar gear, parkas and boots.

There was a cushion between this world and the one we had been witnessing. The price of gasoline was 10 kopecks a liter, about 55 cents a gallon, the cheapest I had seen in Europe. Electricity was 4 kopecks a kilowatt-hour, say 5.5 cents; it had been at that price for more years than anyone could remember. A ruble is worth $1.30 at the bank in Moscow but these figures are hard to convert into some sort of understandable terms for the American or the West European consumer; the purchasing power of the kopeck and the ruble depends on what one is purchasing. There is still a 5-kopeck fare, one-twentieth of a ruble, on the Moscow subway, but a Fiat-sized car costs in the neighborhood of 6,000 rubles. Rent for a two-room apartment runs around 10 rubles a month, making it a lot cheaper to stay put than to go for a drive. There are no queues of cars to be seen at service stations, there are almost no service stations to be seen in a random tour of Moscow. A selection of literature in an Intourist office yields a magazine interview with the minister of energy and electrification who speaks of tapping the hydropower of Siberia or mining untouched deposits of lignite to fire up fifty thousand megawatts' worth of new thermal plants. Breakfast at the hotel turns into a chance meeting with three executives from a New York chemical concern, closing a contract for an ammonia plant on the Volga to run on natural gas. They see good prospects for the future, the resources are there.

One could almost forgive the Soviets if they lectured to their less-fortunate contemporaries as the ant gloated to the grasshopper. They do not; they realize the good times will not go on forever. In 1972, a member of the U.S.S.R. Academy of Sciences, Dr. N. N. Semenov, was forecasting that fossil fuels would last only eighty years at the prevalent rate of growth in their use. Even if new sources were found, the accumulation of carbon dioxide in the atmosphere would limit their consumption. Fusion power would bump up against that other ceiling, the capacity of the earth to absorb heat. Semenov calculated that if man-made power amounted to 5 percent of the earth's intake of solar energy, then the planet's temperature would rise $3.5°$ C. with perhaps irreversible effects on climate and the polar ice caps. A seven-hundred-fold increase in power levels would do the trick and, said Semenov, this is not too far off in the perspective of history.

Still, the Soviet Union can enjoy a breathing spell before it, too, must cast about for new sources of power and heat. This explains why there are no big solar energy projects afoot even though Soviet scientists have been involved in the field for decades. Dr. B. Z. Tarnizhevsky, deputy chairman of the Council on New Sources of Energy set up by the State Committee for Science and Technology of the Council of Ministers of the U.S.S.R., put it tactfully in an interview: "Perhaps the shortage of oil does not stimulate us as much as it does the United States." He and Dr. N. S. Lidorenko, an associate member of the U.S.S.R. Academy of Sciences, have made a study of present and future uses of solar energy in their country. Geography is the principal determinant; the U.S.S.R. has people and sunshine but the sun does not shine where the people are. The possibilities of solar heat in Leningrad are as promising as on Cape Farewell, the southern tip of Greenland; both are at the same latitude. Moscow is not much better off; its parallel runs right through the heart of Labrador. In these regions, no one talks of rooftop solar collection. What the Soviets do talk about is long-distance transmission of power, thermal and hydroelectric, in great quantities, and this is reflected in the thinking of Lidorenko and Tarnizhevsky. Until a feasible way is found to store energy, they suggest that the on-off nature of solar power can be overcome if solar plants are incorporated into a pool that also includes thermal and hydro power. They developed the idea: "An ideal worldwide solution for the use of solar energy would be the construction of solar plants over the planet's equatorial belt within the latitude of a favorable sunny climate, leading to the establishment of a global power pool." It would solve the problem of the Soviet Union where there is plenty of sun but only over the lonely deserts of Central Asia. Tarnizhevsky and another researcher, A. N. Smirnova, have gone into the chances of getting more overall reliability out of a purely solar power system by spreading individual plants over the map. With the help of a computer, they set up a hypothetical group of plants in a wide swath twenty-five hundred miles long on the southern Soviet border. They learned that 75 percent of the total July sunshine in the area would be available for power generation 99 percent of the time, but the best individual site showed only 39 percent reliability.

Imagination appears again in a scheme put forth by Professor Yakov Shefter to lift wind generators by helium balloon up into the jet stream where they could work with never a lull, sending power down the cables that tether them. Lidorenko observed that "the idea presents many difficulties, but perhaps not as many as Glaser's proposal." On the ground, Shefter is examining the feasibility of wind generators in sizes of one, four, ten and fifty kilo-watts as a power supply for agricultural communities.

While they do not see any of this coming to pass overnight, Lidorenko and Tarnizhevsky are quite hopeful about prospects for solar power. They have gone on record as stating that the first large thermal plants should start operating between 1980 and 1985, while photovoltaic power should be coming onto line between 1985 and 1990. Tarnizhevsky said they base their optimism on research and development in the U.S.S.R. and the United States. Under the agreement on scientific and technical cooperation between the two countries, they are to join forces in solar energy. Lidorenko ventured his opinion that the U.S.S.R. could contribute through its plans to manufacture high-performance solar collectors. He also said that work on solar cells is being speeded up in the U.S.S.R., not only for national but international use. He spoke of gallium arsenide solar cells that have achieved efficiencies as high as 20 percent. Even at lower and more customary efficiencies, photovoltaic power looks appealing. The two Soviet scientists have come up with a new answer to the old objection against the amount of land that would have to be set aside for solar plants. They calculate that a photovoltaic station in the deserts of central Asia could turn out the same number of kilowatt hours every year that the Lenin hydro-electric plant on the Volga now produces, but the solar cells would cover only one-seventieth of the area of the basin serving the Lenin plant. They remark that hydroelectric basins in flat country take up land previously used in the economy while the desert land is not good for anything else.

In smaller, much smaller sizes, solar cells are being used as a power source for navigation lights and markers not only in off-shore waters but on remote reservoirs and rivers (there is no lack of remoteness in the Soviet Union). The number of customers for these cells is growing rapidly and demand is much bigger than the supply, we were told by German V. Rosanov, chief of the Batteries and Accumulators Department at the Ministry of the Electrical Engineering Industry. In particular, he was pleased with a way that has been found to keep the cells clean. "It is a simple trick," he said. "The cells are placed inside glass tubes, such as those used in fluorescent lamps. The tubes are hermetically sealed and the cells are protected." Two researchers, the late Dr. A. P. Landsman and N. V. Pulmanov, have remarked that the system is especially helpful for cells using cadmium sulfide films, much more vulnerable than silicon to the elements. Field tests show that the cylindrical tubes shed ice, dust, and snow that might cut output.

A whole range of solar cells has been placed in service to operate lights running from one to thirty watts. The cells can supply power for communications and, at a much higher level, to run irrigation pumps. Lidorenko and Tarnizhevsky describe a unit that puts out five hundred watts with the help of parabolic mirrors to concentrate the sun on the cells and cooling fins to protect them from overheating. In the morning, the unit looks east to catch the rising sun, then the mirrors follow it all day long. As soon as the sunshine is strong enough, an electric pump cuts in, bringing up water from a depth of thirty-nine feet at a rate of six hundred and fifty gallons an hour. In 1969, one of these pumps was put to work at Bakharden in the Kara Kum Desert of Turkmenia on a state farm raising sheep destined for the fur trade as Persian lamb. With this kind of a market, the pump with its solar cells and sun-tracker was able to pay for itself and more have been ordered from the assembly plant in Erevan, the capital of Soviet Armenia. As is always the case with solar pumps, there is no need to worry about storing power as long as one can store the water. Rosanov said that the Soviets are not doing any extensive work on storage of large quantities of power; instead, they have concentrated on an iron-nickel battery for use in electric cars which, he said, is low in cost, easy to service, and aimed to last the life of the car. General Electric has expressed interest in it.

It is in Turkmenia and the other republics of Central Asia that the Soviet finds the most promising market for solar energy. There, it makes no different what the price of energy is at the refinery or the power station; by the time it reaches the customer, any form of centrally distributed power may be very expensive indeed.

Deserts like the Kara Kum are not quite deserted; there are thousands of small settlements flung across them, but not densely enough to make the stringing of power lines worthwhile (electricity demand averages out to only ten watts per square kilometer). Understandably, the solar scientist tends to concentrate on Central Asia with its peculiar structure of a rural population spread thinly over a huge area yet with scientific resources at its disposal.

Solar dreams and schemes in the U.S.S.R. have always been based on the sunshine available here. Turkmenia, running from the Caspian Sea along the frontier that separates the U.S.S.R. from Iran and Afghanistan, receives more than three thousand hours of sun a year, making it the equal of Tunisia and the leader of the Soviet Union. Runners-up include the other Central Asian republics, then Armenia, and finally Georgia, seventh with about two thousand hours. It was in Tashkent, the capital of Uzbekistan, that a research station went into operation during that worldwide period of solar euphoria after the Second World War. It produced solar water heaters (good for a shower bath), crop driers, parabolic collectors made of bent window glass on a concrete base, and a solar furnace that melted tungsten. Tea could be made with a solar samovar. The tradition of practical use of the sun at Tashkent has been carried on by a physical-technical institute there that recently developed machinery to dry fruit or vegetables and a "heat box" to harden concrete.

The sunshine of Central Asia inspired scientists who had left the wintry gloom of Russia. V. B. Weinberg, one of the founders of Soviet solar science, worked out a scheme for a power plant and published it in 1959. He wanted to use a large number of steam boilers, each heated by a parabolic trough five hundred and forty square feet in area. The steam would be piped first to superheaters and thence to a turbine. Like other proposals for extensive solar farming, it carried a heavy penalty that must be paid when heat is transferred along pipes. Lidorenko and Tarnizhevsky comment that one can always hook up a turbine and dynamo to each boiler, but the low efficiency of such small plants can make matters even worse.

It was to get around this loss in heat transfer that Baum and his laboratory came up with their new approach that preceded current interest in the "power tower" that we have described earlier: a boiler high off the ground, catching the rays of mirrors mounted on cars moving along concentric tracks to follow the sun. In other words, the sun's heat could be moved optically. The plant was never built, but it has never been forgotten. One of Baum's collaborators, Dr. Boris Garf, recalled it when we met in the Krzhizhanovsky Power Institute on Lenin Prospekt in Moscow. The institute was once the sole solar energy laboratory in the Soviet Union. It sent out cadres who started others; the most prominent was Baum himself, its former chief, who had moved to Ashkhabad, the capital of Turkmenia. The institute's deputy director, Dr. Ivan T. Aladiev, was about to leave for the United States to discuss solar plants with the National Science Foundation. This led Garf back to Baum's scheme of twenty years before to collect the sun and generate power. "The station was never built, but not for reasons that had anything to do with its scientific merits," Garf said. It was a technological inspiration: the boiler one hundred and thirty feet off the ground only had to rotate once a day to stay in focus while thirteen hundred flat mirrors, each ten by sixteen feet, moved on their tracks. In the most advanced version that had been projected onto paper, the boiler would have produced between eight and nine tons of steam per hour. Some of this would be fed into a turbine to generate 1.2 megawatts, the rest would be used as heat, for example, to distill brackish water. If all the steam had been sent into a turbine, the plant could have generated 2.5 megawatts, but heat was wanted as well as power. A prospective site in Armenia near Erevan had even been chosen for the plant before it was shelved. Nevertheless, the laboratory kept on with its scientific studies and concluded that a boiler on a tower surrounded by flat mirrors was still the most realistic of all possible schemes. "Our estimates showed that if we tried to use a single parabolic reflector to produce the same power, it would have to measure nearly one thousand feet in diameter," Garf said. But some of the original assumptions were changed; in particular, researchers scrapped the idea of trains and decided that the mirrors should be planted in a full circle around the boiler. "This leads to a better use of the sun falling on the site and produces more power, too. Since there are no trains, there is no jerkiness in the movement of the mirrors. With mirrors on a number of cars starting and stopping, there is always the danger that something might go wrong." Garf was pleased to see that American researchers, working independently, had arrived at a similar approach but with a major innovation: they increased the height of the tower elevenfold.

His institute has weighed the merits of storage techniques, whether through water, bins of hot rocks, or molten salts, and concluded, as others have, that everything is a matter of costs. Storage of low-temperature heat for houses has been taken up, but not pushed very far. "Our climate just does not favor solar heating. Even in Turkmenia at a subtropical latitude, there are not many sunny days in winter but, on certain days, the temperature can fall to five degrees Fahrenheit. We have not given up on solar heating in Central Asia, but we think that it must be linked to refrigeration so that we can cover the costs of the collectors."

Research into all solar branches is carried on at the Krzhizhanovsky Institute, which makes up in versatility for the lack of ambient sunshine on Lenin Prospekt. Scientists there had been trying to raise the efficiency of flat collectors, and they had found that a tin oxide coating on glass cut the amount of energy radiated back to the sky. One researcher whom we met, Dr. Marina D. Kudriashova, has concentrated on selective metal surfaces and she is hopeful that they could be mass-produced in the future by applying a "selective" layer to aluminum foil, then using the material to cover the surface of a collector in any desired shape.

We spoke English with Dr. Kudriashova; otherwise, Garf talked French. He was a man of sixty-seven, lithe and surprisingly tall when he unfolded himself from his desk, a mountain-climber by avocation. In the office of Dr. I. N. Malevsky, the head of his laboratory, he demonstrated to us a model of a small electric generator. It looked like an array of floodlights, something reminiscent of Francia's approach to thermal power. In the center of each of eight reflectors set up on a rack, there appeared to be a small light bulb. This light bulb, in fact, was a thermoelectric generator. The Soviets have never turned away from the principle of the thermocouple as a source of electricity. They maintain that it is much less expensive than solar cells and they have been able to use it successfully. The device was a prototype section of a large generator that was being set up at Baku by his institute and the Azerbaijan Academy of Sciencies. Each bulb could produce 1.5 watts at noon, transforming into electricity not the light of the sun (as in the case of the solar cell) but its heat. A splatter of metal appeared at the top of the bulb; Garf said it consisted of bismuth, stibium, and tellurium. In the full-sized array there were not eight bulbs but one hundred and fifty, putting out two hundred and twenty-five watts to drive a small motor and pump water for livestock. The reflectors around the bulb served to concentrate the sun's rays on the generators, and Garf had been able to get them off-the-shelf from the lighting industry. The model also incorporated an automatic sun-following system: first, a quick-acting device to pick up the sun in the morning or after a cloudy spell, then a sun-tracker that worked all day long. Costs could vary by a factor of five or ten, depending on the length of the production run, but Garf maintained that the generator could vie with a diesel-driven pump in the sort of spot miles from anywhere that one finds so often in the deep south of the Soviet Union. Like every solar scientist we met in the U.S.S.R., he was cost-conscious. Competition from oil and natural gas, still in ample supply there, was such that the sun stood a chance only in remote areas or under conditions where pollution could not be tolerated at any level, particularly in health resorts or vacation centers.

It was this motivation that sent Baum, whose name is known throughout the helio-centered world, to Ashkhabad, one of the most distant capitals of the Soviet republics. It is certainly off the beaten path to anywhere else; it lies eighteen hundred miles southeast of Moscow with its back against the Iranian border. We arrived there after four and a half hours of flying time from Domodedevo, the Moscow airport used for domestic flights. The trip was

as boring as air trips always are. Halfway out, the pilot announced
(and a Hungarian student we had befriended translated) that we
were above Volgograd but, for all that we could see through the
clouds, we might have been stacked over La Guardia. By far the
most adventurous part of the journey had been the landing that our
Intourist driver made on the icy roads outside the air terminal thirty
miles from Moscow. On his first right turn off the highway, he went
into a spin and round and round we whirled like the wheel of
fortune. Once through the airport gate, he headed for the international hall and then repeated the same trick with a truck bearing
down on us from one side and a parked Ilyushin on the other. That
man will win the figure-skating championship at the next Winter
Olympics if they let him bring his car. The rest of the trip to
Ashkhabad was an inevitable letdown. At one point, through a
haze, we saw the Caspian Sea and its shores, brown and dry; but
mostly we saw the trials of our airsick fellow passengers and shared
their relief when we started our landing approach into Ashkhabad.
We were over a range of bare mountains, interspersed by puddles
of sand, the edge of the Kara Kum Desert of Turkmenia. Tracks
ran from one puddle to another over the mountains; one could not
imagine what might have jumped those puddles — perhaps a train
of camels. As we lost altitude, a spray of vegetation appeared on
the slopes; then the scene shifted quickly. Fields turned green, irrigation canals glittered on their borders; settlements appeared,
streets on a grid, low-rise housing; from the air, it looked like rural
western America or, at least, more so than anything we had seen
on this side of the Atlantic. On the ground, the impression held:
the dry heat and light air of Fort Collins, not nearly as many cars
but small houses, trees along die-straight avenues. The airport was
only a mile and a half from our Hotel Ashkhabad, at seven stories
the tallest building in town. In 1948, Ashkhabad had been flattened by an earthquake that left eleven thousand dead and only
four buildings standing. It had been rebuilt low and spread out,
neighborhoods tied together by the trolleybuses that sped on
Svoboda Prospekt (Freedom Avenue) at all hours of the day and
night beneath our balcony where the washing dried almost before

we could hang it. The room in Ashkhabad was slightly cheaper than our quarters at the Metropole in Moscow but acres smaller; it could have fitted into the bathroom of the czarist troïkotel without more than a few drops overflowing. But the restaurant was modern, the service smiling, and the action fast in the evening as the local talent went to work on the girls sitting by themselves at the table next to ours. The champagnski flowed like borscht, the band never stopped playing, the girls never sat out a waltz. Then a Turkmeni boy asked me if he could have the next dance with Madeleine. Round and round they went, Madeleine had not danced for months (she did not count my bear tramp beneath the chandeliers of the restaurant in the Metropole). I do not know what the music was but her Breton sabots moved as if the sap was still flowing in them and, at the end, it was the Turkmeni who mopped his brow as he escorted Madeleine back to her chair and returned to his friends' table for a rest after his strenuous recreation.

We had ample time over a weekend to see the setting where Baum had established himself. Ashkhabad was like so many of the other solar centers we visited; there were no monuments, no vestiges of glory, just people living and working. The city had a quarter of a million of them, more than a tenth of the entire population of Turkmenia, which is twice the size of Great Britain but consists mostly of the Kara Kum Desert. Intourist dutifully took us around the sights, from Karl Marx Square downtown to the Kara Kum Canal that runs to the Caspian Sea and, some day, will make Ashkhabad a port. There weren't all that many sights; soon we were visiting a kindergarten and a farm which gave Madeleine a chance to zero her Nikons in on young children and young camels. The city left an impression of elbowroom; industry was sited on the edges while the center was filled with parks for reading, sitting, walking, volunteer tree-planting, and al fresco beer drinking. The snows of Moscow and the rains of Brittany were far from sight and mind as we walked out in the streets of Ashkhabad, through the gardens and the back alleys. A girl of twelve started talking to us, she showed us her school; a family struck up a conversation with Madeleine and soon we were sitting down to dinner with them, sitting all the way down on their carpet, eating pilaf from a bowl and toasting Franco-American-Soviet friendship with Russian vodka and Armenian cognac, while the family's two sons, light as sprites, romped around us and relatives popped in and out, young women in their long Turkmeni robes, each more beautiful than the other, with their quiet babies. We were among geophysicists, doctors, geographers, librarians; in some ways, this was a place like Fort Collins or Lexington where you could walk down a street and end up in the intimacy of a family. Springtime in the Kopetdag was a happy time; the Kopetdag were the range that barred the horizon to the south, their name means "many mountains" in Turkmeni. Summertime was less heavenly, the temperature was more like hell. There were no air-conditioners in our hotel room nor in the homes that we visited. We saw some in a bank and in the local carpet factory where girls sat at hand looms, tying half a million knots to make a square meter of rug. Friends said that the movie houses were air-conditioned, too, making them the sort of summer attraction they were in my youth in Manhattan. If all that heat could be turned into cold, a solar energizer would have a good thing going. Otherwise, Ashkhabad did not appear to offer much of a market for new sources of energy; it was doing too well with the old ones. During our city tour, we were told that Turkmenia was third among the Soviet republics in the production of oil and natural gas, outranked only by Russia and Azerbaijan. There was plenty of the stuff around; the wind backed the haze from factory chimneys against the "Many Mountains"; down on a collective farm, a truck driver pointed to the wall-to-ceiling wood stove that heated two rooms of his home and remarked that he would soon replace it with a gas furnace. Trying to push solar heating in Turk-menia was something like selling power windows to camel drivers.

At the climax of his long career, Baum has set out to meet the present needs of Turkmenia with the sun. He and his collaborators are based at the Physical-Technical Institute of the Turkmeni Academy of Sciences on Gogol Street in Ashkhabad, but they spread out. At Bakharden, the state farm where solar cells supply electricity to a pump, brackish water must be made fit for sheep to drink once it reaches the surface. Baum spoke first of water when we met him and two of his co-workers at the Academy. At seventy-one, he was as forthright as he had been twenty-two years before when I first saw him at a symposium that Trombe had organized at Mont-Louis. The intervening time seemed to have improved his English without slowing him perceptibly.

"We have worked twelve years on solar distillation," he said. "We have regions in the deserts of Turkmenia where water can cost a hundred rubles per cubic meter [one hundred and thirty dollars for two hundred and sixty-four gallons] although the price usually runs between ten and twenty rubles per cubic meter for water brought in by truck. We know now that we can produce drinking water with solar stills at a cost of two or three rubles per cubic meter. At this price, it can be used not only for drinking but it also makes sheep-raising profitable." In 1968, Dr. Rejep Baïramov, a laboratory director at the Physical-Technical Institute, set up a solar distillation plant at Bakharden. The first stage amounted to 6,500 square feet of concrete, steel, polyethylene, and window glass, all of which could be tested under actual desert conditions. Over a year of measurements, the still produced five hundred tons of distilled water and a great deal of experience in running it. Baïramov found there was no real need to clean the window glass on the collectors; one merely had to make sure that all joints stayed watertight and salt deposits were not allowed to accumulate. One man could take care of a still three times the size of this first stage. Armed with these findings, Baïramov has gone ahead with a second unit twice as big. This will give him a total of 19,500 square feet of solar still, enough to water two flocks of sheep with eight hundred head in each flock. "We hope to build many of these plants," Baum said. The distances of the desert are on the side of the sun. It has been estimated that if drinking water must be trucked more than thirty miles, solar distillation will always be cheaper, no matter how brackish the supply at hand.

Cooling is a different matter. Here, Baum has taken on the opposition on its home grounds. He has set out to build a better air-conditioner that would be cheaper to own and operate in

Ashkhabad itself whose hot, dry climate reminds Baum of Arizona, the state he saw for the first time in 1965 when he spent a month visiting the solar capitals of the United States. "The temperature here can rise to 115 degrees Fahrenheit in the shade," said Baum. "This sort of weather lasts four or five months, starting in May. We have studied the performance of different types of buildings; we have sought to improve indoor climate by covering windows and creating shade. This helps, but we cannot achieve the proper conditions. In daytime, the indoor temperature can still go up to 102."

This led Baum and his colleagues to turn to artificial cooling, using all the properties of Turkmenia's climate, the dryness as well as the heat. As Baïramov put it in the terms of a planned economy, "the production schedule coincides with the demand graph": the hotter the day, the more power there is to drive the air-conditioner. This does not always work in every climate; the economics of solar cooling have not looked favorable in the past when a flat-plate collector or a concentrator was used as the power source for a conventional absorption type of cooling system. Baum has written:

> The major flaw in all these installations has been their inefficient use of solar energy. Experimental solar cooling installations that have been constructed up to the present use only from 4 to 9 per cent of the sun's energy when their working temperature is below freezing and from 10 to 15 per cent when it is above freezing, as in air-conditioning. Consequently, the amount of solar collecting surface needed for each refrigeration unit is very large, leading to an inflation of size and costs. This is one of the main obstacles to the widespread use of solar cooling equipment.

It is instructive to look at the history of the approach taken by Baum and his principal collaborator, Dr. Annageldi Kakabaev. In the early 1950s, there was a tendency to try brute force: set up a solar boiler and run a refrigerator at full steam. But now things are done more subtly. Throughout the Central Asian republics of the Soviet Union, the summers are as dry as they are hot. This led naturally into an attempt to cool by evaporation and also by that flow of heat away from the earth under clear night skies. By 1971, Baum's laboratory had developed a system in which a twelve-inch layer of water was chilled to about 63° F. then stored in an underground tank. During the day, the water was piped through a cooling panel that kept a room quite liveable.

Then the laboratory introduced the principle of evaporation into a much more sophisticated air-conditioner based on the absorption refrigeration system. In Baum's case, the working fluid that must be "regenerated" after it has gone through the cooler is a lithium-chloride solution. He does it not by heating it with a flat or parabolic collector but by running the solution over an inclined surface, say a roof, open to the sun and the air, something like the way water is heated in the simplest of flat collectors. The sun not only heats the solution, but evaporates water out and brings it back to its original strength so that it can return to work in the cooling unit. Water is automatically added to replace what is lost during the evaporation process.

Baum said that this approach has cut the costs of solar cooling drastically. It does not require a solar collector to regenerate the weak solution, just a cheap rugged roofing material over which the solution flows. Its efficiency in the use of the sun's heat is high: 60 percent in the hottest and driest parts of Turkmenia towards the east; 45 percent around Ashkhabad; 25 percent near the Caspian Sea.

The laboratory started its test by cooling only one room in 1970. Then a larger plant was constructed, big enough to handle two rooms, and, on the basis of results obtained, a house was built at Bykrova, six miles outside Ashkhabad. This is a brick building containing nine small apartments, three floors and a basement, a total of 4,300 square feet of living space. The rooftop "regenerator" covers 1,900 square feet but, Baum assured us, it is enough to take care of two more floors. For three summers, his researchers were able to watch the house in action. Outside temperatures varied from 104° to 114° F. as they do in these parts; the rooms, used as a laboratory rather than living quarters, stayed between 75° and 78°

F. As his next step, Baum wants to try a really big house with one hundred and eighty sun-cooled apartments in Ashkhabad itself. "We are sure it will work. It is the cheapest and the most reliable system. If we can build it, then the government will be able to judge if this is worth doing on a major scale."

Baum said the purchase price of his system is about the same as that of conventional air-conditioners in Ashkhabad, but the solar cooler does not use electricity at four kopecks per kilowatt hour. It should add no more than 3 to 5 percent to the cost of a square foot of living space, a figure that is not high in Ashkhabad where housing costs are low, much lower in terms of local purchasing power than the price of electricity. He hesitated to draw parallels with economic and solar conditions elsewhere, but he did say that his laboratory was working on ways to use solar cooling in damper climates.

Baum has his eye on solar heating for Central Asia in the future and he has never quite abandoned his power plant which, he thinks, could be used to produce heat for really massive water distillation. He agreed with what one hears elsewhere when he remarked that the era of cheap energy will never return, but he did not go along with all the solutions that have been suggested. "I do not believe in the use of photosynthesis to provide energy. We must use it to feed people; nature does this wisely and precisely to provide us with food. We should not burn crops to obtain the heat and energy that we can get from another source." And Kakabaev remarked, as many Soviets have, that coal, oil, and gas should be regarded more as raw materials than as fuel.

Baum is a physicist by training with a doctorate in technology. Ever since 1926, he said, he has been interested in solar energy. "I heard agronomists talk of the sun. They were saying even then that if man used it, he would not have to fear for the future. And I heard that the sun was free." Fifty years later, Baum knows better: "In nature, everything is free. But if you use it, you must pay for it."

XVII.

Getting the Sun to Market

IT HAD BEEN a warm damp winter in Brittany and most of Western Europe to the delight of householders and the despair of fuel dealers. Supertankers idled on the high seas if they were lucky; the others were laid up and their crews laid off. Snow was sparse even in the north. Elsewhere, when it did arrive, it came so late that it was wiped away by early spring and flushed into the sodden earth. Winter was more a matter of staying dry than keeping warm. Doors swelled and stuck, the washing dripped from Monday to Sunday, and the rain kept coming. If the whole North Sea had been diverted onto our heads, the result would not have been dissimilar. My own feelings were mixed. Personally, I was only too glad to brave the worst of January and February with a bit of chestnut kindling and a few splinters of oak; professionally, I found it hard at times to keep a proper focus on BTUs when wool socks met most of my own needs. Solar heating had taken the sting out of the winter of 1974–75; if nature could do it, why couldn't we?

More and more people were asking this sort of question. They agreed that solar energy did not soil the atmosphere or boil the hydrosphere; now they wanted to learn what it did to an oil bill.

The only way to find out was to try it, to build something and see if it worked. Walter A. Meisen has the title of Assistant Commissioner for Construction Management of the Public Building Service of the General Services Administration in Washington, but he does not talk the way that sounds. At the solar energy roundtable in New York, he said aloud what many must have been wondering:

I really think the problem is that the use of solar energy is being discussed by scientists and we have a passion for being a technological society. As such, we almost frown a little when someone finds a very simple economic way of using solar heat. That is much too simple for us. And so we have to find something very exotic, so that we can earn the name of scientist or engineer.

I am just an architect. If we start talking about BTUs, I get a little lost. I get lost when we say as a nation, with 6 per cent of the world's population, we use 30 per cent of the available energy. I keep thinking to myself: something does not sound right. It is always prefaced with: if present trends continue. Well, present trends cannot continue. And unless we start looking at why they cannot, and what we can do about them, then I am afraid we are going to keep refining solar energy to where it is perfect and you know we will never reach that.

I keep thinking about the internal combustion engine, the reciprocating one mainly, and if we had tried to perfect it to the point where it is today, we would never have had a Model T or anything else. Solar energy is here now and the only thing that is stopping us is that we don't think it is really sophisticated enough in its present state. So we are not offering it to our clients. We are not offering it to the people who would probably welcome it. We don't insist that we maintain 74 degrees the year round. We really insist on some reasonable measure of comfort, and human beings appear to be much more adaptable to the ranges of climate than our thermometers. We sort of reject the human being and say, well, we had better rely on the thermometer. I don't think this is the right way to approach it. I think we can build collectors today and the only thing that is stopping it is people not going into production.

The only other thing that is stopping it is that you're waiting for the government to do it and God forbid that we have to rely on the government to do it for us.

Despite this injunction, Meisen's General Services Administration tried some solar pump-priming as far back as 1972. In that year, it contracted for a new post office and federal building in Saginaw, Michigan, as an environmental demonstration project with an eight-thousand-square-foot collector to supply all of its hot water and part of its heat. However, it has been delayed by budget trouble. Then the GSA designated a federal building to be put up in Manchester, New Hampshire, as its energy conservation demonstration project. It was a bigger undertaking and the GSA took on the New York firm of Dubin-Mindell-Bloome as its energy consultants, thereby adding to its ranks Fred Dubin, who never hesitates to spend his own energy to convince his clients, audiences, and readers to spare theirs. Dubin sees eye to eye with the GSA; he has always maintained that solar energy must come into the market in large public or private buildings before it can be brought within the reach of the individual homeowner. The sun plays well in the orchestra of energy conservation measures that Dubin has put together over the years. It is so much cheaper to save energy than supply it. Trying to heat and cool a glass box with the sun is like trying to pedal an El Dorado the day the gas pumps close. Neither was ever intended to run that way.

Dubin was among the more active figures at the Paris solar energy congress in 1973. It was hard to miss him; he moved in a flurry of statements and reprints, all preaching thrift in the use of power. Later, during our trip through the United States, we were in constant telephone touch (a contradiction in terms, despite the Bell ads), but the best we could do was a long-distance interview between Denver and Manhattan. Only in Paris did we catch up with Dubin or, to be fair, did Dubin catch up with us. The conversation started in his hotel room and moved to a café in the Tuileries Gardens where we could talk in the shade of a parasol while the sun played over a Paris Sunday in the park. Dubin was in Europe on a State Department grant and he was traveling through a number of cities — Athens and Paris, Amsterdam and Copenhagen, Oslo and Stockholm — to talk about energy conservation. He enjoyed it; Dubin looks every inch the American abroad when he is in Europe and delights in everything he sees and hears. He was at ease in the Tuileries as he talked about his life and how it led him to the sun.

Dubin was born in 1914, just the right year to ensure that he would enter the job market at the perigee of the Depression. He got a degree in engineering from Carnegie Tech in 1935 and took it with him to a steel company, but left after a few months. He did not like big industry then and he has tried to keep his firm a manageable size since. He went back to his native Connecticut and joined forces with his father who had become a heating contractor after driving racing cars, Stutz Bearcats and Cords.

Air-conditioning looks to be the climax of Baum's career; it was the start of Dubin's. "We were in at the beginning. We bought ice outside and put it in a bucket, then ran a fan over it. Or we pumped chilled water through a sprayer. We had to make things up as we went along. There were no courses in air-conditioning. What is funny is that we are going back to this approach. Recently, I visited a big hydraulics laboratory where they were blowing air through ice-water sprays. It's a way of accumulating energy and getting around that four o'clock peak in air-conditioning. You make the ice with off-peak electricity."

Dubin followed refrigeration out of the ice-bucket days and into conservation. He found that he could heat a big Boston supermarket with what was being given off by its refrigeration system. That led him into waste heat. Soon he was using the exhaust air from a Boston school to melt snow in its driveway. Even when there was no snow, the exhaust could be used to preheat incoming fresh air. "We built the heat exchanger by hand out of sheet metal. It was bad engineering, but it worked."

Dubin's firm was no stranger to the energy feast of the 1950s and 1960s when intricate control systems had to be installed to cope with buildings' reactions to the sun. It handled one of the largest American air-conditioning systems, that of the State University of New York in Buffalo where fifteen million square feet of buildings spread over fourteen hundred acres had to be cooled. Even on these jobs, Dubin kept his eye on the electric meter. He was among the first architect-engineers to suggest such revolutionary innovations as "operable" windows or lower light levels. In lighting, the customer is hit twice. He must pay for light which, Dubin writes, he doesn't need:

> Consideration should be given to lowering the overall average lighting levels by about 50 per cent. Better quality of illumination, rather than higher intensities, should be the goal. Many recent experiments confirm that lighting levels between 10 and 40 foot candles are sufficient for visual acuity and physiological needs, where levels of 60 to 150 foot candles are now being provided.

That is the first hit; then comes the rebound in summer when the air-conditioners must be turned up to get rid of the heat of the lights. Dubin wants to move heat from one place to another inside a building instead of getting rid of it here and adding it there.

Such wrinkles and shortcuts were helpful, but they left Dubin dissatisfied. "Here we were, fooling around with storm windows and lighting fixtures when the electric utilities were throwing away 65 percent of the heat produced by their generating plants." In 1960, Dubin became interested in total energy systems. If a large building or housing project could generate its own power with a steam or gas turbine on the site, then it could use the waste to supply heat and hot water. Now his ideas about moving energy were scaled up from buildings to communities. Such systems have been incorporated into housing projects (one of the biggest is Rochdale Village in New York with 5,680 dwellings) but they never really caught on in the United States. "They were too capital-intensive," Dubin remarked. "They didn't pay back fast enough for investors." He did not abandon the approach but went beyond it to solar energy. At first, he looked at it from a passive viewpoint, the way

Baum had done in Ashkhabad. He and his firm tried to reduce the load on air-conditioners. They worked with colors, overhangs, roof sprays, and ponds. In 1972, they decided to use the sun instead of merely fighting it. "There was no solar business in those days. We were motivated by intellectual curiosity. It seemed to be a sensible way to use energy."

Their first opportunity to see what all their ideas might add up to came when the government gave them the contract for the federal building in Manchester. "We were asked to save 20 percent of the energy costs. We said we could save 35 percent with off-the-shelf hardware. In the end, we cut the annual energy load of a seven-story office building from ten billion to six billion BTUs a year." Dubin has translated this saving into national terms: "If all residential, commercial, institutional and industrial buildings to be constructed in the United States next year were designed and operated in accordance with the energy conservation practices sug gested for the New Hampshire office building, more than six hundred billion cubic feet of gas, or four billion gallons of oil, or their heat equivalent in other fuels, could be saved yearly. If we project the energy conservation measures to existing buildings and save only 15 percent of the energy consumed now, the amount of oil which could be saved could exceed twelve billion gallons per year."

Dubin and his office analyzed the energy appetite of the proposed building and saw that heating amounted to more than 60 percent of the total. Much of the demand came at night and on weekends when the building was empty, or in the mornings when people came to work. The engineers decided to try to use heat built up during the day — and which normally had to be evacuated — to run the place at night. Like so many others, they turned to the heat pump. By day, it could keep the building comfortable, transferring heat from lights and office machines into a hot water storage tank for use at night and during the first hours of the morning. "Happily, the same storage system that was used with the heat pump system could be used for solar energy," said Dubin. "So we found the start of a fit between the mechanical energy conservation system and the solar energy system."

Once they had whittled away the heating requirements, they asked their computer how much of the remainder could be met with solar energy. Dubin worked with Professor Everett Barber, who teaches an energy conservation course at Yale and runs Sunworks, Inc., a small company specializing in solar engineering and equipment, at Guilford, Connecticut. They determined that a fifteen-thousand-square-foot collector could handle most of the heating and cooling for three of the building's seven floors, allowing for between fourteen thousand and eighteen thousand square feet per floor. They conceived their collector as a sawtooth affair facing south on the roof, each tooth of the saw protecting its neighbor from the wind. Dubin wants to be able to change the tilt of the collector with the seasons so as to obtain the best angles for winter heating and summer cooling. Construction of the building started in mid-1974 and was scheduled for completion two years later.

On the basis of his studies, Dubin has concluded that "there is sufficient solar insulation in New England for a solar energy heating and cooling system." If the collector were enlarged to 40 percent of the building's square footage (this would mean panels over the parking lot as well as the roof), it would cover 70 percent of the heating and 95 percent of the cooling. Dubin states: "The life cycle costs of the solar heating and cooling system are estimated to be less than the costs of straight electric resistance heating and electric compression cooling by about 25 per cent." This may be bad for the electric companies, but Dubin does not think that what is good for them need always be good for the country. "Utility companies have an obligation to society to control their growth. Building up peak demands with nonessential loads, and then offering concessions for off-peak use is not in the best interests of society, given the nature of our impending crises. Utility companies have been guilty of slanting engineering and economic feasibility studies in order to promote more use of electricity. Contrary to some of their studies, electric heat is not cleaner. Central fossil fueled heating plants have seasonal efficiencies greater than 25 or 30 per cent. Subsidies for the use of electric heat are not in the best interests of the public at large."

Designing the Manchester building turned out to be the first of a dozen-odd solar projects involving Dubin's firm. In April 1975, ground was broken for a solar-heated administration and research building by the New York Botanical Gardens on the site of its Cary Arboretum at Millbrook, east of Poughkeepsie in New York State. The building, intended to accommodate eight scientists and their staffs, was designed by Malcolm G. Wells with Dubin once again as energy consultant. This is at least a $2.3 million project and, like the Manchester design, uses every trick in the trade to cut losses of heat, water, and electricity. First, there are natural defenses: the Arboretum building will have sod on its roof to keep the weather out and merge with its surroundings. More earth is piled up against the north and west walls, trees provide a lee when the north wind blows, and the walls themselves are of masonry, a foot thick and insulated outside and in. Windows are double-glazed, oversized to the south to let the sun in but down to little more than slits on the north. On nights and weekends, the windows are shuttered.

The building is planned for thirty-five thousand square feet of offices and laboratories on two levels, one below ground. On the roof, an eight-thousand-square-foot collector will be spread over seven panels facing south and mounted again in sawtooth form. The collector is to heat water, protected by antifreeze, for storage in basement tanks which, in turn, will serve as the source for the heat pumps that will warm the building. In the event of a long sunless cold spell, deep wells on the eighteen-hundred-acre estate will be tapped by the heat pumps. In summer, well water will cool the building and, if necessary, run a small air-conditioner. Rain water will be stored in cisterns, then used to flush toilets and water the arboretum's plants. Another botanical touch is an attached greenhouse. Carbon dioxide breathed out by the staff will be pumped into the greenhouse and breathed in by the plants. It should be a profitable two-way relationship: the greenhouse atmosphere will be available to help heat and humidify the building which is replacing two converted farmhouses serving as temporary quarters.

The Cary Arboretum building was one of the biggest solar designs in the northeastern United States at the time it was announced. Dubin was ready to go further north with another project near Brookline, Vermont, which he planned for a developer. Here, Dubin was able to express some of his ideas on energy-conservant housing. The first is that the individual free-standing house has seen its day, at least in this latitude. The project is to consist of a "village" of twenty houses in two ten-house clusters, huddled for mutual warmth and comfort, each house about one thousand square feet. Dubin likes to group the houses because he can then set up his solar collector independently on the south slope of a hill without trying to incorporate it into the design of a house. He plans forty-five hundred square feet of collectors feeding heat into three tanks with a capacity of twenty thousand gallons. It's best not to put all one's water into a single tank. Otherwise, there are times on overcast days when the water coming down from the collector may actually be colder than the water stored from a previous sunny stretch. When this water falls below 110° F., a heat pump is to boost it up where it belongs. According to plans, and if financing is completed, the village is to be built in two phases: first, a group of ten houses getting power from the local utility and then, after further research, a wind generator will be set up to serve all twenty houses.

Solar buildings have been as hard to finance as any others. A wild-sounding scheme to run a new community of ten thousand on sun and wind in the British Virgin Islands was as wild as it sounded because the promoter went bankrupt before Dubin could do more than make a few suggestions. A shortage of money has also shelved, at least temporarily, a Dubin proposal to use a solar collector to heat the swimming pool of a club for inner-city boys in Wilmington, Delaware, which was to be built in an old quarry. But for every project dropped, another crops up. Dubin has received letters from a Texas developer who wants to use solar energy in a twenty-story office building and another in Virginia who is thinking about it to service three hundred houses. While the queries come in, Dubin keeps working for a mixed bag of solar-minded clients, among them a NASA base in the California Desert where an existing air-conditioner is to be revamped to work on sunshine. A house engineered for Gifford Pinchot in Guilford, Connecticut, has gone into operation with copper solar collectors provided by Barber.

Dubin is convinced that sunshine will work, but he is less certain that it will pay the individual homeowner to put it to work. Even with the rise in the cost of heating oil, he estimates that fifteen or twenty years are needed to win back the costs of solar heat in a private house. This is the kind of foresightedness that institutions or corporations are better able to afford. As Dubin predicted, there has been if not a spate then a minor outbreak of solar plans for large buildings. One reads an announcement that the Community College of Denver intends to heat an eleven-million-dollar building serving thirty-five hundred students with a fifty-thousand-square-foot collector (larger than a football field, we are told). A new science museum in Richmond, Virginia, also hopes to become the world's largest solar-heated-and-cooled building; the state of Connecticut plans a million-dollar housing project for the elderly near New Haven with the sun heating half of forty apartments; New Mexico State University will put a crown of seven thousand square feet of collectors on its contender for the title of the nation's solar biggest, an agricultural department building. A university press release says that "thermal storage will be sufficient to provide necessary heating and cooling even in the unlikely event of three consecutive cloudy days."

The greater likelihood of such an event in the New York City area has not deterred poetic speculation over the possibility of solar skyscrapers. The real estate section of the *New York Times* was quite optimistic at first over a study undertaken for Con Ed by the Energy Laboratory at MIT to see if the sun could be used to dehumidify the top twenty-five floors of the Citicorp Center, a nine-hundred-and-ten-foot-high addition to the Manhattan skyline that was being built for the First National City Bank. The proposal was for a twenty-thousand-square-foot collector to cover the Citicorp roof, sloping at an angle of 45 degrees. The idea was to dry incoming air by passing it through a spray of water-absorbing liquid which, once it had done its work, would be "regenerated" by solar heat. The system was described as more economical than the normal way to get dampness out of the air by chilling it, then heating the air all over again so that it does not deep-freeze the building's occupants. But further study showed that if the flow of cool air was simply reduced, then the system would save not $50,000 a year but only $3,000. That was the end of the project. The *Times* has also reported on a purely hypothetical Encon Building described as "the work of a task force assembled by Julien J. Studley, a real estate broker." Here, collectors would cover not only a slanted roof but the east and south walls. The designers were saying that the sun could handle 20 percent of summer cooling and 30 percent of winter heating. Along with energy savers like double windows and automatic switches reacting to natural light, the collectors would put the cost of the Encon Building up to $44 a square foot, according to Studley who was thinking in terms of 1.1 million square feet piled high over Third Avenue in New York. That is $4 more a square foot than in an ordinary skyscraper that eats up $1.90 a square foot every year in energy bills: $1.50 for electricity and 40 cents for steam. Studley claims that a square foot in the Encon would need only 80 cents for electricity and 10 cents for steam, thereby saving 90 cents a square foot or $1 million a year. In less than five years, the extra investment would be recouped. That is one of the quickest payoffs that has been promised in solar energy.

The first tiny step towards Encon may have been taken early in 1975 when Gump Glass, a Denver firm, announced they were buying a $60,000 solar heating system for part of their headquarters: sixteen hundred square feet of collector to heat seven thousand square feet of office and showroom space. The supplier is another Denver firm, Solaron, Inc., whose vice-president and technical director is George Löf. The system they are installing is similar to the one in Löf's home, hot air supplemented by gas, but it should save 77 percent of the fuel costs for the area heated. According to a newsletter, *Solar Energy Intelligence Report,* Solaron states that it is the "nation's first publicly held company organized for the sole purposes of developing, manufacturing and installing solar air

heating and cooling systems for residential and commercial buildings." The newsletter quotes the president of Gump Glass, Jerry
Sigman: "In no way are we going into this for experimental or
prestige purposes. We are profit-oriented and economy-minded,
and we view this as an eminently practical way to assure ourselves
of an unlimited source of fuel for as long as we are in business."

About the same time, a much bigger plant, intended for test purposes, was put into operation by General Electric at its Valley
Forge Space Center in Bala-Cynwyd, Pennsylvania. Its claim was
that this represented the first private solar system on an industrial
scale in the United States. On the roof, forty-nine hundred square
feet of collectors manufactured by GE provided 75 percent of the
heat and hot water needed by the plant's cafeteria covering twenty
thousand square feet. Cost of the system, admittedly hand-made,
was over one hundred thousand dollars and it had been instrumented at one hundred different points.

Slowly, we may be getting there. While making no attempt to
compete in cost with conventional systems, the National Science
Foundation began its major solar heating experiments with installations in four schools, two of which (in Warrenton and Osseo) have
been mentioned here. Then it decided to combine heating and cooling at the George A. Towns Elementary School in Atlanta,
Georgia. This school covers thirty-two thousand square feet and
accommodates five hundred pupils. A ten-thousand-square-foot
collector has been mounted on its roof to take care of 60 percent
of its heating, cooling, and hot water. The cost of the solar system
is $450,000 and it involves industry: Westinghouse is the prime
contractor, PPG (formerly Pittsburgh Plate Glass) is supplying the
collectors and Barber-Coleman the controls. The absorption cooling
system that will run on solar heat comes from Arkla Engineering,
and Dubin has done the mechanical design work. Georgia Institute
of Technology is monitoring the performance of the equipment,
while the overall solar design was in the hands of a firm of architects from the Pittsburgh area, Burt, Hill and Associates.

Like Dubin's company, Burt, Hill has made a reputation in trying to cut the use of energy in all its forms except solar. P. Richard

Rittelmann, at thirty-five a partner at Burt, Hill, is almost as familiar a figure as Dubin to congressional committees and energy
roundtables. "Long-standing government controls," he has said,
"have allowed the prices of energy to reflect only cost and not the
value of a diminishing resource. If somehow energy had been permitted to seek its true market value, energy conservation would
now be a way of life, and alternate energy opportunities could
quite conceivably have followed an evolutionary development
process rather than the revolutionary processes to which the nation
must commit itself."

One of the things I like about Rittelmann, whom I have never
met, is that he does this sort of thinking not in some center for
advanced studies in the sun but in the firm's offices at Butler, Pennsylvania, north of Pittsburgh. Like Sheffield, Butler is a steel town;
the mill is all that matters and it was a change from the campuses
we had been touring. On the advice of Tom Ainscough, who was
working on solar energy at Burt, Hill and took charge of us in
Rittelmann's absence, we stopped for the night on the edge of
Butler at a Holiday Inn where we fitted our yellow Nova into a
parking space next to a Mack tractor on a holiday without its
trailer. The Holiday Inn was a temple of the Gas and Oil Age; not
only did every guest park his Nova or his Mack outside his room
but the lobby was devoted to an art show of hand-painted automobiles. Madeleine was struck by a masterpiece: "On the trunk
lid, a devil's head leered above a sky of pink and blue clouds; but-
terflies and stars floated over a field strewn with flowers along the
doors while, on the hood, the silhouette of a necking couple could
be seen against the setting sun. Then there was a wonderful old
Packard, gleaming and shiny, tan and brown like a scarab, head-
lights as big as the engines of a 747."

The next morning, we met Ainscough in the firm's offices that
occupied the sixth floor of the Mellon Bank Building in downtown
Butler, a red brick structure going back no doubt to the height of
the Iron and Coal Age. Ainscough did not know when it went up;
all he knew was that the building was a good deal older than he
was at thirty-seven. He had been an art teacher, then he studied

civil engineering at night, and finally did his advanced work in architecture at Carnegie-Mellon. At that point, he left the educational world for good, first as an industrial designer with Pittsburgh Plate Glass and then, in 1966, as an architect with Burt, Hill, where he remained until 1974 when a subsequent move took him to North Carolina as director of an energy conservation program. Madeleine sketched him: "Blond and curly-headed with a little round nose, funny, charming, surrounded by photos of the houses, solar and otherwise, that he had designed." Burt, Hill's was a busy office with clients running from American Hardware to the Zion United Methodist Church. Even when the architecture was not solar, it was energy-minded. "Dick Rittelmann and I could see the so-called crisis coming," Ainscough said. "In 1968 or 1969, we stopped building glass boxes and worked the other way. We used overhangs to shade windows — draperies don't help, you know; once the heat is in, it can't get out. This naturally led us into solar energy. It hit us like that, it looked like it might be an answer. We wrote articles and we began to spread the word. Soon, we realized that solar energy will not tolerate an ethic of waste. We can't go on doing what we do today in air-conditioning where we cool all the air coming into a building to 55 degrees, then reheat it to 72 degrees. We realized that what we need is not an evolution but a revolution."

They had a chance to try their ideas when they were asked to cut the fuel bill of a Pittsburgh elementary school built in 1937. "We lowered the ceilings, we blanked out some windows with brick piers and reduced the size of others with canopies that let in the sun in winter and keep it out in summer. According to a computer estimate, this should save 80 percent of the gas bill. Under the old system, outside air was heated, circulated through the building, and exhausted. We have recirculated as much of the heated air as possible but code restrictions still require us to bring in a percentage of fresh air. In Pittsburgh, outside air is not fresh air." Heat must be used and reused, it must not fly out the window. In a pamphlet on solar energy architecture, the firm notes: "The careful extraction of heat generated within the building, a judicious selection of building materials and efficient system design has produced an office building in Toronto, Canada, which requires no outside energy for heating until the outside air temperature drops below 12 degrees F. There are other projects across the U.S. which claim equally impressive results."

Ainscough and Rittelmann think that solar architecture will enter the market the way other innovations have done: from the top down, from the large building to the private home and from the rich to the rest of us. Rittelmann once shared his ideas on the subject with a congressional committee:

> There is a very definite diffusion process within the architectural profession itself and within the building industry in general. . . . The much used and abused "split-level" entry in today's housing market had as its beginning an innovative architectural solution to accommodate the hilly sites around Los Angeles, to take advantage of land that would otherwise have been unsuitable for building. The so-called "split-level" entry has had such diffusion throughout the housing industry that it is quite often seen in homes on perfectly flat lots where its use has no logic other than to satisfy aspiration.
>
> The air-conditioning industry had its beginnings with individually designed, rather expensive, rather cumbersome built-up systems which have had extensive progression through the building industry to the point where it is possible today to purchase a completely packaged air-conditioning unit that can be installed without the services of an architect or engineer and in some cases by the buyer himself.
>
> The recently commercialized trash compactor had its beginning in large individually designed and constructed commercial units. We must admit to ignorance, however, regarding what market force motivates a consumer to spend $200 for an appliance which makes 40 pounds of trash out of 40 pounds of trash. We suspect, however, that aspiration is a stronger motivating force in this case than need.

Ainscough insists that aspiration must come into play to motivate the future solar buyer. "That is why we don't want solar energy on low-cost housing. You can't hide five or six hundred square feet of collector. The tenants wouldn't like it; solar energy would be a symbol of their plight. Everyone else wouldn't like it, either; it would be a symbol of low-cost housing. People aspire upward. I think that if solar energy started in low-cost housing, it would go like a lead balloon."

As pioneer solar architects, Rittelmann and Ainscough ran into a dearth of equipment to specify and started to make their own. "Dick builds furniture, he likes contemporary woodworking. I race sports cars and I build racing engines on the side myself. Between us, we can handle anything." They devised a solar heating module based on a water storage tank built from a paper tube four feet in diameter and lined with two inches of plastic. Rittelmann did the woodwork and Ainscough welded the frame. They put a collector with an Alcoa selective surface on the module and took it up to the roof of the Mellon Bank Building for trials. That was where Ainscough showed it to us with the Butler County Courthouse in the background and the module guyed by ropes so that it would stay on the roof until a few hundred gallons of water could be poured in to keep it down. Ainscough likes this sort of thing; he later became involved in a demonstration project using his own home and wrote:

What I am trying to do is show economic feasibility for fitting a solar-assisted heating system to existing homes — that is, something the average person could afford and even install himself. For instance, you can use a 1,500 gallon septic tank for storage. It has more than enough capacity for the average house and the cost is only $185, delivered. Frankly, I was surprised that you could buy six and a half tons of anything for that little.

Cost has delayed the start of Burt, Hill's showpiece project in Shanghai, West Virginia, the Wilson Solar House designed for Mrs. A. N. Wilson, a retired National Park Service official. It is not a big house as living space goes, something like 1,400 square feet with 588 square feet of solar collector. It looks like a large glass tent connected by a greenhouse to a smaller tent, the garage. To keep the heat in, the architects have used triple-glazing: a double layer of glass outside, then a yard of air space occupied by plants and another sheet of glass inside that can be opened or closed. The greenhouse serves as an entrance hall, cutting down the chill when the front door is opened. The house was planned to run not only with a flat collector to supply heat and hot water but with solar cells and a wind generator for electricity and a composter to handle household wastes. It was conceived as a forty-thousand-dollar home with the solar and wind equipment to be added with the help of government funds encouraging solar research. When the aid was not forthcoming, the house stayed on paper but it has served to attract industrial interest and enable Rittelmann to go more deeply into the ground rules that govern the building and selling of solar homes. Some have left him in despair. He has found thirty thousand separate code bodies in the United States with jurisdiction over the use of a solar collector, but "most of them certainly don't understand what a solar collector is." Tax laws work against the principle of life-cycle costing under which a solar house has some hope of catching up with a conventional home where fuel-using appliances cost less to buy but much more to run over their working lifetimes. "We had a situation where a large industrial client was considering various roofing systems for some new buildings that we were doing. As it turned out, the best roofing system on a life-cycle basis was the one that was next to the most expensive. The worst was the cheapest one that we expected would fail or need replacement in five years. The decision was to put on the cheap roof, which shocked us, and then make arrangements with the contractor to come back in five years and put on the expensive roof.

"The reason was the difference in taxing between a capital outlay for the cheap roof, and replacement by the expensive roof which comes out of an operational and maintenance budget. The same thing is happening in energy conservation. Right now, we have many conditions where we have to choose the least efficient equipment."

Under these conditions, perhaps one should be surprised not that there are so few solar-heated homes in the United States but that there are so many. By the end of 1975, William A. Shurcliff had counted one hundred and fifty-one there and thirty-six more elsewhere, making a grand world total of one hundred and eighty-seven. Shurcliff is a Harvard research fellow who had put out the eleventh edition of his *Solar Heated Buildings: A Brief Survey*, listing all that "did exist, do exist or are expected to exist very soon." The majority were built following the shift in the balance of petroleum power; seventy-four were completed in 1975 alone in the United States. Shurcliff's book is sobering. Scanning scientific publications, clipping the daily press, writing letters, he has put between two yellow covers and on 172 pages all the buildings on earth that rely on the sun to any extent for their comfort. There are not many; I see more in the village where I write and so do all of us at any moment of the day. Yet it is instructive in the range of options that it shows. There are the well-known names: Thomason with no fewer than six houses and nine patents; Hay in California at the bottom of the price bracket with an estimated cost of only thirty thousand dollars for a home with his roof-pond system; the Löf houses in Denver and Fort Collins, the old series of MIT experiments. Builders seem to have sought not only sun but also a favorable intellectual climate. This would explain the large cluster of houses in New England where MIT did its pioneering, in Colorado around Löf and others, in Arizona near the Meinels. The biggest and most comfortable solar homes are in southern California where a small collector is enough to cope with the mild winter. Then there is a house being planned in Toronto with a sixty-thousand-gallon concrete tank to store summer heat all winter long. Shurcliff gives a wealth of engineering details and the impression that solar builders, at least the ones he lists, are far more than backyard tinkerers. Many are backed by developers, aerospace firms or realtors with an eye to bringing out something that could be sold widely. Some are expensive; in Tucson, a price of more than two hundred thousand dollars has been estimated for the Decade 80 Solar House that is being built and financed by the Copper Development Association, truly a palace for a sun king with thirty-two hundred square feet of living space. At the other end of the scale are the experimenters who have built their own houses with their own money.

They go ahead with little reward save the recognition they get from a newspaper looking for copy on a quiet day. The *Boston Globe* keeps track of inventors in that MIT sphere of influence even when it might appear only occult. One story tells of an insurance salesman in Vermont who is using a hundred and forty cases of empty beer cans — to be filled with water and antifreeze — as his storage medium. An engineer in New Hampshire has resorted, instead, to one hundred tons of rocks in a sixty-thousand-dollar home where his solar heating system, built mostly with his own labor, accounted for an extra ten thousand dollars which he hoped to pay off in seven years by saving on electric heat. Many of these houses are the fruit of personal or voluntary labor. Such is the home of Richard Davis, an instructor at the College of the Atlantic in Bar Harbor, Maine, who moved into a "zero energy" house designed by a fellow instructor, Ernest McMullen, and constructed with the help of students. His collector is something like Thomason's: water runs over five hundred and forty square feet of south roof into a two-thousand-gallon basement water tank surrounded by thirty tons of rocks. Davis estimates he can get through three or four sunless days before using his wood furnace, a back-up system favored by many who really want to cut the apron strings tying them to a utility or a fuel dealer. Electric power comes from two wind generators built by Davis and friends. It is stored in sixty-five batteries rescued from a junkyard and then an inverter converts it from DC to AC.

Every paper has its running solar stories. In 1974 the *New York Times Sunday Magazine* described the start of a solar-heated home by Barber in Connecticut but the *Wall Street Journal* had him worried a year later because Guilford would not permit him to put up a sixty-foot tower for his wind generator. Space given these efforts is far out of proportion to their economic import. It is as if editors sensed that their readers look at the world and look for a way out.

These reports add to the conflicting impressions that go with any discussion of solar energy. It is so hard to get a consensus; each authority has its own references, many of them involving such subjective notions as "a comfortable temperature" or "an adequate home." It does seem that the worst problems are encountered by the system that tries to duplicate all the comforts to which most Americans and some Western Europeans have grown accustomed. Some have been related to a Senate committee by Meinel and his wife. Not only have they advocated vast solar farms to produce power, but they have lived in a solar home. They stated:

> We recently built two different types of solar heaters for our home in Tucson. One used water as the working fluid and the other air. The water system gave us much trouble. After a few weeks, it began leaking at its joints because of the daily thermal cycling of the piping. Also on cold nights, even in the desert, it froze up. Air in the lines from small leaks further blocks the automatic pumping. Each of these problems can be solved with appropriate changes to the system, but they all tend to drive up the cost.
>
> The second system, using hot air, gave better performance, even though the absolute efficiency was somewhat lower. It was better adapted to the design of our house and to its forced-air heating and cooling system. A thermal storage bin was constructed behind an adobe wall below the bank of solar collectors.
>
> When you get a solar unit, you quickly know that it requires a rather complicated set of controls in addition to the wall thermostat in the house. . . . The cost of the collector, the point on which most people focus their attention, is only a minor part of adding such a system. The collector is about one-third, the thermal storage one-third and the balance divided between the logic unit, ducting controls and modifications to the house. Our unit satisfied us, but may be too expensive for the average home owner.
>
> In looking at a number of houses in Arizona where the owner was interested in solar energy, we were surprised to see how few existing homes meet the requirements for a successful solar in-

stallation. Problems include too many trees, the wrong orientation of the house, poor roof shape, etc. New homes and new subdivisions offer better chances for successful solar installations.

It is our opinion that solar heating and cooling for individual homes is not going to be widely successful. People who like gadgets may find them interesting and satisfactory. Most people will not be bothered with them. Some who get them may be distinctly unhappy, feeling they were the victims of oversell. We think that solar installations will be far more successful for apartments, condominiums and commercial businesses. The larger units required will be supported by more sophisticated money management by the owner and by better maintenance. The units might be large enough to encourage operation by a utility.

The Meinels have a company, Helio Associates, which has worked with the Tucson Gas and Electric Company on the design of an installation for commercial buildings under the acronym of GAP (Gas Augmentation Program). A happier-sounding title is the one adopted by the Environmental Quality Laboratory at California Institute of Technology and the University of California in Los Angeles in their Project SAGE (Solar Assisted Gas Energy) which is being carried out with the Southern California Gas Company. The purpose is to examine the economics of combining sun and gas to heat water. The gas company has put four hundred thousand dollars of its own money into SAGE with another four hundred thousand dollars coming from government sources. An apartment house in El Toro, with thirty-two units, is being fitted with solar heaters to see how they can be adapted to buildings already up and another system has been planned for a new building.

Yet where does this leave the man with his own house who is not yet ready — and may never be ready — to move into a condominium in Tucson or a complex in La Jolla? In a difficult spot, no doubt about it, if he wants to live with the sun the way he has always lived. As a rule, solar heating costs so much at present that anyone who can afford it need not worry about his fuel bills anyway. Yet there are glimmers coming through the clouds. The Trombe systems, applicable principally to the sort of concrete houses built all over Europe, will take just about a zero off the figures that are quoted in the United States. It is sad that, as of this writing, they have not been introduced on a large scale in a variety of climates. Something along this line has been tried in Manchester, New Hampshire, that Yankee capital of energy thrift, by a group called Total Environmental Action. As Trombe does, they use a foot-thick concrete south wall with two layers of fiberglass sheets in front of it to trap the sun. The fiberglass sheets are a "Beadwall" devised for Steve Baer who has started a company called Zome-works in Albuquerque. This sounds like a godsend: a small electric motor blows a million styrofoam beads into the space between the two fiberglass sheets so they will not radiate all their heat away at night, then blows them out again in the morning when the collector can go to work again.

Baer himself has used the Beadwall to go with his drum wall, a south wall consisting of fifty-five-gallon oil drums filled with water and stacked behind a window. The sun shines through the window onto the drums, heating the water; then at night the window is protected by the Beadwall or shuttered to keep the heat in. At the price of the drums — from two to four dollars apiece secondhand — this wall can be built for less than five dollars a square foot, according to Baer. He states that it works at latitudes between 30 to 45 degrees (on the east coast of the U.S., this runs from Bangor to Jacksonville). In summer, the drums can cool a home if the shutter is opened at night and closed by day. They have been installed in Baer's house at Corrales outside Albuquerque and they provide 90 percent of the heat needed by four rooms if, according to Shurcliff, one accepts a temperature range from $55°$ to $80°$ F.

Baer is also the inventor of a skylid, an insulated shutter that can be used to cover skylights, ordinary windows, or small collectors. It works with two canisters of freon, one inside, the other outside. When the outside canister is the warmer of the two, the lid opens; when the inside is warmer, it closes, thereby offering an automatic shuttering system without any power source. In 1974, Baer said that three panels of his skylids, each measuring ten feet two inches by five feet eight inches, could be had for three hundred and forty dollars, including the freon drive mechanism. He belongs on the solar roll of honor for having marketed a device that does not cost more than spending the heating season in the Bahamas.

Baer is honest, he has no alluring sell. With ingenuity and fortitude, his devices can be incorporated into existing houses. They certainly are not the answer to the prayers of the average homeowner who wants his old comfort without his new energy bill. The world is crying for a handy solar heater that can warm hearth and hearts. Here, the human capacity for wish fulfillment is infinite.

As solar energy gained momentum, solar swindlers appeared on the scene. They left some of the scientists we saw in helpless rage. One spoke of a group in a nearby town that claimed, according to newspaper reports, to be heating a seven-room house with one hundred square feet of collector. "We see this in the papers. It's terrible, it gives solar energy a black eye. It violates the basic principles of physics and it violates the solar constant. A square foot simply does not receive that much sunshine. Either these people are using a gas burner or they're living at 40 degrees Fahrenheit in winter."

I had the opportunity to get some information about one such system and some opinions on its worth even though the company producing it did not answer my request for material. Since my sources ask that I do not use their names, I will not mention — nor have I mentioned anywhere — the name of the company promoting the system. It is probably just as well; in forthcoming years, such operations are likely to appear under different names, all basking in the glow of the sun.

In the case that I have tried to run to earth, the company claims that it has achieved a breakthrough that enables it to get around the old rule that one square foot of flat collector is needed to heat two square feet of living space. It speaks of a new surface that enables it to heat five times as much area. One study of the system noted that the company had based its estimate on a house whose actual heating demand was three times higher than had been stated. Besides that, the occupants would die of asphyxiation long before freezing to death because no allowance had been made for ventila-

tion. In one statement, the company announced 90 percent efficiency for its solar collector — but with an outside air temperature
of 60° F. In another, it said that its system would supply 90 percent of the house's heat; the study showed that it could yield only
50 percent with a collector twice the size. A prospective licensee
spent three months traveling fifteen thousand miles to report that
he had been unable to find a single organization, institution, or individual with an iota of confidence in the firm. The best he had
heard was that its claims were "highly improbable," the worst was:
"they should be in jail for mail fraud."

What is discouraging is that the company receives publicity,
articles in well-known newspapers which it distributes as reprints
that generate more articles. I do not know how much it obtains
from the sun, but it gets plenty from the press. True, no newspaper
is responsible for statements made by a source that it quotes, but
not every reader knows this and many take a newspaper account as
a reference. With a boom in solar energy quite likely, it would be
wise for newspapers to cross-check some of the good stories they
are bound to pick up, just to make sure they are not tall stories.

XVIII.

〜〜〜〜〜〜〜〜〜〜

A Solar World

〜〜〜〜〜〜〜〜〜〜

THIS BOOK has been written existentially. A year elapsed between its start and finish; by the end, it never could have been begun the same way. The first chapter has become a reference point from which one can measure the rate at which solar energy has progressed and is likely to go on progressing. In this year, it achieved more public and political acceptance than in the twenty years that had gone before. It is hard to say where and how far it will go from now on. The contradictions in this book, the conflicting verdicts and forecasts, reflect a science that has not yet been tested.

The year was one of sudden accelerations and grinding halts, another of those uncertain stretches in our journey through time. Our own experiences have been related in the accounts of the places we visited while gathering material. At first, I introduced them as a transparent device to sweeten such heavy doses of flat collectors and photovoltaics; then I began to realize that they belonged to the year that all of us shared. They speak of the way the world lived and looked on life not so long ago. They show, too, that life is still with us to be savored in great tasty chunks, no matter what the room temperature or the price of hi-test.

Their nostalgia, their confusion in time go with this reawakening science reliving a past and a world where the present offers but few clues to the future. During this year, the unthinkable became accepted, if not acceptable. From the viewpoint of energy policy, as good an observation post as any, the use of the sun and wind is no longer derisory although it does not seem to enter into present calculations. While nuclear power will stay with us, it must shoulder more and more of a new burden of proof as its future is debated. The debate should not stop; too much has been done in the past with insufficient information in hand. The notion is starting to appear that there may not be a single big quick fix after all, a petroleum-surrogate to heat our homes, nourish our industries, and run our public and private transportation systems. It is hard to find another integrated money-spinner like oil, with its cascade of payoffs and concomitant control over the economic and hence the political process. With nuclear energy, the political control must come first and the cascade of by-products is an economic burden. It is best, therefore, to spread the risks, to hedge the bets rather than put them all on a single horse that just might be scratched.

Solar energy is definitely a starter. In the United States, government support is nowhere near the level devoted to fission and fusion power but it has come beyond the point of no return. At a time when the country achieved what neologists called negative growth, solar energy was in a state of positive shrinkage. As we may recall, the total solar budget for the federal government in the 1974 fiscal year just missed $15 million. The next year, it was up to $55 million; then for fiscal 1976, it spurted to $114.7 million in obligations and $86 million in actual outlays with virtually all the money channeled through the new Energy Research and Development Administration. Lloyd Herwig, whom we had met when he was director of solar research at the National Science Foundation, had become acting scientific advisor to and acting deputy director of ERDA's solar energy division in Washington. Although his office was still temporarily in the NSF quarters on G Street when we saw him again, he had acquired in his new incarnation the assurance that goes with a budget more than sextupled.

By the end of fiscal 1976, said Herwig, ERDA planned to have solar heating in at least one hundred buildings, 80 percent residential and the rest commercial, thereby almost doubling the world population of solar houses. A facility to test thermal power systems is being built at Albuquerque, a "power tower" with a field of flat mirrors concentrating the image of the sun onto a central receiver, something like Odeillo without the big dish. It will turn out five thermal megawatts (that is, megawatts of heat, not electricity) and it will be five times the size of Odeillo, making it by far the biggest solar plant ever built. But by 1979, according to ERDA's plans, the United States will have a plant rated at between thirty and thirty-five thermal megawatts and generating ten megawatts of electricity.

Photovoltaics will move ahead just as quickly. ERDA is making its main thrust through the Department of Defense which is responsible for a host of faraway places in odd corners where electricity is needed in quantities of a few kilowatts, not just the flicker that runs a navigation buoy. Given such a market, manufacturers should be able to go into production and bring the price of solar cells down from twenty thousand to two thousand dollars per peak kilowatt, merely by using silicon of a quality lower than that demanded by the semiconductor industry. This figure, Herwig said, does not take into account savings expected to come from new ways of making solar cells.

Everywhere, ERDA is moving out of R and D and into pilot plants. One such plant will go up to convert urban wastes into methane, precisely the technology needed to get methane from forest wastes and from the energy plantations of the future. The components of ocean thermal gradient plants, the deep-sea pipes and the heat exchangers, are being built and tested. The type of hundred-kilowatt wind generator that NASA has erected in Cleveland is to be uprated to two hundred and three hundred kilowatts with blades still one hundred and twenty-five feet in diameter but capable of working in higher winds. "The solar energy technology to produce electricity that appears to be nearest to economic viability is wind conversion," Herwig has stated. "Assuming success of present research projects and program plans, practical wind energy systems ranging up to megawatts of electric power per unit should become commercially available by about 1980."

What excites Herwig is the amount that American industry spends on solar energy. ERDA is putting out a catalog of the hundreds of companies designing and offering products. "Industry is at least matching expenditures by the federal government," he said. "In the heating and cooling of buildings, it is spending twice as much." He concludes that the sun will be able to furnish 50 percent of the energy demand of the American home. This is an average; the figure runs from 25 to 75 percent depending on where the house is located. It should take five years to develop and improve the components of solar heating units, then put them together into something that will work. "By 1985, optimized solar energy systems with back-up energy can be competitive with other energy systems for homes. Without a doubt, they will be competitive with electricity." According to ERDA's timetable, solar heating and cooling will be competitive in some regions of the U.S. by 1980, then in most regions five years later.

Herwig predicted that solar energy will make its first big impact when it is introduced with ERDA's help into existing commercial buildings. Next, it will enter the energy market for new housing and, sad to say, it will appear last of all in the old homes where most people live. In this category, it will be vying first with electric heating.

According to one ERDA projection, 1 percent of the buildings going up in the United States will be solar-equipped by 1980 and at least twenty-five hundred homes and two hundred commercial buildings already up will get solar heating and cooling each year. This means that twenty thousand new solar houses and two thousand new solar commercial buildings are to be built in 1980. They will save eight hundred and fifty thousand barrels of oil that year and twenty times more in 1985 when, according to ERDA's forecasts, there should be five hundred and ninety-six thousand solar homes and fifty-five thousand commercial buildings. This is bold talk in Washington, especially when measured alongside the esti-

mate by the Federal Energy Administration that heat production
by high-temperature solar collectors in the United States in 1975
was enough to save but one hundred and eighty-eight barrels out
of a daily use of seventeen million barrels.

These forecasts are as good as the money behind them. Like
every other major change in our lives, solar energy will have to be
heavily subsidized by the public sector (all the more so because it
has no hope of multibillion-dollar military programs to pick up its
tab for research and development) before it can enter our private
world. The support has finally come. It is not yet a sure thing —
projects amounting to far more than $100 million a year have gone
down without a trace in the past — but it is assured of a good run.
For the first time, the profit motive has appeared: there is money
to be made from solar energy, even if most of it is government
money for the time being. Vested interests have appeared, pressure
groups are being created to keep the money coming. There is popular — one could even say populist — pressure behind solar energy,
as evidenced in the battle waged by several states to become the
home of the Solar Energy Research Institute. Solar houses will
reach the demonstration stage where, almost anywhere in the country, anyone will be able to walk into them, look around, ask the
price and decide whether to buy or not. Solar generation of electricity — whether through the collector farm, the power tower, the
windmill, the photovoltaic cell, or the thermal reserves of the sea —
will be carried out on a scale large enough to let us decide whether
or not we want more of it. We are going to reach this point, at
least in the United States, in the next ten years, barring any ghastly
failures, scandals, or cost overruns. The political decision to go
there has been made.

The next test will come when all this technology arrives at the
stage of the marketable innovation, the Volkswagen . . . or the
Edsel. By that time, government money will not be enough to assure continued growth. Either solar energy will sell or it will go
back onto the shelf, to come down at a future date so remote that
it is not likely to concern anyone of an age to think about it today.

By that time, too, we should be much closer to the answers to two other questions: the economic and political cost of a reliable long-term supply of energy from oil, coal, or uranium, and the environmental acceptability of nuclear power based on a long period of widespread service. This moment would arrive much more quickly if these two questions were suddenly answered, whether by another sharp rise in fuel prices or spectacular nuclear incidents. Otherwise, the ERDA schedule should hold, with the United States well in the lead because, as we have seen, it is by far the sunniest of the advanced industrial nations. Immediate benefits elsewhere are likely to accrue first to developing countries in the solar belts.

Other U.S. agencies are involved. NASA not only revived wind-power but, at its Marshall Space Flight Center in Huntsville, Alabama, it undertook a major solar heating and cooling test with three trailers parked together to represent a 2,500-square-foot home. Its big contribution should come in the development of solar cells. The Jet Propulsion Laboratory, operated by NASA in Pasadena, is managing a program to cut the cost of solar arrays down to five hundred dollars per kilowatt by 1985 when five hundred thousand kilowatts are to be generated by the sun. Ten years earlier, Varian Associates had generated a 55 percent gain in its stock in two days when it announced a new solar cell. In itself, this represented a breakthrough: for the first time, solar energy became a stock market leader.

Legislation has been introduced to offer low-interest government loans for house owners and builders who want to install solar devices; a bill in Michigan has proposed property tax exemptions on windmills and solar collectors; Iowa legislators bravely passed a bill to try solar heating and cooling in part of the state capitol; Home Federal Savings in San Diego became the first American bank to announce private loans for solar heating and cooling. Building codes are starting to take notice of solar collectors. One state requires all new one-family homes to be built so that solar water heaters may be added later; the city manager of Santa Clara, California, has announced plans to set up a municipally owned solar utility with the city putting systems into new buildings for customers who will pay a monthly fee to cover amortization and maintenance. Ingenuity is appearing in law as well as in science.

Much of this information has been gleaned from the *Solar Energy Intelligence Report,* a Washington newsletter which, in itself, is an indubitable sign of health. The newsletter's first issue appeared in March 1975 replete with the sort of items needed by firms looking for new sources of business, whether in the commercial market or with the government. Previously, the sole newsletter in the field had been the *Solar Energy Digest,* published in San Diego and written more for the individual determined to change his own way of living.

Private as well as public money is being invested. Not so long ago, the manager of any solar energy project was delighted if he could announce that a recognizable name in industry was on his side, if only in the cheering section. Now the shoe seems to be on the other foot; it is industry that seeks solar associations. *Solar Energy Intelligence Report* announced in its first issue that the Electric Power Research Institute planned to spend three million dollars in 1975 on solar energy and was working on a five-year program in solar heating, photovoltaic power, thermal power on a one-megawatt scale, and conversion of kelp to methanol. In that first issue, too, one learned that Bell Laboratories was making progress in cheap cadmium sulfide solar cells; IBM had determined that tungsten coated with a new surface could absorb and hold solar heat at temperatures near 1,000° F.; the Boston Edison Company was planning to team up with Heronemus, that old utility-baiter, for a wind system to supply power to a hospital; and six companies had joined forces to see if the sun could heat and cool high-rise office buildings. They were, to drop a few names, Phelps Dodge Brass, Sun Oil, Alcoa, Oliver Tyrone Corporation, PPG Industries, and Standard Oil Company (Ohio).

Arthur D. Little, Inc., in Cambridge was advancing on its way to a "solar climate control industry" and issued a status report listing six firms that had made solar equipment available in the United States by April 1975. PPG Industries was producing a flat-plate collector; so were Reynolds and Revere Copper and Brass, using aluminum and copper absorber surfaces respectively and expectedly. Also mentioned were Hitachi in hot water, Solaron in hot air, and an Australian company, Beasley Industries, exporting water heaters to the American market. One of my favorite items of industrial intelligence came from the Hsinhua agency to the effect that "the Shanghai No. 15 Radio Factory recently turned out 1,000 solar energy stoves for peasants on the outskirts of the city and in other parts of China." The dispatch, published in August 1974, went on to describe the stove as resembling an upside-down umbrella that can be easily taken apart and quoted users who said that it took twenty minutes to boil three liters of water and fifteen minutes to cook a kilogram of rice.

Solar energy began to gain literary and social acceptance. *The New Yorker*'s book review pages carried an ad for *Energy for the Home* published by Garden Way in Charlotte, Vermont. On a macro-level, the sun was turning out to be the answer to any serious questions raised about the future of the human race at its present address. A number of such exercises were carried out; one of the most ambitious was *Mankind at the Turning Point,* the Second Report to the Club of Rome, by Mihajlo Mesarovic and Eduard Pestel. One might call it a sequel to the first report, *The Limits of Growth,* which used the technique of running our behavior through a computer to see where it might lead. Mesarovic and Pestel regard the all-nuclear economy as a "Faustian bargain — and worse, for we would be selling not merely *our* soul to satisfy our immediate comfort needs, but the well-being and perhaps the very existence of *generations still unborn.*" This has been heard before; what is new are the authors' projections of what such an economy would be like, assuming that technological kinks can be ironed out. They state that the job can be done, that all energy can be nuclear, but it will take twenty-four thousand breeder reactors throughout the world. To get there, we must start building four per week as of now; if the reactors last thirty years, we shall have to build two a day merely to replace the ones that wear out.

As an alternative, they come up with three strategies: short, intermediate, and long-range. For the next ten years, they think the flow of oil from producing to consuming regions must be kept up, if only because nuclear power has little to offer by way of a solution in so short a time. During the intermediate transition period, coal can be introduced as a supplement. They maintain that supplies will last through the year 2100 if world annual per capita energy use can be held to four kilowatts, the present level in France as compared to ten kilowatts in the United States.

In the meanwhile, solar energy should be brought in. They say: "If the solar energy farms are built, at first, within the oil-producing regions and jointly financed by the producers and the developed regions, the revenue from oil exportation would be reinvested in energy; the world monetary inflation, too, would be helped by this joint venture. Even more importantly, the energy-producing regions would remain energy-producing regions, even if and after their oil runs short. Thus would the economic stability of the world system be served leading also gradually to a limited worldwide decentralization of energy production."

They give their computer its head, they drive on:

To cover millions of acres with solar energy farms would constitute the most extensive engineering undertaking of all times. One per cent of the world's land surface would be required, and the land, pipelines and accessory equipment would probably cost between $20,000 and $50,000 billion. Solar energy farms would be situated far away from industrial centers and therefore pipelines will have to join energy producer and user (as is the case with oil). . . .

As we propose the implementation of a program which our world computer model suggests is feasible and desirable, we take certain things for granted. . . . Mostly, we assume — or at least hope — that our life-styles will change, that we, the consumers, will understand that our frivolous use of energy takes food from the mouths of children. . . .

It will take more than hope to change these life-styles; the use of energy that they imply is not all that frivolous. It is in societies where energy and resources are used most abundantly that wealth appears to be distributed most evenly. A study published by *Electrical World* shows the United States to be leading the world in getting the most gross national product per capita in terms of kilowatt-hours per capita, with Sweden as the runner-up. The author, Fremont Felix, remarks that he is often asked why the United States could not maintain this figure while using energy the way West Germany does. "The answer is that this could be done only if the GNP per capita in the U.S. was the same as that of West Germany — instead of being two-thirds higher. And this could be done only if the pay scale in the U.S. were the same as it is in Germany." Felix is an out-and-out proponent of nuclear power which "will not only ensure the U.S. of an adequate supply of electricity from domestic sources but, over the life of each nuclear plant installed, will result in costs equivalent to those of a plant firing oil at no more than $5 per barrel." This can be and has been argued; what is more difficult to counter is his linking of economic justice to high per capita use of energy.

I think that a shift to massive sun and wind power would involve an even more massive shift in the way that people live and earn their living. Concentrations of power demanded by large industries cannot be achieved with the energy falling onto the roof of a plant. This might heat the cafeteria or cool the offices but it is not going to run industrial processes. Here is the start of a conflict of interests. As things are done now, the same network serves domestic and industrial users. If the individual were supplying a proportion of his own needs by capturing the sun, then industry would be obliged to fend more for itself as it did in the early days of mills run by their own waterwheels or steam plants. The industrial mix might change if this citizen subsidy were reduced and the present premium on high-energy, low-employment activities might be lessened.

Those were not halcyon days in the early industrial era. Political power as well as energy was concentrated in the hands of a few. Systems of government that had evolved when land was the main if not the sole resource continued to rule. Replacing them by systems with a broad popular base was a long and bloody task that is far from finished. And yet, the curves of increasing population and depleting resources indicate that we are headed back towards a lower per capita distribution of energy. Somehow this must be rendered acceptable to an enfranchised people.

Perhaps it can be done. Our visits to solar scientists led us to societies living at various levels with happiness and equilibrium not necessarily correlated with use of power. There is a sense of societal security in places where a way of life, though frugal, is based on permanent resources. There is less collective anxiety or *après-moi-le-déluge*. I had the occasion to see two oil-producing countries in South America. In one, the president had decreed that all automatic elevators had to be manned to alleviate unemployment; in the other, petroleum was being exported from a harbor where laborers pushed loads about in carts rolling on home-made wooden wheels, planks nailed side-by-side and rounded off. Oil and nuclear power represent a multiplication of human energy whose fruits tend to rise to the top if a state of political maturity has not been reached. No doubt those on top will not climb down without a fight or, at least, without trying to stay there as long as they can.

That prospect is not appealing; barring catastrophe, it need not come about. The alternative of diffuse solar energy with diffuse political power is now plausible. Slowly, not through any sudden wave of abnegation but through market and industrial acceptance, the sun should make its entry into our lives. Ushering it in is a task worthy of the mettle of science and technology: to get more from the same, from the solar constant, instead of getting more from more and more. I love the square-rigged ship and the horse-powered farm; I know they cannot run a population of four billion that will go on rising like a badly ballasted airship before it comes down. This means more than conservation or just cutting out the waste. A study in *Technology Review* published by MIT shows that if the average weight of the American car were brought down from thirty-six hundred to two thousand pounds by 1980, then U.S. oil reserves would last five more years to 1998 instead of 1993. Is it really worth the trouble?

Now the prospects of a solar world do not appear so outrageous; we find it easier to live with them. Perhaps this world will come because, unlike petroleum, the sun and wind are widely distributed beyond boundaries. We have reached the point where we can hardly live with the political distribution of fossil fuels and, a little later on, uranium deposits. Solar energy may be a political necessity. All the far-flung plants in the sun belt could be built by an international consortium of producers and users. After all, the resource is spread so widely that it is cheaper to try to collect it than to fight for it. There can be no absolute monopoly at all levels. The man who does not want to heat his house by buying hydrogen made on the coastal deserts of South America can always say to hell with it and run up a collector on his roof. He will learn a lesson in self-reliance and he will teach it to his neighbors.

If we can live this way, if we can ever achieve an identity again as producers, not consumers, if only through the hot water or the odd kilowatt we can manufacture on a rooftop, then we will start to return to our old function as fashioners of the green earth. We will cease to regard it as expendable, we will cease to be expendable ourselves. We will regain the freedom we had when we could fill up and pull out; we will be beholden to no central despot. Liberty can come from independence just as it has come from abundance. All this will mean going a long way. I think the sun can start to take us there.

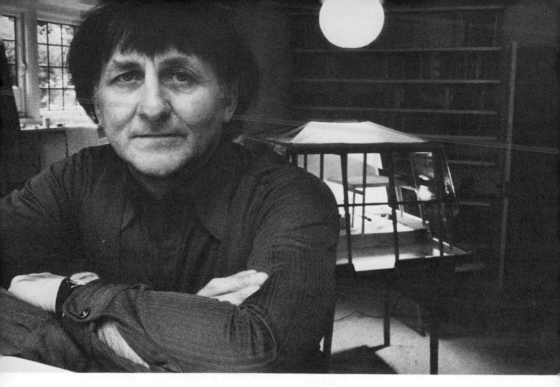

Alexander Pike and his conception of the Cambridge house.

James Thring demonstrates the rotor of a vertical wind generator on his version of the Cambridge house.

DAVID WALKER

John Page in his office.

When the vertical wall in this model of the Trombe house is heated by an electric lamp, smoke rises and starts to flow along the ceiling, showing how heat moves in the full-sized house.

To illustrate the greenhouse effect, Page plunges a heated metal cube inside another cube made of treated glass. His instruments will detect no heat loss because the glass is opaque to long-range radiation.

Here, Page models the Hay house heated and cooled by water tanks on its roof. The sun is represented by the bank of lamps while the big flat plastic frame contains water and ice to simulate the clear cold sky over the actual house in California.

M. ALI KETTANI

Jean-Pierre Girardier and the motor of one of his solar pumps.

Boris Garf and part of a thermoelectric generator.

MARINA D. KUDRIASHOVA

VALENTIN A. BAUM

ANNAGELDI KAKABAEV

REJEP BAÏRAMOV

FRED DUBIN

LLOYD HERWIG

Index